Technology

Markets

Regulation

Public policy

communications
the next decade

A collection of essays prepared for the UK Office of Communications

Edited by Ed Richards, Robin Foster, Tom Kiedrowski

November 2006

ISBN: 1–905774–08–7
978–1–905774–08–1

Published by Ofcom

Foreword

Viviane Reding

Member of the European Commission,
responsible for Information Society and Media

For many years, experts from all over the world talked about the digital convergence of communication networks, media content and devices. Today, we see that technological convergence actually happening.

Traditionally separate markets – such as telephony, internet and television – are changing fast and converging while market players quickly have to adjust their strategies to this new reality. This is, for example, why telecoms and cable operators are moving into each other's markets.

The European Commission's initiative *i2010: European Information Society 2010* is a comprehensive strategy for modernising EU policy and deploying regulatory instruments to encourage the development of the digital economy. As the European Commissioner responsible for Information Society and Media, it is my task to ensure that an appropriate framework for the digital economy and for investment in ICT is in place. In practice, this means two things: creating an open and competitive single market for information society and media services within the EU, and creating the right regulatory environment for business and citizens. This is essential for delivering sustained growth and skilled jobs in Europe.

Today's regulatory framework for electronic communications builds on principles of competition law and technology neutrality. It takes account of the convergence of telecommunications, data transmission and broadcasting services, by dealing with all communications networks and services (fixed, mobile, satellite, broadcasting and so forth) in a consistent way. The relevant Directives are a balanced and well-considered attempt to provide a coherent framework for the delivery of different services which traditionally have been subject to different regulatory regimes. However, it needs attention in a number of areas to remain effective for the coming decade.

Regulation fit for the future

In June 2006, the Commission launched a public consultation on the review of the rules that govern electronic communications. The revised framework could enter into force around 2009-2010.

Therefore, the timing of this book seems to be just right, providing food for thought for possible improvements to the regulatory processes and contributing to the policy debate. The forward-looking focus – which includes trends and challenges for the ICT sector going beyond 2010 – fits well with that of the Commission's review, which aims at ensuring that the framework remains valid for that timeframe also. So many changes in technology, markets and consumer expectations will have occurred by then – a century of the ICT industry! – that there will be a need for comprehensive reflection on possible evolutions and scenarios from a diverse set of authors.

Globalisation raises challenges and new opportunities to all economies, whether for industrialised or developing countries. The revamped Lisbon strategy aims to make the European economy more flexible so that it is better prepared for such challenges.

On the one hand, technology has contributed to globalisation *par excellence* by shortening distances and reducing the cost-of-use of most information society services. On the other, we should be aware of the positive impact of technologies, products and services developed by the ICT industry for the rest of the economy. These technologies have largely contributed to the growth of productivity of the European economy.

Technology has especially changed some key parameters of the e-communication industry. Digital convergence has brought together the world of audio, video, data and voice communications, changing forever the distinction between the different types of services. Borders have become increasingly artificial and, thanks to the internet, certain services are accessible from any place in the world.

The next generation of upgraded networks (the so-called NGNs) are likely to be based on the IP protocol, leaving the historic differences between different network infrastructures far behind. We may find that this ultimately has significant effects on business models: in particular, that the services consumers receive over communications networks will be offered separately from the connection to the network. This is likely to have a significant impact on the traditional vertically-integrated structure that has characterised telecoms incumbents since their creation. The price of voice calls is likely to decrease towards zero, as is already the case for some PC-to-PC calls.

Against this background, policy makers, regulators, legislators and market players can no longer confine themselves to narrow and sector-specific approaches. Convergence requires new ways of coping with this reality. The EU regulatory framework of 2002 already responds to such challenges by relying on two main: technological neutrality and a market–based approach to economic regulation.

Regulation that is based on specific technology, or which is unrelated to market dynamics, is destined to become quickly outdated. In contrast, the principle of technological neutrality delivers coherent regulation of all the services provided over such networks whatever the communication network used for its delivery. Regulation aimed at promoting a specific sector, company or technology can introduce significant distortion in the markets, leading to inefficient investment by market players and an uneven playing field for those players.

This will only dampen the interest of innovators and entrepreneurs to invest in Europe. Technological neutrality and a market-based approach to regulation mean that we devote the right resources to the objectives of regulation (promotion of innovation, fostering competition and protecting consumers) while making Europe an attractive place for investment and innovation.

But what does a good regulator look like?

Regulation requires regulators. From an institutional point of view, the question is whether a 'converged' regulator is the most suitable model for dealing with future challenges? Member States have followed different approaches: in some, (e.g. Italy and Finland) the same regulator is responsible for regulating both content and transmission services. In the UK, Ofcom is the independent regulator and competition authority for the communications industries, with responsibilities across television, radio, telecommunications and wireless communications services.

Under existing European law, the scope of activity of national regulators is part of the institutional autonomy of Member States; it is not dealt with by Community policy. Moreover, regulation of infrastructure (i.e. the EU regulatory framework for e-communications networks and services) is separate from the regulation of broadcasting content (i.e. the *TV Without Frontiers* Directive) as they respond to different policy objectives. These frameworks will continue to remain separate in the near future, though the two sets of rules will have to be consistent as digitalisation and technological progress go on.

One point I often stress – and bear this in mind when we design and implement policies – is that technology is not about a better 'lifestyle'; rather, it is about a better life for everyone through better services for consumers and citizens and new opportunities for growth and jobs. This is the thinking that has guided the i2010 initiative for a European information society.

For consumers, convergence means the possibility of accessing the same services and content (email, music, television) using different terminals over different types of networks. In a convergent world, consumers will be in control of their entertainment and media content.

Here, Europe has great potential as regards the development of innovative services (e.g. mobile TV) and the provision of diverse offerings of content, including content production. The global marketplace of the internet creates new economics which allow content producers to address even very specialised niche demands. This means new offers and choice for consumers – wherever they are in the world – as long as these products are accessible. Current proposals in the area of spectrum are aimed at facilitating the development of innovative products and wireless services.

Competition: the catalyst

Effective competition in the markets benefits consumers in terms of increased choice and lower prices. The framework provides a basis both to control the exercise of significant market power and the means to allow national regulators to take steps to allow self-sustaining competition to emerge. As markets become effectively competitive, the framework provides that sector-specific regulation will be removed. Thereafter, competitive markets will be subject only to competition law and consumer protection regulation. As market dynamics are difficult to predict (notably in the long-term), the EU model of regulation based on market analysis is more sustainable than pre-fixing a date for the phasing out of *ex ante* regulation.

One question is often asked: to what extent does the framework incite or inhibit strong and sustainable competition and investment? Undoubtedly, investments can flourish in a variety of situations and it is difficult to isolate the impact on investment of such a specific factor as the framework. Many factors have a positive influence on investment (e.g. the level of GDP per capita, population density, size of the company, and so on).

However, the available evidence suggests that competition remains its main driving force and that the countries that have applied the EU regulatory framework in an effective and pro-competitive manner have attracted more investment. For example, the highest penetration of broadband rates are seen where there is facility-based competition, with head-to-head competition between cable and telephone networks.

Some argue that we need a regulatory forbearance to encourage investment in new network infrastructure. But in my view there is no evidence to show that 'regulatory holidays' would generate more investment than competition. We have to be clear on this point: regulation should promote competition, and should not favour monopolies. Investment in new and competing infrastructure will accelerate the day when transitional access obligations can be further relaxed – or even removed.

Efficient spectrum management is key to the European ICT industry. We have started to implement a new strategy within the 2006 review of the electronic communications framework. Flexibility can facilitate the use of spectrum by a multiple number of wireless technologies, greatly reducing inefficiency and, therefore, spectrum scarcity. Greater flexibility in the use of spectrum will release the economic potential of this resource, stimulate the development of innovative products and increase competition in services and technologies. This in turn will improve our overall macro-economic competitiveness, will create jobs and increase the range of services available to citizens.

The Commission's plan includes giving spectrum users more flexibility in deciding how to use radio resources. Firstly, this would be achieved by strengthening the use of unlicensed spectrum. Secondly, owners of spectrum usage rights would be enabled to apply the principles of technology and service neutrality where possible. Thirdly, using criteria based on economic efficiency, selected bands agreed at EU level could be subject to secondary trading across the EU as one way to facilitate access to spectrum. Fourthly, in specific cases to be agreed at EU level, undertakings who develop services with a wider reach than a single country should be able to obtain a single common authorisation, or be granted the right of use for frequencies enabling them to provide services on a European basis.

Putting the new course in place will require the collaboration of the Council and the European Parliament at first, and of all stakeholders in general. We need to put together all the tools and instruments available to the EU and the Member States to promote and accelerate a favourable development of the communications and media industries.

In the meantime, I am confident that this book will be a very useful resource for all those concerned with policy in the field of electronic communications and media and will facilitate stakeholders to better understand the challenges faced by the actors of the ICT industry – whether independent regulators, ministries, undertakings, content and media companies, or citizens.

Contents

SECTION 2

THE CHANGING NATURE OF REGULATION IN THE PUBLIC INTEREST

SECTION 3

ECONOMIC REGULATION BEYOND 2010

SECTION 4

UTILISING THE AIRWAVES

SECTION 5

GLOBAL AREAS OF FOCUS

All contributors have been asked to write in a personal capacity. Views expressed are those of the authors, not those of Ofcom.

Biographies

M.H. Au

Mr M. H. Au has been the Director-General of Telecommunications of the Government of the Hong Kong Special Administrative Region since August 2003. He is also a member of the Hong Kong Broadcasting Authority. He was appointed an Assistant Director of OFTA upon its establishment in 1993. In 1997, he was appointed Senior Assistant Director (Regulatory), OFTA, in charge of the economic and technical regulation of all public telecommunications services. Mr Au joined the Telecommunications Branch of the Hong Kong Post Office in 1985 as Chief Telecommunications Engineer and became an Assistant Postmaster General in 1991. Mr Au holds the degrees of Bachelor of Science in Electrical Engineering (University of Hong Kong), Bachelor of Laws (University of London), Master of Laws (University of Hong Kong) and Master of Finance (Curtin University of Technology, Australia).

Philip Booth

Philip Booth is Editorial and Programme Director at the Institute of Economic Affairs and Professor of Insurance and Risk Management at Cass Business School, City University. He has previously worked at the Bank of England as a special adviser. Philip Booth is a Fellow of the Institute of Actuaries and of the Royal Statistical Society. He has published widely on investment, finance and pensions and is co-author of the books, *Investment Mathematics* and *Modern Actuarial Theory and Practice*. Other edited books include *Towards a Liberal Utopia* and *The Road to Economic Freedom*. He is Editor of the journal *Economic Affairs* and Associate Editor of the *British Actuarial Journal* and *Annals of Actuarial Science*.

Steve Burdon

Professor Steve Burdon has been associated with the University of Technology Sydney for over 25 years and is currently a visiting Professor. He has also held a number of positions at other Australian universities, including Associate Professor and Fellow at Melbourne Business School.

Prior to this, he held a number of senior executive positions including MD Asia-Pacific for British Telecom, CEO of OTC and Group MD for Telstra. Professor Burdon holds a number of non-executive board positions in public and private companies and has extensive experience advising leading Australian organisations. He has wide teaching and research experience in various areas including outsourcing and alliances, e-business, national ICT policy, innovation and entrepreneurship.

Jonathan Cave

Jonathan Cave is a game theorist and applied economist who trained at Yale, Cambridge and Stanford. A reformed antitrust official, he is currently a member of the Economics Department at Warwick University and Senior Economist (Senior Research Fellow) at RAND Europe, specialising in regulation, law and economics, policy analysis and evaluation.

He has worked on telecoms and sectoral, economic and/or social regulation; Information Society Policy (legal issues, sustainability, indicator development, IPR and scientific information, eGovernment and eDemocracy; trust and biometrics); R&D policy, procurement, etc. His teaching and academic work includes corporate finance, networks games and complex systems.

Martin Cave

Martin Cave is Professor and Director of the Centre for Management under Regulation, Warwick Business School and President of the international telecommunications think tank, ThinkTel. Until 2001 he was Professor of Economics at Brunel University. He specialises in regulatory economics, especially of the communications sector. He has advised a number of regulatory agencies in telecommunications and broadcasting. As well as his academic work he has also undertaken studies for the European Commission and advised regulatory agencies and the European Commission. He was a member of the Competition Commission from 1996 to 2002. He is the author of the Independent Review of Spectrum Management (2002) for the UK Government, co-author of Understanding Regulation (1999) and co-editor of the Handbook of Telecommunications Economics (Vol. 1, 2002, Vol. 2, 2005).

Peter Cochrane

Peter Cochrane was Head of BT Research from 1993-99, in 1999 he was appointed Chief Technologist. In November 2000 Peter retired from BT to join his own startup company – ConceptLabs – which he founded with a group out of Apple Computers in 1998 at Campbell CA, in Silicon Valley. A graduate of Trent Polytechnic and Essex University, he was the Collier Chair for The Public Understanding of Science & Technology at The University of Bristol from 1999 to 2000. He is a Fellow of the IEE, IEEE, Royal Academy of Engineering, and a Member of the New York Academy of Sciences. He has published and lectured widely on technology and the implications of IT and was awarded an OBE in 1999 for his contribution to international communications, the IEEE Millennium Medal in 2000 and The City & Guilds Prince Philip Medal in 2001.

Philip Cullum

Philip is Deputy Chief Executive of the National Consumer Council. He is author of *Consumers and regulation* and *The stupid company: how British businesses throw away money by alienating consumers* and is a leading commentator on issues of regulation. He represents the consumer interest on bodies such as the DTI Better Regulation Ministerial Challenge Group and the OFT Payment Systems Task Force. Before joining the NCC, Philip was an Associate Partner at Accenture, responsible for creating new thinking aimed at top CEOs and policymakers, and then Executive Director at Opinion Leader Research.

David Currie

Chairman of Ofcom since August 2002 and, since January 2001, Dean of Cass Business School (formerly City University Business School) and Professor of Business Economics. Formerly Professor of Economics at London Business School (1988-2000), and Deputy Dean (1992-1995, 1999-2000), Director of the Centre of Economic Forecasting (1988-1995) and Director of the Regulation Initiative (1995-2000) at London Business School.

A well-known figure in international economic policy, David Currie has published widely on government regulation of industry and on international macroeconomics and policy. He also serves (since September 2004) on the board of the Dubai Financial Services Authority. He was previously non-executive Chairman of Coredeal MTS and a non-executive director of the Abbey National, and has also served on a variety of UK government bodies including the Gas and Electricity Markets Authority (Ofgem, the UK energy regulator) and the Treasury's Panel of Independent Forecasters (the 'Wise Men').

He sits on the cross-benches in the House of Lords as Lord Currie of Marylebone.

Carolyn Fairbairn

Carolyn Fairbairn is a partner in London's McKinsey office, where she leads McKinsey's UK media practice. Before joining McKinsey, Carolyn spent eight years at the BBC, where she was Director of Strategy & Distribution from 2000-05 and a member of the BBC's main executive board. Prior to that, Carolyn was a member of the Number 10 Policy Unit, and has also worked as a financial journalist for *The Economist* and as a development economist for the World Bank. Carolyn holds an MBA from INSEAD and a BA in economics from Cambridge.

Robin Foster

Robin Foster is an independent adviser on economic, policy and strategic issues in the communications sector. He is chief adviser to the Board of Human Capital, media strategy consultants, and Associate Director of the Global Communications Consortium, a research programme at the London Business School. Until August 2005, he was Partner, Strategy and Market Developments at Ofcom. Previous positions include director of strategy at both the Independent Television Commission and the BBC, and head of the telecoms and broadcasting consulting division at consultants NERA.

Robin has published a wide range of speeches and papers on communications sector policy and regulation, including *Future Reflections* – a detailed scenario analysis of the future of television in the UK, published by Bournemouth University Media School in 2002.

Chris Giles

Chris Giles has been the economics editor of the *Financial Times* since Autumn 2004. He writes on micro and macroeconomic issues affecting the UK and other leading economies. Mr Giles was a Policy Manager in Ofcom working on the Public Sector Broadcasting Review. Previously at the *Financial Times*, he was the economics editorial writer. Before joining the *FT* in 2000, Mr Giles was an economics correspondent at the BBC. He started his career in research, spending seven years as an economist for the Institute for Fiscal Studies.

Isolde Goggin

Isolde Goggin is a Member of the Commission for Communications Regulation (ComReg), the statutory body responsible for regulation of all aspects of communications services in Ireland including telecommunications, broadcasting, postal services and spectrum management. Prior to her appointment as Commissioner, Ms Goggin was Director of Regulated Markets Division at the Competition Authority from 1996 to 2002.

Isolde Goggin worked for nine years with Telecom Éireann (now Eircom), the then State-owned telecommunications incumbent, initially in a variety of engineering areas and later in the commercial area in the international Division.

From 1989 to 1991 she worked in DGXIII (now DG Information Society) of the European Commission where she acted as an expert consultant to the Satellite Policy Unit concerned with EU policy for future satellite-based fixed, mobile, broadcasting and position-fixing services, including issues such as frequency planning and orbital positions. From 1991 to 1996 she was a business manager with Ericsson Systems Expertise Ltd., a subsidiary of the Swedish multinational telecommunications company, based in Dun Laoghaire, near Dublin.

Ms Goggin qualified from Trinity College, Dublin, with an engineering degree and obtained a Masters in Business Administration from University College, Dublin. She also holds a Postgraduate Diploma in European Competition law and is a Chartered Member of the Institution of Electrical Engineers.

Thomas W. Hazlett

Thomas W. Hazlett is Professor of Law and Economics at the George Mason University School of Law. In addition to his duties as a professor, he directs the Information Economy Project at the National Center for Technology and Law. Professor Hazlett is an internationally recognised expert on telecommunications policy. He has published widely in economics journals, law reviews, and popular periodicals, and he is currently a columnist for the *Financial Times' New Technology Policy Forum*. He frequently provides expert testimony to the courts, government agencies, and the U.S. Congress, and has served as a consultant to public and private organisations throughout the world. Professor Hazlett received his PhD in Economics from U.C.L.A. and has held faculty positions at the University of California, Davis, Columbia University, and the Wharton School of the University of Pennsylvania. In 1991-92, he served as Chief Economist of the Federal Communications Commission in Washington, D.C.

Ruth Hieronymi

Ruth Hieronymi has been a Member of the European Parliament since 1999 within the group of the European People's Party and European Democrats (EPP-ED) the North Rhine-Westphalia. Within the European Parliament, Mrs Hieronymi is a Member of the Committee on Culture and Education and the Delegation for relations with Canada. She is the media policy spokeswoman for the EPP-ED Group, chairwoman of the Audiovisual Policy Intergroup and Parliamentary rapporteur for the directive on Audiovisual Media Services (AVMS). She has been a member of the WDR broadcasting council since 1991. Mrs Hieronymi received her Master's (MA) from University of Cologne in 1977.

Reed Hundt

Reed Hundt served as Chairman of the Federal Communications Commission (FCC) from 1993 to 1997. Prior to heading the FCC, he was a partner in the Washington, D.C. office of Latham & Watkins. Mr. Hundt currently serves on the board of directors of Intel, Pronto Networks, Tropos Networks, Polyserve, Entrisphere, and Access Spectrum, and as an advisor to The Blackstone Group, and McKinsey & Company. He is a member of the advisory committee at the Yale School of Management, and Co-Chairman of The Forum on Communications and Society at The Aspen Institute. Mr. Hundt is the author of *You Say You Want A Revolution: A Story of Information Age Politics* (Yale University Press, 2000) and *In China's Shadow: The Crisis of American Entrepreneurship* (Yale University Press, 2006). He graduated *magna cum laude* from Yale College (1969) with exceptional distinction in History. He is also a graduate of Yale Law School (1974), and is a member of the

District of Columbia, Maryland, and California bars.

Tom Kiedrowski

Tom Kiedrowski is a Policy Manager within the Strategy and Market Developments Group at Ofcom where he works on a variety of projects including the Approaches to Regulation project that resulted in the production of this book. Prior to joining Ofcom, he was the Regulatory Affairs Manager for the European Competitive Telecommunications Association (ECTA) in Brussels. Between 1997-2004, Tom held a number of managerial positions within BT Group Regulatory Affairs. Before entering the communications sector, he worked at the London Investment Banking Association (LIBA). He is member of the Royal Institute of International Affairs (Chatham House) and the International Telecommunications Society (ITS). Tom was educated at Kent and Warwick Universities and holds an MA in International Political Economy.

Charles Leadbeater

Charles Leadbeater has advised companies, cities and governments around the world on innovation strategy and drawn on that experience in writing his soon-to-be published book *We-think: why mass creativity is the next big thing*. His previous books include *Living on Thin Air* and *Up the Down Escalator*. In 2005 he was ranked by Accenture as one of the top management thinkers in the world. Mr Leadbeater is a senior research associate with the think-tank Demos, a visiting fellow at Oxford University's Said Business School and a Senior Fellow at the National Endowment for Science Technology and the Arts. He is a former winner of the David Watt prize for journalism. In 1994 he moved to the *Independent* as assistant editor in charge of features and became an independent author and advisor in 1996. Mr Leadbeater spent ten years working for the *Financial Times* where he was Labour Editor, Industrial Editor and Tokyo Bureau Chief before becoming the Features Editor.

Graham Mather

Graham Mather is a solicitor and President of the European Policy Forum, an independent international research institute. He is a member of the Ofcom Consumer Panel. Mr Mather is a member of Competition Appeal Tribunal in the UK. From 1989 to 1994 he served two terms as a Member of the Monopolies and Mergers Commission. He was MEP for Hampshire North and Oxford from 1994 to 1999. Mr Mather has been a Visiting Fellow of Nuffield College, Oxford, which specialises in economics, sociology, politics and European affairs. He has also been General Director of the Institute of Economic Affairs and Head of the Policy Unit of the Institute of Directors.

Ed Mayo

Ed Mayo is Chief Executive of the National Consumer Council. Ed Mayo has been named a Young Global Leader by the World Economic Forum and one of the UK 100 most influential social policy thinkers by *The Guardian*. Ed is author of a series of reports on the economy and society, including *Shopping Generation* and *A Playlist for Public Services*. Most recently, he has chaired a commission on affordable housing and co-chaired the UK Sustainable Consumption Roundtable. Ed has a long track record of social entrepreneurship, including co-starting the Fairtrade Mark and London Rebuilding Society, the inner city financial co-operative.

John Naughton

John Naughton is a control engineer with a strong interest in systems analysis and computer networks. He is both an academic and a journalist. He has been Professor of the Public Understanding of Technology at the Open University since 1972. In 2000, Professor Naughton was short listed for the Aventiz Prize for Science Books for his *Brief History of the Future*. He runs the Wolfson Press Fellowship Programme at Cambridge University. He was educated at University College, Cork and Emmanuel College, Cambridge.

Eli M. Noam

Eli M. Noam is Professor of Economics and Finance at the Columbia University Business School since 1976. Director of the Columbia Institute for Tele-Information, a research center focusing on strategy, management, and policy in communications, mass media, and IT. Served for three years as a Commissioner for Public Services of New York State. Appointed in 2003 by the White House to the President's IT Advisory Committee. Noam a regular columnist for the *Financial Times* online edition. He has published about 400 articles and 27 books. His forthcoming books include *Media Concentration in the United States* (Oxford, 2007). His academic board memberships include the Oxford Internet Institute. He received the degrees of BA, MA, Ph.D (Economics) and JD from Harvard University, and an honorary doctorate from the University of Munich.

David Puttnam

David Puttnam spent 30 years as an independent film producer receiving accolades for films including *Chariots of Fire, The Killing Fields* and *The Mission*. He was Chair and CEO of Columbia Pictures from 1986 to 1988. David retired from film production in 1998 and now focuses on his work principally in education. He was the founder and Chair of Trustees for the National Teaching Awards. He has served on a variety of public bodies. In July 2002 he was appointed President of UNICEF UK. David was awarded a CBE in 1982, a knighthood in 1995 and created a life peer in 1997. He chaired the joint scrutiny committee on the Communications Bill that subsequently became an Act of Parliament that established Ofcom.

Viviane Reding

Viviane Reding is a Member of the European Commission responsible for Information Society and Media since 2004. Between 1999 and 2004 Mrs Reding was a Member of the European Commission responsible for Education, Culture, Youth, Media, Sport. She was a Member of the European Parliament between 1989 and 1999 within the group of the European People's Party (EPP). From 1979-1989 Mrs Reding was a Member of the Luxembourg Parliament, Member of the Office of the Chamber of Deputies, Member of the Benelux Parliament and Member of the North Atlantic Assembly leading the Christian Democrat/Conservative Group. Before entering politics, Viviane Reding enjoyed a distinguished carrier as a journalist and was President of the Luxembourg Union of Journalists between 1986-1998. She is a Doctor of human sciences from the Sorbonne in Paris.

Ed Richards

Ed Richards was appointed Ofcom Chief Executive in October 2006. He was previously Chief Operating Officer where his responsibilities included strategy, research, consumer policy, business planning, finance, human resources and Ofcom's functions in the Nations and Regions.

Prior to Ofcom Ed was Senior Policy Advisor to the Prime Minister for Media, telecoms, internet and e-govt. Before that he was Controller of Corporate Strategy at the BBC. He also worked in consulting at London Economics Ltd, as an advisor to Gordon Brown MP and began his career as a researcher with Diverse Production Ltd, where he worked on programmes for Channel 4.

Irwin Stelzer

Irwin Stelzer is a Senior Fellow and Director of the Hudson Institute's Economic Policy Studies Group. Prior to joining the Hudson Institute in 1998, Dr Stelzer was resident scholar and director of regulatory policy studies at the American Enterprise Institute (AEI). He is a member of the Board of the Regulatory Policy Institute (Oxford); a Visiting Fellow at Nuffield College, Oxford; and honorary senior research fellow, The Smith Institute.

Dr Stelzer is also a consultant to News International and News Corporation, a wind farm developer, and companies with interests in the direction of competition policy and financial services regulation. In the past he has served as a consultant to many electric and gas utilities in the United States, and to BSkyB. In 1961 he founded National Economic Research Associates (NERA) and served as its president until a few years after its sale in 1983 to Marsh & McLennan. He also has served as a managing director of the investment banking firm of Rothschild Inc.

Damian Tambini

Damian Tambini is a lecturer at the London School of Economics. Previously he was Head of the Programme in Comparative Media Law and Policy at Oxford University. He is also an Associate Fellow of the Institute for Public Policy Research (IPPR), The Oxford Internet Institute, and at Oxford University Said Business School. He is a fellow of the Royal Society of Arts and serves on the Advisory Groups of: The Oxford Media Convention, the Creative Archive License Group and POLIS.

Damian's research interests include media and telecommunications policy and democratic communication. He has published numerous books, articles and pamphlets on telecommunications and media policy. His next book, *Codifying Cyberspace* will be published early in 2007.

Leonard Waverman

Leonard Waverman is Professor and Chair of Economics, as well as Director of the Regulation Initiative and the Global Communications Consortium at the London Business School. Current research is on the growth and productivity impacts of the rollout of telecommunications and computers and was the subject of the *Economic Focus* section of *The Economist*, March 12, 2005. His most cited current publication is *Telecommunications Infrastructure and Economic Development* joint with Lars Hendrik Roeller, *American Economic Review*, Sept 2001. He is currently finishing a book (joint with Melvyn Fuss), entitled *The Networked Computer*, to be published in 2007 by Cambridge University Press.

Dr. Waverman is a non-executive Board member of GEMA – the UK's Electricity and Gas market Authority. He is member of the Scientific

Advisory Board of the German Institute for Economic Research in Berlin (DIW) and a Fellow of Columbia University's Centre for Tele-Information. He is on Vodafone's Advisory Board on the Social Importance of Mobile, and is a Director of the Nexus Mundi Foundation.

He was on the Advisory Committee introducing Competition in Ontario's Electricity system (1995-1996), a part-time Board Member of the Ontario Energy Board, as well as of the Ontario Telephone Service Commission, and a member of the US National Association of Regulatory Utility Commissioners (NARUC) for six years. He edited the major Journal in energy economics – *The Energy Journal* for six years.

Professor Waverman is a citizen of Canada and of France and has received the honour of *Chevalier dans l'Ordre des Palmes Academiques* from the Government of France.

Jonathan Zittrain

Jonathan Zittrain holds the Chair in Internet Governance and Regulation at Oxford University and is a principal of the Oxford Internet Institute. He is also the Jack N. & Lillian R. Berkman Visiting Professor for Entrepreneurial Legal Studies at Harvard Law School, where he co-founded its Berkman Center for Internet & Society. His research interests include battles for control of digital property and content, cryptography, electronic privacy, the roles of intermediaries within internet architecture, and the useful and unobtrusive deployment of technology in education. He has recently co-authored a study of internet filtering by national governments, and is writing a book about the future of the now-intertwined internet and PC. Papers may be found at www.jz.org.

Introduction

David Currie

Member of the House of Lords and Chairman of Ofcom

Ed Richards

Chief Executive Officer of Ofcom

The reality of convergence – and the sweeping transition, from analogue to digital technologies – is radically changing the communications sector.

New corporations, which have come into being over the past decade, are already global, multi-billion pound businesses. They're offering new services over new platforms to meet previously hidden demand, requiring longer-established players to reassess and adapt their business models to meet consumer demands in today's rapidly changing marketplace.

These changes can bring great benefits to customers, both individual and corporate, and are driving growth and innovation in the wider economy. But, just as existing businesses need to reassess their models, so public policymakers and regulators need to assess how their role and activities need to evolve to keep pace. How far are traditional public interest goals still valid or relevant, or are they now being met by the market? And in cases where they do remain relevant as goals, how far do the specific public policy and regulatory interventions still contribute effectively to meeting those goals? What changes do public policymakers and regulators need to make to ensure that their actions help, and not hinder, innovation, investment and competition to serve the consumer and public interest?

The debate around these questions is being played out in many countries. It is to help inform that debate that Ofcom has commissioned this book of essays from a wide range of commentators, experts and participants in the communications sector. Both severally and collectively, the essays represent the personal views of their authors, not those of Ofcom. They are designed to stimulate debate and so contain a diverse range both of intellectual and geographic viewpoints. The latter recognises that there are validly different national responses to common challenges which reflect the different national market circumstances.

In New Europe and the major emerging economies, fixed wireline infrastructures are being leap-frogged in favour of a greater predominance of mobile and wireline infrastructure. In Hong Kong, as M. H. Au's essay reminds us, 80 per cent of its households are encompassed in just 8,000 buildings, thus providing the market circumstances for the availability of multiple, competing high-bandwidth networks to most of Hong Kong's consumers. In the USA, intense market-led competition between a limited number of platform operators has led to high bandwidth network roll-out in the main urban centres, and an emphasis on wireless development elsewhere to prevent rural or small town dependence on monopoly supply. It has also led to a withdrawal from *ex ante* regulation, but also new concerns about access expressed through the net neutrality debate. And in the European Union, the legal frameworks for both platforms and audiovisual content are simultaneously, if separately, being revised.

Despite the difference in national market circumstances, four broad themes emerge from these essays as common challenges for policymakers and regulators.

- The first is how to encourage and support sustainable inter- (and intra-) platform competition as technology-change substantially alters the underlying architecture of networks over the next three or four years? What is the best framework to encourage innovation and investment by all players without giving the established operators too much leeway to leverage their established position into new markets? By contrast, when does the moment come when we find that we are still regulating those parts of the value chain (broadcasters/channel packagers and network operators) because that is where the traditional levers lie, when the real power has moved to other parts of the value chain – and where today's intervention to yesterday's problem hinders tomorrow's innovation?

- The second theme concerns access. In the days of stable technology and a limited range of one-size-fits-all services – basic voice telephony, internet access, television and radio reception, and access services for disabled people – universal service was a straightforward concept and not unduly expensive to impose, often through implicit cross-subsidy requirements on a limited number of incumbents. In today's and tomorrow's much richer and more diverse communications markets, the task of determining which services should be universally available becomes a more complex issue.

  The emergence of multiple geographic markets within a single national market also begs a question: namely, must a universal service be at a universal price? It is not even just the matter of availability but also of the ability to take up and use new communications services that is prompting public policy concern and debate about digital inclusion and the digital divide.

  Quite how sharp those social concerns will be in reality, and how far the market will address them, must be moot. What is evident is that great clarity will be needed in the criteria for determining what should be universal services in the future. And, as the (declining) monopoly rents that enabled the implicit cross-subsidies which supported universal services are fully competed away, the question "who pays?" comes into much sharper focus.

- The third theme is wireless spectrum, which is a resource of increasing importance to the communications sector and an engine for innovation in the sector. As device technology advances to allow more and more spectrum to become usable, the issue ceases to be whether there is a 'shortage of spectrum' but a shortage of rights to use the spectrum. Several countries – the USA, Australia and the UK included – are increasingly moving towards market mechanisms to allocate and assign spectrum. They're recognising that traditional methods of allocation both substitute a regulator's judgement for that of the market, and are simply too slow to keep up with the rapid pace of development in the communications sector. Just as a functioning market in land and property rights took time to develop and raised knotty practical and transition issues, so too is the case in spectrum with the added complication of multinational co-ordination. Several of the essays which follow address those practical and transitional issues, and all concur that heading more towards market mechanisms is the right direction of travel.

- The fourth theme is content. The great increase in bandwidth has enabled much more content creation, wider choice, greater user-control and increasing user-participation. These developments are welcome, but they also increasingly question the traditional public policy goals of, and the ability of the traditional players to deliver, regulation in broadcasting. The three traditional public policy goals have been quality, plurality and standards. There is rightly a debate about how far, with the proliferation of content and delivery systems, public policy and regulatory intervention remains necessary to deliver those objectives; and, if so, what the right form of intervention should be and where it should apply. In the USA, the free-to-air networks are becoming an oasis for regulated standards; elsewhere the first amendment applies. In other markets, broadcasting is controlled and the internet is unregulated (recognising, as someone has remarked, that on the internet even the First Amendment is merely a local ordinance). In Europe, the debate has focused around the concept of platform neutrality, at least as regards standards, but continuing with more direct intervention in broadcasting alone to achieve plurality and quality objectives.

  Several of the essays test each of these approaches. Perhaps unsurprisingly, there is less consensus around these content issues than there is about the competition, economic and spectrum resource issues tackled in other chapters.

What is clear across all these issues is that they raise questions that need to be tackled appropriately at different levels: some local, some national, some regional, some global. There is no one-size-fits-all answer and, as Ed Mayo and Philip Cullum remind us, no uniform consumer either. Philip Booth also advocates that public policymakers and regulators should have the trust and the humility to intervene at the most local level possible – subsidiarity – so that the intervention is closest to the multifaceted individual consumers that it is designed to serve.

We hope that this set of essays will contribute constructively to the debate, helping public policymakers and regulators towards answers and processes that bring out the best of the communications revolution, both culturally and economically, for citizens and consumers alike.

overview

Overview

Robin Foster

Chief Adviser to the Board, Human Capital
Associate Director, London Business School Global Communications Consortium

Tom Kiedrowski

Policy Manager, Ofcom

Introduction

In her foreword to this book, Viviane Reding highlights the fact that the challenges posed by convergence in communications are felt globally, not just locally. The different authors in this book are all, one way or another, seeking to present a view of policy thinking in the shadow of the uncertainty that convergence brings. Although our contributors and research bring out differing perspectives and ideas, one thing is clear: the ongoing nature of the change underway will require a fundamental review of the approaches to regulation across a range of issues, and now is the time to prepare the groundwork for that review.

This chapter brings together some of the themes that the authors raise, and the questions they pose, and places them in a wider context, drawing on research carried out by Ofcom during 2005 and 2006 about the approaches being developed by communications sector regulators around the world. Our contributors describe a world in which communications markets are once again being transformed. A new wave of technological change, entrenching the next generation of internet-based technology in consumer life – digital audiovisual services, new internet protocol (IP) based, high-bandwidth networks, high capacity and low-cost storage of data – is creating new business models and destroying old ones. Market structures are changing, bringing into commercial competition firms that have been undisputed masters of their 'own patch', sometimes for decades. Consumers should benefit from more choice and lower prices. New media and communications services will bring the potential for important social as well as consumer value. Jonathan Zittrain points to the world of opportunities that now exists and the promise of much more to come with 'netizens' contributing to a two-way system, rather than just existing as consumers. These trends, our authors argue, will render obsolete much of the established regulatory framework, and require new models for the future to secure continuing citizen and consumer benefits. As one of our contributors, John Naughton, asserts: "Companies that do not adapt will vanish from the earth. The big question is whether the same holds for regulators."

Trends in convergence

Digitisation has made it possible to deliver the same communications services – voice, video and data – over a number of different electronic communications platforms. Our contributors now see a future in which continuing innovation in services and network architecture is designed to facilitate mobility, convenience and flexibility of use. A decade ago, mobile, broadband, digital TV and radio, wireless networks and WiFi hotspots were niche products or non-existent. Today, they are all part of the mass market.

With the adoption of IP-based next generation networks (NGNs), new services and applications will be developed more quickly and at a lower cost than that associated with legacy networks. NGNs will allow for services that offer downloads and streaming of broadcast-quality images, with the scope for the uploading of consumer-generated content carried on a single, converged, core network platform in combination with a variety of fixed and mobile access networks. Even wireless technologies, be they 3G and 4G mobile or WiMAX, will be capable of higher bandwidth delivery, although to varying degrees.

A dramatic improvement in compression, peer-to-peer capabilities, and cheap high capacity storage will transform the capacity of distribution networks, the cost of distribution, and the ability of consumers to store data and content in the home. Intelligent user interfaces – including navigation and search tools – will help users find their way around the huge volume of content they will have access to over the high-bandwidth networks.

Taken as a whole these technology changes result in important developments in the sector which will have profound consequences for markets, citizens and consumers and, in turn, for policy-makers and regulators. The authors of this volume identify or predict a number of distinct trends.

First, there will be a shift from a communications sector characterised by relative stability, high-entry barriers and monopoly, to one which is fast changing, subject to disruptive competition and increasingly open. Market boundaries will be redefined, and there will be increasingly blurred boundaries between what is considered broadcasting, entertainment and publishing; fixed or mobile; a supplier or a consumer. But although competition will increase, we may also see new bottlenecks emerging at different parts of the value chain. For example, control of digital rights management, search tools, or premium content could become bottlenecks and have increasing importance over time.

Second, there will be a transformation in the relationship between consumers and the providers of content and services to them. Consumers will have access to the infrastructure, information and tools which will allow them to search more widely for content, switch between alternative voice and other service providers, and participate more in their consumption of communications services. The impacts on both the competitive environment and on public interest objectives traditionally pursued in the media sector are potentially significant.

Third, there will be a change in the relative positioning of network providers and applications service providers. Currently, intelligence resides predominantly in the network; however, in a converged environment, it is likely that some intelligence will move to the edges of the physical network, which could shift some control from the underlying network to end-users and application and content providers, altering the balance of power between players in the sector.

Finally, there is likely to be a shift from narrow nationally focused policies and regulation, to an environment where the implications of global technology adoption as well as cross-border policy and business decisions become more important. This will require policymakers not only to take an increasingly international perspective in their decision-making, but also to make choices about the appropriate level of engagement with regional and global institutions.

Market transformation

Inter-platform competition

The future should therefore bring increasing inter-platform competition: platforms will be able to deliver multiple services, and any given service will be able to be delivered by multiple, competing platforms. The most obvious example of the resulting increase in competition between platforms is in voice services. First mobile offered an alternative to the traditional PSTN for voice services, then Voice over Internet Protocol (VoIP) on the public internet offered a further alternative. A similar trend is evident in audiovisual, where services are now delivered via analogue and digital terrestrial, satellite, cable and now IPTV, all in competition with one another.

These trends in services and technologies, as well as regulatory innovations, have stimulated competition between platforms as well as more intra-platform competition with the result of increasing the challenges to traditional business models in the sector: few established businesses can be confident in their continued market success. Incumbent market shares are falling in radio, television, fixed and mobile telecoms, and price competition is generally increasing. As a result the major players in existing markets (who develop converged network platforms) are likely to seek to exploit their potential, and preserve their established market positions, by expanding their range of activities. One of the key changes over the next ten years is likely to be providers in once disparate parts of the communications sector increasingly moving into each other's territory. There is the possibility of powerful media companies entering telecommunications markets with bundled service offers, and vice versa, as well as internet application service providers potentially moving into all three realms.

Governing the airwaves: importance of wireless and the need to optimise spectrum policy

Many of the new platforms that have emerged in the last decade, and many of the prospective new ones, are wireless of one form or another and will be designed to facilitate mobility. Thus the supply of spectrum is a key enabler to facilitating choice and competition as Peter Cochrane, Martin Cave, Isolde Goggin and Tom Hazlett all identify. Spectrum decisions will have a major impact on the evolution of the communications sector, and are therefore seen as a critical building block in any new regulatory framework.

The growth of wireless has been brought about both by consumers' demand for mobility, and by a range of technical advances in compression technologies, battery life, processing power and miniaturisation. Importantly, wireless technologies now benefit from global economies of scale in device and network equipment manufacturing, which rely on similar radio spectrum frequency bands being used for similar applications in different countries.

In the past the problem of scarcity has been dealt with by a process in which spectrum has been assigned to users on the basis of administrative judgment. Negative externalities have been dealt with by granting exclusive rights of use for particular blocks of spectrum.

Such spectrum planning, relying heavily on central planning and expertise, was seen as the best way of addressing potential problems of interference and the need for international harmonisation of spectrum use. Isolde Goggin, in her chapter, remarks that the result has been a system which is good at safeguarding national rights (for example, for television broadcasting) and at avoiding interference. However, as Peter Cochrane and Isolde Goggin articulate, a key problem is the under-utilisation of large chunks of spectrum under the 'current command and control' arrangements. In a fast changing world, it becomes increasingly difficult for central planners to identify and determine the best use of each part of the spectrum. It is better, all three contributors argue, to rely on market mechanisms, including auctions and tradable spectrum rights.

Martin Cave sets out the case for a market-based spectrum management regime in building on the argument that spectrum should be treated like any other input. He discusses how licence-exempt spectrum or commons can be fitted into this market-based framework and how spectrum management can be adapted to achieve flexibility in spectrum use in a richer context of adaptive technologies. In order to respond both to current and future needs for the period following 2010, Cave calls for a radical rewriting of the EU regulatory framework that would allow significant harmonised spectrum trading and liberalisation in the European Union.

The compatibility of spectrum property rights with a spectrum commons approach is covered by Tom Hazlett. Just as a public park is best provided when the costs of the park are known and competitive land-use alternatives are permitted, Hazlett argues that exclusive spectrum ownership facilitates efficiency. He argues that liberalising spectrum policy empowers competitive spectrum owners to experiment with alternative network architectures or access models and that it also permits governments to evaluate whether particular forms of spectrum access need to be subsidised.

As regulators make progress with liberalisation, and as awareness of spectrum trading increases, secondary spectrum markets are likely to become increasingly effective at allocating spectrum efficiently. In time, the characteristics of market maturity – liquidity, due diligence and market intermediaries – are likely to become increasingly evident. As the remaining primary assignments are complete, the role for regulators in spectrum assignment is likely to decline; their role in licensing, enforcement and international co-ordination will continue.

Our contributors also identify some of the problems that have to be addressed when moving to a more market-orientated spectrum system. In particular, spectrum authorities often face complaints from incumbent users when they try to free up spectrum for competitive purposes – especially if incumbents have committed to large scale investments on the assumption that they will have access to a scarce resource. Isolde Goggin notes that the debate on digital switchover has focused minds on how to balance market mechanisms with the need to ensure the availability of spectrum for services of general economic or social interest, and argues that there is "no right answer" to the balance between spectrum markets and a more policy-driven approach to spectrum allocation. Much depends on the nature of the public policy goals and the scope that markets might have for providing the right incentives for any particular group of spectrum users. Broadcasting may be a special case, she argues – especially where public service broadcasting is concerned. But even here there may be scope for introducing economic incentives for broadcast users of the spectrum – perhaps via spectrum pricing to encourage efficient use of the spectrum they have been allocated.

One thing is clear: even in a market-based approach, policy-makers and regulators affect the outcomes of spectrum assignment, for example through the sequencing and packaging of assignments and through their role in international spectrum negotiations.

New approaches to regulation

The trends described in this book are exposing the very different frameworks under which radio, television, fixed and mobile telecoms have historically been regulated. Much of the regulatory framework for different platforms is enshrined in legislation. Policy-makers will need to decide whether regulators should labour under the constraints of existing regulatory frameworks, or make changes to the legislative structure to take account of convergence.

Economic regulation

Changing market boundaries

Much of the focus of economic regulation in communications to date has been on promoting competition to monopolies, or, where this has not been feasible, replicating the effects of competition through regulatory pressure on costs and prices. In most countries where *ex ante* regulation is applied to a network operator a finding of dominance, or market power, in a defined relevant market is required first. In light of convergence, it is expected that boundaries between products and services will continue to blur; new forms of mobile and portable devices are likely to appear with interactive and broadcasting features. It is probable that this will have some implications for the market definitions that underpin *ex ante* regulations. Some new markets will emerge (e.g. for hybrid fixed and mobile services), and it is possible that cluster markets will develop for new service bundles. In particular, it is likely that there will be increased availability of multi-play service provision, with providers offering service bundles which may cover fixed and mobile telephony, broadband and broadcasting services. Convergence will inevitably lead to new market definitions, so the analysis will need to be updated.

Dynamic approaches

Against this background, regulators around the world are focusing increasingly on the dynamic benefits from the competitive process. In his chapter, Philip Booth argues that it is important for regulators to concentrate on the competitive process and less on trying to recreate the hypothetical outcome that would result from perfect competition. Expressed in another way, this means a different approach to balancing the risks inherent in any regulatory decision – worrying more about the risks that regulatory intervention might stifle innovation or distort market decisions than the risk that competition proves not to be as effective as hoped. Jonathan Cave identifies options for using market mechanisms more readily to achieve any given regulatory goal. Len Waverman argues what is required is a re-evaluation of the costs and benefits of *ex ante* regulatory intervention compared against the costs and benefits of relying instead on *ex post* competition policy framework. He stresses the importance of investment in advanced electronic communications services as a key determinant of economic growth. Even if there are some enduring bottlenecks, for example, regulators could still benefit from changing the focus of the way in which they approach regulatory decisions. As Irwin Stelzer argues, "less is more", especially in periods of rapid technological change, and the presumption should be in favour of market forces to operate with technology that best satisfies consumer needs and avoids regulatory 'mission creep'.

Infrastructure competition

How this debate plays out partly depends on where you are starting from. The approach needed in pursuing the twin aims of competition and investment is clearly determined to a large degree by the legacy of past infrastructure developments and the existing state of competition. The likely impact of inter-modal competition, especially on competition in the local loop/access network, depends on how much chance there is of it emerging (or threatening to emerge) in a way that can exercise a reasonable degree of constraint on the actions of a dominant incumbent operator. Even if there is effective competition in urban areas, there may be only one dominant supplier in less densely populated areas. M.H. Au, in his chapter on Hong Kong, explains how the particular conditions in that market have enabled huge progress to be made in promoting competition and withdrawing from some aspects of regulation – such as the planned end to local loop unbundling requirements, now that effective broadband competition has been established.

In the United States, where there is extensive access to both cable and fixed-line telecom networks, there is arguably greater scope to allow the market to take its course than in countries which have relatively limited cable competition, and where there is still a dominant incumbent telecommunications provider.

Investment in next generation infrastructure

Reed Hundt, in his chapter, raises the issue of communications sector infrastructure, and the role that it plays in providing wider competitive advantages for nations and their citizens. Electronic communications networks are not – for the most part – tradable across national boundaries, but they are an essential input into business and society in each country. Therefore, a premium should be placed on high-quality, cutting-edge and affordable telecommunications networks in order to improve efficiency and secure economic growth and jobs. Hundt argues that governments have a responsibility to ensure that their citizens do not come off worst in the competitive battle between countries, which means support is required for the creation of high-quality, modern and efficient broadband networks – and regulators have a role to play in ensuring the environment is conducive to this goal. But, as Hundt points out, the removal of unbundling rules in the United States and the reliance on inter-platform competition has not ended the debate about competition and access.

Net neutrality

Prompted partly by developments in the US, there is an emerging international debate about the issue of 'network neutrality'. Although used by different people to mean different things, the real issue is about the extent to which network operators should be free to offer (and charge for) differentiated levels of service to access their networks. As different qualities of service become easier to implement in the long-term, this debate may intensify.

Eli Noam points to internet content and applications providers, as well as some traditional media companies that, he argues, have banded together in order to seek protection from the network companies' power over access prices, quality, price discrimination and the perceived likelihood that they will favour their own content subsidiaries.

However, Reed Hundt points out that there are good economic reasons to allow such differentiation, in order to ensure that those who value enhanced service features and quality can attain them (for a suitable price) and that those who do not, don't need to pay for things that they do not value. He argues that price and service quality differentiation will also provide the best environment for investment in network capacity and service levels.

Potential for new bottlenecks

Changes in services delivered may mean that new bottlenecks could develop at multiple levels of the value chain, for example in DRM or search engines. As our economic authors have cautioned, *ex ante* regulation at one level could be risky as it could favour one level over another and therefore distort the market, since companies' investment decisions will be influenced by arbitrary regulatory distinctions rather than commercial realities. However, a removal of regulation could also be risky. This would imply that a provider power may be offset by the countervailing buying power of the purchaser. At the company level, this could be adequate as individual companies would be constrained in using their market power to distort competition, but at the consumer level it could be less so because retail price levels may be significantly higher as result. As part of this, regulators will have to balance the need to secure access to bottleneck platforms with the need to ensure incentives for efficient investment and innovation.

This is also important for the discussion on social goals and their implications for regulation. As Charles Leadbeater argues in his chapter, we should encourage the further development of a new highly social and distributed form of media production. The most important thing for regulators here, he suggests, is to abstain from interventions which unwittingly or deliberately sustain incumbents while squashing the scope for disruptive new entrants to emerge. In particular, established players may try to use extensions to copyright, patents, and other elements of intellectual property to protect their market position – rewarding current rights holders without necessarily providing any incentive for further innovation.

Regulation to support social objectives

Continuing social objectives

Our contributors argue that the blurring of product and service boundaries has significant implications not only for economic regulation, but also for the complex framework of social or 'citizenship' regulation – a myriad of rules and obligations which has evolved over the years to sustain what are seen to be important societal outcomes. Such regulation aims to deliver outcomes that even an efficient market would not guarantee. Interventions to secure such 'societal outcomes' include:

- securing the widespread accessibility of fixed telecommunications services through universal service obligations
- securing through the application of licence conditions to broadcast TV and radio channels:
  - standards for the protection of audiences against exposure to harmful or offensive content
  - standards to preserve the distinction between advertising and editorial content
  - range, quality, plurality and broad availability of broadcast audio visual and audio programming (achieved through

measures including output quotas, news impartiality requirements, advertising restrictions, independent production quotas, and radio format specifications).

The need for new models

In many cases, the increase in competition within and between platforms represents the success of economic regulation. Yet, it can often undermine the mechanisms through which these societal outcomes have been delivered to date. These mechanisms can be thought of as implicit 'compacts': for example, access to valuable scarce broadcasting spectrum in exchange for Public Service Broadcasting (PSB) or commitments to protect against harmful or offensive content; or universal access to telephony at a uniform price in exchange for historic incumbency.

Competition erodes these 'compacts'. For example, competition in telecoms has focused on high-value customers in low-cost locations, taking market share and eroding the margins that to date have cross-subsidised lower value customers in high-cost locations to an incumbent. In television, the scarcity value of terrestrial analogue spectrum is declining as multichannel penetration increases. As the value of one half of these 'compacts' declines, their effectiveness in securing social outcomes is likely also to decline. The origins of PSB of course lie very much in the analogue age. A key economic feature of broadcasting in the analogue world is its public good characteristics. In the age of satellite, cable and IP-based networks this will all change.

Eli Noam fears that as the internet becomes the main platform for most media uses, there will be pressure to extend existing telecoms regulatory rules to much of the media system as a whole. He argues that as the regulatory rules for the internet move towards those of telecom, because the issues that led to telecom regulation – discrimination, market power, national security, consumer protection – are the same issues being raised by the internet, it then follows that the telecom rules become the rules affecting all media.

Of course, some historic 'social outcomes' may instead be delivered by the market. Chris Giles makes some important practical points about the need to recast our public policy objectives and to develop new mechanisms, such as contestable funding, to deliver them.

Similarly, a number of universal service regulations currently ensure that blind or deaf people have access to special services allowing them to use telephones. The arrival of the internet, and of Next Generation Networks (NGNs) using open standards, allows products designed specifically for this market to be profitable, which may change how we need to consider such regulation.

Around the world, local and regional governments are examining different options to secure universal broadband provision – typically with access to some form of public funding, made available on a competitive basis. As Steve Burdon notes in his survey of developments in Asia, many regulators are adopting what he terms a "proactive" approach to managing emerging competition, which involves setting national interest goals – such as investment in modern infrastructures – as part of the overall regulatory framework.

Changing social needs

The new models will need to reflect changing social needs. In the future, consumers will face greater choice, and, given improved information and tools and a much better understanding of media issues (media literacy), will be able to exercise their own individual decisions about what to consume. John Naughton describes the waning influence of the established broadcasters and, like Charles Leadbeater, notes the importance of new media as a tool for increasing social democracy. Naughton argues that the new media ecosystem will be richer, more diverse, and immeasurably more complex than the old system. Blogging, for example, helps individuals to bypass the gatekeepers of the old media world, in which "freedom of the press was available to anyone who was rich enough to afford one".

Nevertheless, there is still seen by many to be a strong and continuing rationale for intervening in the market to deliver desired social outcomes. David Puttnam argues, for example, that the citizens' interest is still important in the digital world and needs to take precedence over consumer interest. In particular he notes the continuing importance of ensuring diversity and plurality in digital media, and the key roles that media literacy and an effective IP policy will play in that environment. Ruth Hieronymi in her essay argues strongly for the continuation of sector-specific media law in the EU, and explains the intentions of the current proposals for extending content regulation to on-demand TV services.

Others see even more radical change – often for the better. Damian Tambini, in his chapter, sets out the case for a more holistic approach to role of communications in citizenship, which would encompass access to networks and services, content and public participation. He suggests that policy should develop the notion of citizens' rights, including access to key information services – regulators may have a role in protecting some groups from an unacceptable level of information exclusion and certainly have a role in monitoring the extent and understanding the impact of such exclusion – particularly as an input into any debate about universal service obligations. Carolyn Fairbairn proposes some criteria by which we might decide whether new media markets need to be tightly regulated or otherwise, including the extent to which new services might exercise a powerful influence over our lives, and the risk that plurality may suffer.

Sustainability of national content regimes

The internet, especially with higher bandwidth services, will make it increasingly difficult to sustain national content rules and standards, although there should be scope for building a more effective self-regulatory system which meets public expectations. Even if continuing intervention to secure certain standards were desirable in the long run, the advent of broadband and the internet increases the practical challenge of finding effective means of so doing. Our contributors paint a picture of a world in which there are many thousands of content providers – with individual internet users posting audiovisual content alongside more established professional companies – very different from the numbers covered by the current licensing regime. Content will be available from outside a particular country, making enforcement difficult. There is also a transitional risk that – to meet the need to be seen to be doing something – regulation is either increasingly focused on those it can still affect (the traditional broadcasters), while allowing many other content suppliers to escape attention, or applied so widely that it hinders the development of innovative content propositions from the many possible new suppliers who could access the market.

Regulators around the world are grappling with similar challenges. They appear to be pursuing two very different responses. One approach is to tackle regulation of the internet head-on, with all the implications for intrusive regulation and impact on freedom of expression that entails, which is perhaps most starkly demonstrated in China. However this is not the only approach.

The other is to step back, and allow individuals themselves to exercise responsibility for what they and their families consume, aided by the latest in content filtering and rating devices, by better information about the nature of the content to which they have access, and by clear and enforceable laws to prevent the distribution and use of illegal content (such as child pornography). This latter approach could also involve some form of 'opt-in' regulation. That is, rather than determining via legislation who should and who should not be part of the regulated arena, it could be left to the individual industry players to decide voluntarily and perhaps encouraged by certain incentives. Jonathan Cave points out that new forms of regulation, including self- and co-regulation, make greater use of commercial interactions to address governance issues.

Across the majority of countries, Ofcom has already found that a hybrid model of direct and self-regulation has been used to implement consumer protection measures in the e-commerce environment. For example quality seal systems are in place in France (*L@belsite*), Germany (*Trustedshops*) and Japan (*Japan DMA*), while the *Global Trustmark Alliance* promotes the use of quality seals at an international level.

There is likely to be an ongoing debate about what social outcomes remain valuable, and the need for regulation to continue to deliver those outcomes. For example, some outcomes, such as localness or ethnicity in programming, may remain desirable but increasingly be delivered by the market. For others, such as access to basic telecoms services, expectations of the minimum acceptable level are likely to be of most concern. There is likely to be a gradual shift, often initiated by regulators, towards more transparent and platform-neutral delivery of such solutions, such as through a provider of public service content. The provider would commission audiovisual content that served stated PSB purposes, to be distributed via a range of technologies and media platforms.

Institutions and processes: 2010 and beyond

Consumer empowerment

Regulators will need to consider their response to increasing consumer empowerment, and also perhaps actively encourage its development. New media developments, for example, may bring a much more democratic and participative relationship between users and content providers ('user power', 'pull not push'), and a shift in the balance between passive consumption and active content creation and conversation. This might reduce the need for content regulation in its traditional form. However, whereas some consumers are able to keep abreast of these developments and benefit from them, many may be baffled by their complexity. In her chapter, Carolyn Fairbairn highlights the potential that new media can offer content providers and public interest content provision in particular. A heightened role for citizens/consumers in exercising individual choice and responsibility for what they consume should be supported by effective, tough and well-targeted consumer protection measures. In other words, the challenge is to encourage and help those who have grown accustomed to traditional forms of communications to move into the converged world – for example, via media literacy initiatives, promotional activity and other incentives.

As Ed Mayo and Philip Cullum argue, the consumer interest needs to be close to the heart of regulatory decision-making. Their specific 'menu' of proposals includes that regulators need to start with a focus that sees the public not just as prime beneficiaries of regulation, but as its co-producers. They call for methodologies to be devised for action that goes beyond addressing imperfect markets, which they argue is the case at present. Above all, they argue that regulators should do far more to engage consumers in decision-making and they must move away from a regulator 'knows best' mindset.

Institutional Design

What does all this mean for regulatory institutions? At national level many argue that the case for converged regulators is clear. David Puttnam reminds us of the long gestation period for the Communications Act 2003 and is rightly proud of the important safeguards built into that Act. Viviane Reding suggests that the model of converged regulation could be adopted more widely, even though there are significant practical barriers to this in many countries. Italy, Finland and Australia all now have converged regulators of one design or another.

A more difficult question is what balance should be struck among national, regional and global regulation in this new era. No national policy-maker or regulator, in either the developing or the developed world, is able to act in isolation to the developments that are taking place around the world. This is especially true for some aspects of spectrum policy. Graham Mather sheds light on the current debate about the future structure of European telecoms regulation by highlighting parallels with other EU markets, and the case for intelligent co-operation between national institutions rather than a major shift of power to Brussels. Philip Booth also advocates that public policy-makers and regulators should intervene at the most local level so that the intervention is closest to the multifaceted individual consumers that it is designed to serve.

In his chapter, Graham Mather also talks about regulatory effectiveness and identifies the importance of impact assessments in good regulatory decision-making. He also notes, though, that there is a risk that such processes lose meaning over time, and therefore need to be subject to periodic audits.

In his discussion of internet governance, Jonathan Zittrain believes that if the current and future benefits are not to be lost, the internet needs its own specific type of regulation that is neither an intrusive state-led model nor a libertarian free-for-all. Instead he looks to such institutions as the Internet Engineering Task Force (IETF) and Wikipedia as potential guides to future internet governance.

Conclusions

Both Ofcom's own international research, and the many thoughtful contributions to this book underline the changes that are likely to affect the communications sector over the next few years, and the challenges for regulation. Although many different and fascinating ideas are contained in the chapters, the common theme is that regulators and policy-makers cannot stand still. Indeed there is a risk that out-of-date regulation could hinder the development of the sector and – to use John Naughton's biological analogy – destroy the organic changes which would otherwise occur.

The emergence of new wireless platforms has, and will, contribute to increasing inter-platform competition able to deliver multiple services, and any given service will able to be delivered by multiple, competing platforms. As stated at the outset of this chapter, core networks are increasingly similar: they use IP and increasingly modular device components, and they are based on open standards.

These trends of more inter- and intra-platform competition and changing societal attitudes will affect the way that regulation delivers social outcomes in future. As competition erodes the value of one half of these implicit arrangements, they will become less sustainable and less effective. In the future, more transparent mechanisms will need to replace them. Policy-makers and regulators will have to look again at what societal outcomes are desirable, and to determine what can be delivered by the market and what requires new mechanisms for intervention.

These trends will simultaneously increase competition and choice, but also undermine the effectiveness of many of the old regulatory tools used to achieve public interest objectives – for example, public interest content provided by commercial broadcasters, and programme standards. They change the way we need to think about potential and actual competition, (temporary) monopoly power, risk and rewards, and the need for economic regulation. They also call for a reassessment of our public interest objectives in the sector, and the tools we have at our disposal to achieve them. As our contributors argue, they require further tough questioning of the continuing role and aims of regulation in the sector.

Many of our authors call for a radical rethink of the approach to regulation over the next five to ten years. They argue for a focus on investment and innovation, and for the reinvention of approaches to public interest content and other social objectives. They see the scope for a market-based spectrum policy to transform the competitive landscape. They are optimistic about the opportunities presented by new media and communications services for individual consumers and for society as a whole. But they identify concerns that will need to be addressed if the full benefits are to be realised and some worry about the emergence of an entrenched digital divide or 'information deficit'. They see an imperative for less regulation in some areas, and also discuss how regulation needs to evolve and become smarter as it responds to the new challenges ahead.

Clearly regulatory innovation will continue to take place across the world as policy-makers and regulators respond and prepare for the challenges and the new opportunities that convergence and changes in market structure will offer. What will be important to factor in will be the need to engage with consumers and industry to understand these changes and developments. It will be necessary to reflect the highly dynamic nature of industry and technological evolution and avoid an excessively static view of markets.

The debate engendered by the chapters in the book will, we hope, help result in better answers to the big questions in due course. But, in summarising the contributions, we are impressed by the overall optimism about the potential for the communications sector to be a driver of economic growth and a generator of great cultural benefits – and the confidence that policy and regulation can evolve fully to realise those benefits.

section 1

Trends and challenges

Our changing media ecosystem

John Naughton

Professor of the Public Understanding of Technology, the Open University, and internet columnist, *The Observer*

Introduction

It's a truism that our communications environment is changing. It was ever thus: all 'old' media were new media once[1]. But there is something special about our present situation at the beginning of the 21st century. The combination of digital convergence, personal computing and global networking seems to have ratcheted up the pace of development and is giving rise to radical shifts in the environment.

Because we are living through this upheaval, it is difficult to take the long view of it. Our problem is not that we are short of data, or even of information; *au contraire*, we are awash with it, as companies and governments turn to consultants and market researchers for enlightenment or guidance. But the resulting glut of information doesn't seem to be making us much wiser. Indeed our current state might be best described as one of 'informed bewilderment'.

Part of our difficulty is that we lack a discourse that is appropriate to what is happening. Traditionally, we have drawn linguistic and analytical tools from economics, and as a consequence seek to interpret what is going on through the prism of that dismal science.

But economics – at least the economics on which we have relied to date – is the study of the allocation of scarce resources, whereas an important feature of our emerging media environment is abundance, not scarcity.

Besides, much of the cultural production that characterises the new environment is driven largely by non-economic motives and takes place entirely outside market processes. In the words of Yochai Benkler, what we are seeing is the emergence of:

"...a flourishing non-market sector of information, knowledge and cultural production, based on the networked environment, and applied to anything that the many individuals connected to it can imagine. Its outputs, in turn, are not treated as exclusive property. They are instead subject to an increasingly robust ethic of open sharing, open for all others to build on, extend and make their own."[2]

For these and other reasons, a discourse rooted in market-based economic analysis seems unequal to the task of understanding what is going on in our media environment just now. This essay explores the utility of an alternative conceptual framework borrowed from science.

1. Lisa Gitelman and Geoffrey B. Pingree (Eds), *New Media* 1740-1915, MIT Press, 2003

2. Yochai Benkler, *The Wealth of Networks: how social production transforms markets and freedom*, Yale University Press, 2006, page 7

Language

'Media' is the plural of 'medium', a word with an interesting etymology. The conventional, everyday interpretation holds that a medium is a carrier of something. But in science, the word has another, more interesting, connotation. To a biologist, for example, a medium is a mixture of nutrients needed for cell growth. And that's a very interesting interpretation for our purposes.

In biology, media are used to grow tissue cultures – living organisms. The most famous example, I guess, is the mould growing in Alexander Fleming's Petri dishes which eventually led to the discovery of penicillin.

What I want to do is apply that perspective to human society: to treat it as an organism that depends on a media environment for the nutrients it needs to survive and develop. Any change in the environment – in the media that support social and cultural life – will have corresponding effects on the organism. Some things will wither; others may grow; new, mutant, organisms may appear. The key point of the analogy is simple: change the medium, and you change the organism.

This way of looking at our media environment is not new. I picked it up originally from the late Neil Postman, a passionate humanist who taught at New York University for more than 40 years and was an unremitting sceptic about the impact of technology on society. In a series of witty and thought-provoking books – *Teaching as a Subversive Activity, Amusing Ourselves to Death, The Disappearance of Childhood and Technopoly* – he described how our societies are shaped by their prevailing modes of communication, and fretted about the consequences.

Postman deserves to be better known. But he was his own worst enemy, because he was a witty and iconoclastic writer who apparently did not realise that, in academic life, you will never be taken seriously if you make jokes or write clearly. In the academic culture, luxuriant obscurantism is taken as the litmus-test for profundity. The other reason Postman may have been under-rated is that he was a sucker for the Big Idea, the broad sweep across historical periods and disciplinary specialisms. He was therefore regarded with suspicion by scholars whose preferred *modus operandi* is to crawl along the frontiers of knowledge peering through a thick magnifying glass.

Postman's most intriguing book is *The Disappearance of Childhood*.[3] In it, he argues that the concept of 'childhood' – as a special, protected phase in a person's life – is an artefact of communications technology. It was, he claims, a by-product of the evolution of a print-based culture.

Before print, Postman maintained, adulthood began the moment a young person was deemed to be competent in the prevailing communications mode of the society. In the oral culture that pre-dated Gutenberg, a child therefore became, effectively, an adult at around the age of seven. This, he maintains, is why you never see children *per se* in the paintings of Breughel – you merely see small adults; and it is why the Catholic Church defined seven as the 'age of reason', after which an individual could be held accountable for his sins.

But the invention of printing changed all that. Why? Because in a print-based culture, it takes longer (and requires more education, some of it formal) to attain the kind of communicative

3. Neil Postman, *The Disappearance of Childhood*, Vintage, 1994

competence needed to function as an adult. So the concept of childhood was extended to the age of 14 or thereabouts (which, of course, was the original leaving-age for most children in state schools in the UK). And this remained the case from the 18th century to the middle of the 20th.

The title of Postman's book – *The Disappearance of Childhood* – comes from his contention that the arrival of broadcast television represented the first revolutionary transformation of our communications environment since Gutenberg. Just as print had transformed society – undermining the authority of the Catholic Church and stimulating the Reformation, enabling the rise of modern science and the growth of a new intellectual class – Postman argued that the dominance of TV had a correspondingly dramatic impact.

In particular, it had effectively lowered the age of reason. In a society dominated by the idiocies of such a medium, it didn't take long for a child to master the basics. Postman cited research that allegedly showed that American children were 'competent' TV viewers, in the sense that they understood genres and could follow narrative threads by the age of three. This explained, he said, why although there were remedial classes in reading in every American public school, he had never seen a remedial class in TV viewing! (It also explained, he contended, why adults were increasingly dressing like children, and vice versa.)

I'm not sure what to make of Postman's view about education, but his general point – that changes in the communications environment bring about cultural change – is, I think, accurate and profound. He made a convincing case for it in another book, *Amusing Ourselves to Death*. It is in part a devastating analysis of the impact that broadcast television had, and continues to have, on American politics.[4]

So we live in a polity that has been shaped by a single communications medium. Most of us have grown up in such an environment. It seems as natural to us as the air we breathe. And yet it is changing under our noses.

In seeking a language in which to talk about change, I've borrowed another idea from Postman: the notion of *media ecology* – that is to say, the study of media as environments. As with 'medium', the term is borrowed from the sciences, where an ecosystem is defined as a dynamic system in which living organisms interact with one another and with their environment.[5]

These interactions can be very complex and take many forms. Organisms prey on one another; compete for food and other nutrients; have parasitic or symbiotic relationships; wax and wane; prosper and decline. And an ecosystem is never static. The system may be in equilibrium at any given moment, but the balance is precarious. The slightest perturbation may disturb it, resulting in a new set of interactions and movement to another – temporary – point of equilibrium.

4. And not just on US politics, either. Think of the way General Elections have been fought in this country since 1960, or of the way our governance has been affected by the demands of 24-hour 'rolling' television news. The rise of news management and political spin has been the predictable – and understandable – response on the part of politicians. The result is a political climate in which public discussion is often conducted in banal sound-bites; in which there are always only 'two sides' of a question, no matter how complex; in which politicians seek to engineer 'photo opportunities', to dictate the terms on which they will submit themselves to questioning by television presenters and to distance themselves as much as possible from any uncontrolled encounters with actual electors. And so on.

5. Clapham: *Natural Ecosystems*, Macmillan, 1973

This seems to me a more insightful way of viewing our communications environment than the conventional 'market' metaphor more commonly used in public discussion, because it comes closer to capturing the complexity of what actually goes on in real life.

Just to illustrate the point, consider what has happened when new technologies have appeared in the past. When television arrived, it was widely predicted that it would wipe out radio, and perhaps also movies and newspapers. Yet nothing like that happened. When the CD-ROM appeared on the scene, people predicted the demise of the printed book. When the web arrived, people predicted that it would wipe out newsprint. And so on.

These 'wipe-out' scenarios are a product of a mindset that sees the world mainly in terms of markets and market share. Yet the reality is that while new communications technologies may not wipe out earlier ones, they certainly change the ecosystem. The CD-ROM did not eliminate the printed book, for example, but it altered forever the prospects for printed works of reference. Novels and other books continued to thrive.

A vivid illustration of ecological adaptation comes from the interaction between television and newspapers in the UK. There came a point, sometime in the late 1950s, when more people in Britain got their news from broadcast media – especially television – than from newspapers. This created a crisis for the print media. How should they respond to the threat?

Basically, they reacted in two different ways. The popular papers – the ones with mass circulations and readers lower down the social scale – essentially became parasitic feeders on television and the cult of celebrity that it spawned. The broadsheets, for their part, decided that if they could no longer be the first with the news, then they would instead become providers of comment, analysis and, later, of features. In other words, television news did not wipe out British newspapers, but it forced them to adapt and move to a different place in the ecosystem.

The 'organisms' in our media ecosystem include broadcast and narrowcast television, movies, radio, print and the internet (which itself encompasses the web, email and peer-to-peer networking of various kinds). For most of our lives, the dominant organism in this system – the one that grabbed most of the resources, revenue and attention – was broadcast TV.[6]

This ecosystem is the media environment in which most of us grew up. But it's in the process of radical change.

Life after broadcasting

Broadcast TV is in serious – and apparently inexorable – decline. It's haemorrhaging viewers, or at least the viewers who are the most commercially lucrative. And its audience is fragmenting. In particular, it's been eaten from within: the worm in the bud in this case is narrowcast digital television, in which specialist content is aimed at specialised, subscription-based audiences and distributed via digital channels.

The problem is that the business model that supports broadcast is based on its ability to attract and hold mass audiences. Once audiences become fragmented, the commercial logic changes. And, to compound the difficulty, new technologies have emerged such as

6. Note that 'broadcast' implies few-to-many: a relatively small number of broadcasters, transmitting content to large audiences of essentially passive viewers and listeners.

Personal Video Recorders (PVRs), which record onto hard drives rather than tape and are much easier to programme. They're enabling viewers to determine their own viewing schedules and – more significantly – to avoid advertisements.

Note that when I say that broadcast TV is declining, I am *not* saying that it will disappear. That's what the computer scientist John Seely Brown calls 'endism'[7], and it's not the way ecologists think. Broadcast will continue to exist, for the simple and very good reason that some things are best covered using a few-to-many technology. Only a broadcast model can deal with something such as a World Cup final or news of a major terrorist attack – when the attention of the world is focused on a single event or a single place. But broadcast will lose its dominant position in the ecosystem, and that is the change that I think will have really profound consequences for us all.

What will replace it? Simple: the ubiquitous internet.

Note that I do not say the 'web'. The biggest mistake people in the media business make is to think that the net and the web are synonymous.

They're not. Of course the web is enormous[8], but it's just one kind of traffic that runs on the internet's tracks and signalling. And already the web is being eclipsed by other kinds of traffic. According to data gathered by the Cambridge firm Cachelogic, peer-to-peer (P2P) data exceeds web traffic by a factor of between two and ten, depending on the time of day.[9] And I've no doubt that in ten years' time, P2P traffic will be outrun by some other ingenious networking application, as yet undiscovered.

Already, the signs of the net's encroaching centrality are everywhere. We see it in, for example, the remarkable penetration of broadband access in developed countries; the rapid growth of e-commerce; the streaming of audio – and, increasingly, video across the net; the interest of Rupert Murdoch and other broadcasters in acquiring broadband and other internet companies; declining newspaper sales and the growth of online news; the expanding use of the web as a publication medium by public authorities; the spread of public Wi-Fi; and in the stupendous growth of internet telephony – spurred by the realisation that, sooner rather than later, all voice telephony will be done over the net.[10]

The point of all this is that while my (baby boomer) generation grew up and came to maturity in a media ecosystem dominated by broadcast TV, our children and grandchildren will live in an environment dominated by the net. Which begs an interesting question: what will that mean for us, and for them?

7. John Seely Brown and Andrew Duguid, *The Social Life of Information*, Harvard Business School Press, 2000

8. At a recent conference at the Open University, the Head of Research at Yahoo estimated the size of the public web as 40 billion pages. The 'deep' web – the part that lies beyond the reach of search engines has been estimated to be 400 - 550 times larger than the public web.

9. www.cachelogic.com/home/pages/studies/2004_03.php

10. "It is now no longer a question of whether VoIP will wipe out traditional telephony, but a question of how quickly it will do so. People in the industry are already talking about the day, perhaps only five years away, when telephony will be a free service offered as part of a bundle of services as an incentive to buy other things such as broadband access or pay-TV services. VoIP, in short, is completely reshaping the telecoms landscape." *Economist*, 15 September, 2005.

A net-centric world

In thinking about the future, the two most useful words are 'push' and 'pull' because they capture the essence of where we've been and where we seem to be headed.

Broadcast TV is a 'push' medium: a relatively select band of producers (broadcasters) decide what content is to be created, create it and then *push* it down analogue or digital channels at audiences that are assumed to consist of essentially passive recipients.

The couch potato was, *par excellence*, a creature of this world. He did, of course, have *some* freedom of action. He could choose to switch off the TV; but if he decided to leave it on, then essentially his freedom of action was confined to choosing from a menu of options decided for him by others, and to 'consuming' their content at times decided by them. He was, in other words, a human surrogate for one of BF Skinner's pigeons[11] – free to peck at whatever coloured lever took his fancy, but not free at all in comparison with his fellow pigeon perched outside on the roof.

The other essential feature of the world of push media was its fundamental asymmetry. All the creative energy was assumed to be located at one end (the producer/broadcaster). The viewer or listener was assumed to be incapable of, or uninterested in, creating content; and even if it turned out that s/he was capable of creative activity, there was no way in which anything s/he produced could have been published.

Looking back, the most astonishing thing about the broadcast-dominated world was how successful it was for so long in keeping billions of people in thrall. Networks could pull in audiences in the tens of millions for successful and popular broadcasts – and pitch their advertising rates accordingly. Small wonder that one owner of a UK ITV franchise[12] famously described commercial television (in public) as "a licence to print money".

But in fact the dominance of the push model was an artefact of the state of technology. Analogue transmission systems severely limited the number of channels that could be broadcast through the ether, so consumer choice was restricted by the laws of analogue electronics. The advent of (analogue) cable and satellite transmission and, later, digital technology changed all that and began to hollow-out the broadcast model from within.

The web is the opposite of this: it's a *pull* medium. Nothing comes to you unless *you* choose it and click on it to pull it down onto your computer. You're in charge. In the words of Elizabeth Murdoch, the web is a 'sit up' medium, in contrast to TV which is a 'sit back' medium.

So the first implication of the switch from push to pull is a growth in consumer sovereignty. We saw this early on in e-commerce, because it became easy to compare online prices and locate the most competitive suppliers from the comfort of your own armchair.[13] The US automobile industry has discovered, for example, that a majority of prospective customers turn up at dealerships armed not only with information about particular models, but also with detailed data on the prices that

11. B.F. Skinner, *'Superstition' in the pigeon, Journal of Experimental Psychology*, vol.38, 1974

12. Roy (later Lord) Thompson, the Canadian media magnate who also owned *The Times* and *The Sunday Times*.

13. Erik Brynjolfsson and Michael Smith, *Frictionless Commerce? A Comparison of Internet and Conventional Retailers*, Management Science, vol.6, no.4, April 2000, 563-585

dealers elsewhere in the country are charging for the exact same cars.[14]

But the internet doesn't just enable people to become more fickle and choosy consumers. It also makes them much better informed – or at least provides them with formidable resources with which to become more knowledgeable. In an interesting study,[15] John Battelle describes the dramatic effects that powerful search engines such as Google are having on the advertising and marketing industries.

The net is also making it much harder for companies to keep secrets. If one of your products has flaws, or if a service you provide is sub-standard, then the chances are that the news will appear somewhere on a blog or a posting to a newsgroup or email list. And when it does, conventional PR news management techniques are ineffective.[16]

The emergence of a truly sovereign, informed consumer is thus one of the implications of an internet-centric world. This is significant, of course, but it was predictable, given the nature of the technology. And in the end it may turn out to be the least interesting part of the story.

The couch potato bites back

My conjecture is that the most significant consequence of an internet-centric world lies not in the arena of consumption, but in production. In blunt terms, the asymmetry of the old, push-media-dominated ecosystem looks like being replaced by something much more balanced.

The implicit assumption of the broadcast model was that audiences are passive and uncreative. In recent years, what we're discovering is that that passivity and apparent lack of creativity may have been more due to the absence of tools and publication opportunities than to intrinsic defects in human nature.

Take blogging – the practice of keeping an online diary. At the time of writing[17] Technorati, a blog-tracking service, was claiming to be monitoring 47.6 million.[18] Many of them are, no doubt, vanity publishing with no little literary or intellectual merit. But hundreds of thousands of blogs are updated every day or so, and many of them contain writing and thinking of a very high order. In my own areas of professional interest, for example, blogs are often my most trusted online sources, because I know many of the people who write them, and some of them are leading experts in their fields.

What is significant about the blogging phenomenon is its demonstration that the traffic in ideas and cultural products isn't a one-way street, as it was in the old push-media ecosystem. People have always been thoughtful, articulate and well-informed, but up to now relatively few of them ever made it past the gatekeepers who controlled access to publication media. Blogging software and the internet gave them the platform they needed – and they have grasped the opportunity in very large numbers.

14. *Crowned at last*, Economist, 31 March, 2005

15. Battelle, John, *The Search*, Nicholas Brealey Publishing, 2006

16. Companies that have discovered this include Kryptonite, manufacturers of expensive bicycle locks, and the Sony BMG corporation, which used DRM software on CDs which covertly installed a rootkit on the customer's PC, thereby potentially exposing it to malware attacks.

17. July 10, 2006

18. www.technorati.com

The result is a dramatic reversal in the decline of what Jürgen Habermas calls 'the public sphere'[19] – an arena that facilitates the public use of reason in rational-critical debate and that had been steadily narrowing as the power and reach of mass media increased. In recent years, the political implications of this re-energised public sphere have begun to emerge, notably in the debates among Democrats in the US about how to challenge Republican political ascendancy and the Bush presidency.[20]

The explosive growth in blogging has prompted a predictable outburst of 'endism' – as in questions about whether the phenomenon marks the end of journalism. Yet, when one looks at it from an ecological perspective, what one sees is the evolution of an interesting parasitic/symbiotic relationship between blogging and conventional journalism. Several case studies – for example the Harvard study of the Trent Lott case[21], and the *60 Minutes* saga[22] (which led to the premature retirement of TV news anchorman Dan Rather) – have delineated the contours of this relationship. What has happened, in essence, is that a new organism has arrived in the media ecosystem and existing organisms are having to accommodate themselves to the newcomer. And vice versa.

The other remarkable explosion of creativity comes from digital photography. In the last few years sales of digital cameras have grown phenomenally. Many mobile phones also now come with an onboard camera. So every day, millions of digital photographs are taken. Until the advent of services like PhotoBucket and Flickr.com, an understandable response to this statement would have been "so what?" But these services allow people to upload their pictures and display them on the web, each neatly resized and allocated its own unique URL.

I don't know how many photographs Flickr holds, but it must already run into many millions. The most interesting aspect of it is that users are encouraged to attach tags to their pictures, and these tags can be used as the basis for searches of the entire database.

When writing this I searched for images tagged with the word 'Ireland'. The database returned 226,688 photographs. (A few months earlier, the same search had yielded only 85,000 images.) Of course, I didn't sift through them all, but I did look at a few hundred. They were mostly holiday snaps, but here and there were some memorable pictures. What struck me most, though, was what they represented. Ten years ago, those holiday snaps would have wound up in a shoebox and would certainly never have been seen in a public forum. But now they can be – and are being – published, shared with others, made available to the world. And this is something new.[23]

In fact, strictly speaking, it's something that ought not to be possible, at least in terms of the old media ecosystem. The blogging avalanche and Flickr's visual cornucopia are just two examples of *user-generated content*, which to an old-style broadcaster would be an oxymoron, a contradiction in terms. In the previous paradigm, broadcasters *created* the content and users (audiences) merely *consumed* it.

19. Jurgen Habermas, *Structural Transformation of the Public Sphere*, MIT Press, 1989

20. Frank Rich, *Ideas for Democrats?*, New York Review of Books, October 19, 2006

21. *'Big Media' Meets the 'Bloggers': Coverage of Trent Lott's Remarks at Strom Thurmond's Birthday Party*, Kennedy School of Government, Harvard University, Case Program, 2004

22. Dave Eberhart, *How the Blogs Torpedoed Dan Rather*, NewsMax.com, 31 January, 2005. Available online at: www.newsmax.com/archives/articles/2005/1/28/172943.shtml

23. It is also beginning to have a dramatic impact on the market for stock photographs – as the rise of www.istockphoto.com shows.

In the emerging system, broadcasters and conventional media outlets will doubtless continue to create content, but so will a great many others. If present trends continue, there will come a point where more content is being produced annually by users than by the entire output of what the UK Treasury rather quaintly calls the 'creative industries'. And when that crossover point is reached, we will have moved into entirely uncharted territory.

The wealth of networks

The explosion of user-generated content has been made possible by a conjunction of several technologies: the personal computer; inexpensive but powerful software tools; and the open internet. Jonathan Zittrain, a prominent cyber-legal scholar, describes the effects of this combination as 'generativity'. "This denotes", he says:

"a technology's overall capacity to produce unprompted change driven by large, varied, and unco-ordinated audiences. The grid of PCs connected by the internet has developed in such a way that it is consummately generative. From the beginning, the PC has been designed to run almost any program created by the manufacturer, the user, or a remote third party and to make the creation of such programs a relatively easy task. When these highly adaptable machines are connected to a network with little centralised control, the result is a grid that is nearly completely open to the creation and rapid distribution of the innovations of technology-savvy users to a mass audience that can enjoy those innovations without having to know how they work."[24]

The most persuasive narrative to have emerged to date about the significance of generativity is Yochai Benkler's *The Wealth of Networks*.[25] In it, he charts the remorseless industrialisation of the information economy from the early 1800s to the 1960s. In that century and a half, communications technologies tended to concentrate and commercialise the production and exchange of information. He writes:

"High-volume mechanical presses and the telegraph combined with new business practices to change newspapers from small-circulation local efforts into mass media. Newspapers became means of communications intended to reach ever-larger and more dispersed audiences, and their management required substantial capital investment. As the size of the audience and its geographic and social dispersion increased, public discourse developed an increasingly one-way model. Information and opinion that was widely known and formed the shared basis for political conversation and broad social relations flowed from ever more capital-intensive commercial and professional producers to passive, undifferentiated consumers."

This model was readily adopted and amplified by radio, television and – later – cable and satellite communications. But the economics of long-distance mass distribution systems that were needed to reach expanding and geographically dispersed populations were typified by very high up-front costs and low marginal costs of distribution. "These cost characteristics", writes Benkler, "drove cultural production toward delivery to ever-wider audiences of increasingly high production-value goods, whose fixed costs could be

24. Jonathan Zittrain, *The Generative Internet*, Harvard Law Review, vol.119, no.7, May 2006
25. Benkler, op. cit.

spread over ever-larger audiences like television series, recorded music, and movies. Because of these economic characteristics, the mass-media model of information and cultural production and transmission became the dominant form of public communication in the twentieth century."[26]

This was the world in which – as the old joke put it – freedom of the press was available to anyone who was rich enough to afford one.

The combination of technologies that have produced Zittrain's generativity has, Benkler argues, changed all that. The core functionalities needed to create, store and disseminate information, knowledge and culture are now widely available and cheap – at least by Western standards.

"Any person who has information can connect with any other person who wants it, and anyone who wants to make it mean something in some context, can do so. The high capital costs that were a prerequisite to gathering, working, and communicating information, knowledge, and culture, have now been widely distributed in the society. The entry barrier they posed no longer offers a condensation point for the large organisations that once dominated the information environment. Instead, emerging models of information and cultural production, radically decentralised and based on emergent patterns of cooperation and sharing, but also of simple coordinate coexistence, are beginning to take on an ever-larger role in how we produce meaning-information, knowledge, and culture in the networked information economy."[27]

What has happened, in other words, is that ownership of the means of cultural production has passed from those who could afford their high capital costs in the old ecosystem to just about anyone who has a computer, some appropriate software and an internet connection. One doesn't have to be a devout Marxist to realise that such a radical shift in the means of production will, in due course, impact on what Marx called the 'superstructure' – the culture that sits atop the fundamental economic realities of production.

The emerging ecosystem

We can now begin to see the outlines of the media ecosystem that is emerging under the pressure of the developments I have discussed. My guess is that it will be significantly different from the ecosystem that was dominated by broadcasting technology, and in which all of our regulatory apparatuses and many of our business models were designed.

It will be richer, more diverse and immeasurably more complex because of the number of content producers, the density of the interactions between them and their products, the speed with which actors in this space can communicate with one another, and the pace of development made possible by ubiquitous networking.

The problem – or 'challenge', to use the politically-correct term – is whether old regulatory systems and business models can be adapted to work in the new environment. As far as business is concerned, the answer is simple: companies that don't adapt will vanish from the earth. The big question is whether the same holds for regulators.

26. Benkler, op. cit., page 29
27. Benkler, op. cit., page 32

Private is the new public

Jonathan L. Zittrain

Professor of Internet Governance and Regulation, Oxford University

This essay is drawn from the author's inaugural lecture at Oxford University on 25 April 2006. It is published here under a Creative Commons Attribution NonCommercial-ShareAlike 2.5 Licence.

Introduction

'Internet governance and regulation' calls to mind committees and voting, reams of regulation about bandwidth or content or identity, and treaties among states. Those who consider themselves scholars or practitioners of internet governance may well identify with these phenomena. Here, I wish to take a different tack because I do not see the future of the governance of the internet – or the people who use it – there.

To explain what I mean, I want to start from a topic that nearly everyone agrees carries salience in cyberspace – privacy – and then show how the meaning of privacy is itself evolving, and how the means by which we should promote and protect it call for tools that transcend the emerging toolkit relied on for internet governance issues.

Privacy-as-defence

The prevailing view of privacy in cyberspace is that of defence: safeguarding one's information against those who ought not to see it. 'Privacy policies' are found on many websites, consisting of a little-read boilerplate about what the website gathers from a user and what it does with the information. (Typically the answers are 'as much as it can' and 'whatever it wants'.) Security of personal information follows closely on the 'defence' theme, and there have been some valuable policy innovations in this sphere.

For example, a 2003 California law provides that if you expose your customers' private data to others, you must alert customers to the security breach. This has led to a rash of letters drafted by marketing professionals trying to figure out how to send bulk mail to millions of people telling them that a package containing tapes with their credit card and social security numbers has been lost. (It turns out that it helps if you include money-saving coupons with the letter.)

The cutting edge of privacy-as-defence can be found by looking at a hypothetical website, allmuzak.com (figure 1), where two different visitors can be charged two different prices for the same CD.

It is something that might be done as that site records the number of nanoseconds that elapse between the time you view a page about a CD you might want, and the time your trigger finger finally hits 'Buy it Now'. There are those who do not even seem to have time to register the price before they buy, and those who linger and browse, and come back, and finally buy. Perhaps the former should be charged more than the latter because they do not even appear to be paying attention. Is this a good idea or a bad one? As somebody who tends to buy before he thinks, I think it is a bad one, but more rigorously, it causes us to ask how we should conceive of even the simplest of transactions when the underlying technology allows for such nuanced distinctions to be made.

In figure 2, both Geoff and Robert have visited the site and are charged the same sticker price. But with their loyal customer card, they are given an individualised discount resulting in the original price that they were respectively asked to pay in the first example. Is that less offensive than engaging in blatant price discrimination? It is functionally equivalent of course, but it appears to be less offensive.

Privacy-as-defence is a generally well-trod area of inquiry. Here I want to draw from it the lesson that both the problems and opportunities of cyberspace eventually escape the digital realm. The instances of differential treatment on the basis of data mining that we imagine in allmuzak.com will find their way to the real world.

As you use a loyal customer or other kind of card, it may well be that the price of a loaf of bread becomes indeterminate: there is a sticker price that no one pays, and when you take the bread up front, you will find out what your special discount gives you as a result of your relationship with the store. Merchants need not only vary price; they can and will also vary service. Customer cards, soon augmented by radio frequency tags, can serve to identify those who arrive at a home improvement store, monopolise the attention of the attendants and exit without having bought a single nail. With these kinds of cards, the store can discern the good customers from the bad, and appropriately alert the staff to flee from bad customers and approach good ones.

Privacy as strategy

I see privacy moving from defence to strategy, expanding the notion of privacy beyond the mere citizen's plea to have certain secrets or private facts remain under control. The core of privacy transcends this conventional conception.

Part of a more expansive view can be found in the wildly, disproportionately successful iPod. 32 million iPods were sold last year, one every second. It is a $3 billion market. The market for iPod accessories alone is extraordinary: $1 billion. One billion dollars for accessories like a small electronic dog which dances as your iPod plays. A new UK warship, the HMS Daring, was just constructed with built-in iPod docks. (Since the ship cost $1 billion, one might say that it represents a doubling of the market for iPod accessories in a single year.)

The success of the iPod reminds us that we can come to have a certain identity with an object that transcends its purely functional value. If your iPod is stolen, it is somehow not the same to simply get a new one; there was something about *that* iPod. Many people inaccurately believe that their iPods get to know their tastes and in a random shuffle play, start to play the music they like better more often. This is a form of putting one's identity into the artefact, vesting it outwards from one's physical self. This represents an expansion of the private sphere; technology can help people find new ways to express themselves.

Consider Apple's iTunes music store that now includes a section for podcasts. Podcasting, a technology developed several years ago, enables creators to produce a 'channel' of music and video content to which people can subscribe and automatically receive updates. The iTunes music store has since embraced podcasting. Type in 'Harry Potter' there, and you

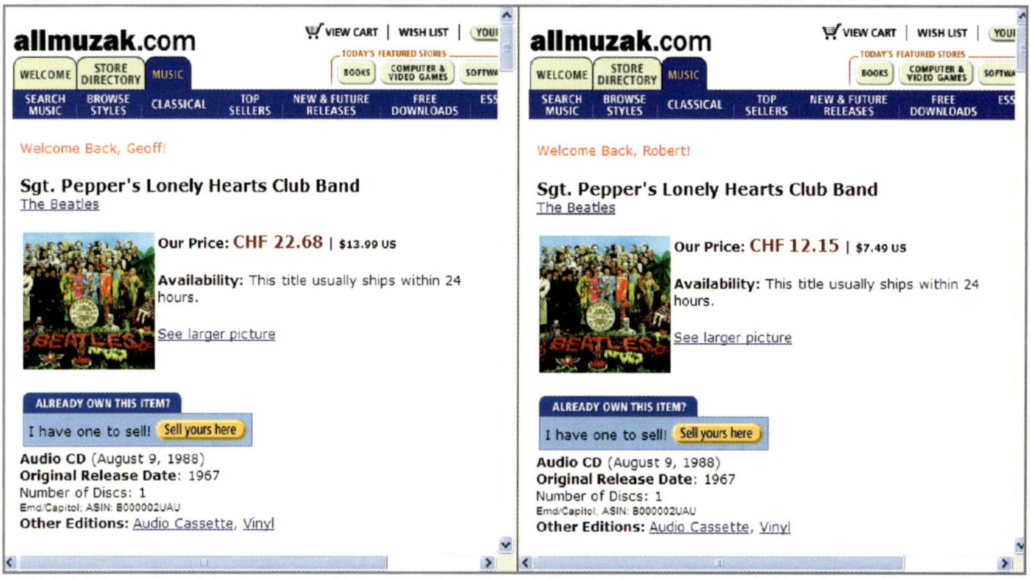

Figure 1

Figure 2

will see multiple featured podcasts, none of which is approved by J. K. Rowling or her agents. Instead, they were made by enthusiasts with a computer and microphone, creating their own Harry Potter shows for others to download and enjoy.

Consider virtual worlds, where more and more people are spending their time. 100 million people worldwide log in each month to play interactive computer games – quite an amazing undertaking, given that my suspicion is that nearly everybody in this room would deny having anything to do with a virtual world. Many invest their identities in those worlds, such that if the world should be shut down or their avatar abruptly eliminated, they would not just feel as if they were thrown out of a movie early; they would feel as if a piece of their very identity had been lost.

Investment of identity can become collective. A site called sorryeverybody.com was put up after the most recent American presidential election. One American decided to say "sorry, we tried" to the world, and invited others to join him in expressing an apology, each with a single photograph. Today there are over 32,000 people who sent in photographs, each with a different message. The website spawned a book spin-off, as well as apologiesaccepted.com, where the rest of the world found a chance to say "thanks anyway" to the 49 per cent, including a wry note from Singapore: "At least you could vote for *your* president". I would be remiss if I did not mention sorryjustisntgood enough.com, the website by someone not ready to accept the apology, and of course, wehavenothingtobesorryfor.com.

Nearly any example in the realm of fast-moving web zeitgeist is, of course, a trinket *du jour*. But these efforts bear further examination. What is it that could cause 32,000 people from around the world to send in photographs that have a political valence to them? The internet audience is so vast, so varied, that any idea with a modicum of appeal might be able to reach a critical mass of takers. When we measure the impact of an idea on a scale more refined than the raw number of page views its web incarnation receives, we see an astounding number of unexpected successes arising from this medium – successes that I say are anchored in a sense of our private sphere finding an opportunity to expand through the very medium.

Private as the new public

In fact, the way these spheres expand and more importantly, interact with one another, suggests that private can in some important respects become the new public. Let me explain, beginning with the National Aeronautics and Space Administration (NASA) click-workers' study described by Yale Law Professor Yochai Benkler in his most recent book, *The Wealth of Networks*.

NASA had a tedious job, the sort that puts Tom Sawyer's white picket fence to shame. It had pictures of the moon and of Mars with craters on them. These were standard bitmap images, and they wanted the craters to be vectorised. To put it even less interestingly, they wanted people to draw circles around circles, and asked for people on the internet at large to undertake this task. As a result, the Crater Identification and Mapping Project found, much to NASA's pleasant surprise, that within a week, the work of a single graduate student for a year could be replicated. (I am not sure whether this put the graduate students out of a job, but it at least put them out of their misery and may very well have provided an opportunity for more enticing work.)

The near-costless aggregation of far-flung work can be applied beyond craters. For example, consider the task of optical character recognition as each book of the Bodleian Library is scanned for a digital, full-text search. Mistakes are made. An automated scanner once came upon 'Tim Berners-Lee' (the name of the founder of the World Wide Web) and rendered it as the 'Timberner's League' which I imagine as a team of lumberjacks.

Today, surfers on the web, truly having run out of things to do, can play a video game where they score points when they highlight such inaccurate translations, and gain even more points when they correct them. Thus in real time, we can build a collective, organic computer of the world to work to improve optical character recognition. These are the kinds of grid applications that suggest a certain emerging public activity from individual private actions.

And we see, again, the movement from virtual to real: 'shot spotters' comprises a series of microphones arrayed in certain neighbourhoods such that if a shot is fired, the microphone picks up on that particular sound wave, triangulates among multiple microphones to it, and summons the police. Augment it with a neighbourhood watch, where we do not just have the police watching from closed-circuit televisions. Instead, the public at large can review CCTV camera feeds and press a button if they see something strange going on through their computer monitors, at which point two other ordinary citizens are asked for verification: "Do you agree that there is something strange going on here?" If the answer is yes, the police can be sent. That is quite a force to be reckoned with.

Today, the central proving ground for individual expression cohering into a single quasi-public whole is Wikipedia. The free online encyclopaedia has over 1 million English-language articles. The idea of an encyclopaedia that anyone can edit live is clearly lunatic. If somebody came up to you in 2001 and said "I am going to build an encyclopaedia to rival Britannica or surpass it, and this is how I am going to build it: by throwing open the doors to the world at large." you would have needed quite a degree of foresight to say anything other than "you are crazy." And yet here is Wikipedia.

What is most impressive to me is that the entries in Wikipedia that we think most controversial and sensitive are typically the ones that reflect the most depth, the most care. Consider the entry on Rachel Corrie, an American protester in the Gaza Strip. She stood in front of a bulldozer to stop it from knocking down a house. The bulldozer struck and killed her. As you might surmise, there is quite a lot of controversy around these circumstances, and the article reflects that debate.

With every article on Wikipedia, there is not only the article itself, but also a corresponding discussion page where people who make changes to the article can justify them to others as they go back and forth. The page's history and discussion page reveal, for example, the participants discussing whether in fact a picture of her burning an American flag in a protest before she went in front of the bulldozer is one that ought to headline the article, rather than a simple profile picture.

These are the kinds of discussions you could imagine taking place in the editing room of *Britannica*, among the handful of editors assigned every interval to write entries like this – though, to be sure, there is not yet an entry on Rachel Corrie in *Britannica*. We are thus able not simply to hear what a source oracularly says, but to see the debates, wrong turns, and dead ends that precede it.

On Wikipedia, these debates tend toward moderation rather than toward an endless edit-war where editors go back and forth on a fact with one person introducing it, another excising it, and a third introducing it again. This is because Wikipedia is not just a technical instrumentality where people edit. A public wiki standing alone rarely works. It is, rather, an entire culture and ethos of self-governance, with rules about debate and editing that transcend the substance of the encyclopaedic topics themselves. For example, on Wikipedia there is a three reversion rule: after the third reverted edit in a back-and-forth, the text is supposed to stay until a third party can participate or until the two warring parties can come to an accommodation.

There are nascent systems of mediation and arbitration on the site – a reinvention of a legal system unmoored from any particular sovereign, and with enforcement powers limited to lockdown of certain pages or exclusion of certain users, actions taken by other users more senior in the Wikipedia hierarchy. These innovations help explain something as inexplicable as the existence of Wikipedia. (I still wouldn't be entirely surprised to find that Wikipedia is a secret project by the Soros Foundation, with paid experts writing the pages.)

And once again, the virtual points the way towards the real. Take Pledgebank, started by the unassuming but brilliant Tom Steinberg, someone who wanted to take on our classic collective action problems. He has set up a very simple website where you can say "on Saturday, I will appear on the banks of the river Isis and pick up garbage as long as 20 people agree to do so as well". Pledgebank keeps track of people making such a commitment, and once the designated critical mass is reached, all parties are alerted.

This is how the Open Rights group was formed in the UK. It used Pledgebank to aggregate commitments to write a £5 monthly standing order for a year, so long as 1,000 other people were willing to make the same commitment. 1,038 people signed up by the 2005 deadline, and by the end of January 2006, 500 people put their money where their clicks were, having issued the standing orders. By July, 650 people had followed through.

The power of the new public

Technologies to overcome collective action are powerful. So powerful that it is worth dwelling for a moment on the possibility of this public going against the 'old' public: sovereign governments. You may be familiar with cartoon icons Jingjing and Chacha. These are the internet police of the Chinese city of Shenzhen. They have their own website and blog. They say things such as "you should all strictly limit your own behaviour on the web and together we will create a healthy internet environment and maintain harmonious internet order. See you often." A somewhat ambiguous salutation.

As we go digital, the prospects for surveillance of what we do *en masse*, not just by corporations looking to charge us a different, maybe lower, price for things, but by governments looking to gather intelligence, is a threat. We see emerging distributed ways of people combating it – for better or worse, depending on whether one has a greater fear of runaway mobs or runaway governments.

For example, a piece of software called Tor asks people to offer up their computer and bandwidth such that someone from far away can shift his or her data through it and then through another and another so that by the time the data gets to its destination, where it came from, or what it says from the point of view of any of the intermediaries, is unidentifiable from an outside source. The success of Tor is directly proportional to the number of people who decide to sign up and become nodes.

This is, I believe, a form of 'netizenship,' inviting people not to look at the internet as something that comes into your house that you use as a utility, but something for which your choice about how you use it can in turn affect the fabric of the network itself. How readily one can be anonymous can then in part reflect the votes counted, one network connection at a time, of those people who want to make a political commitment to that kind of anonymity.

However, even the people behind Tor do not embrace anonymity at all costs. The Tor charter states, in good faith, I think, the value of accountability. Tor's authors are working on technical methods whereby you can remain anonymous but still have a persistent identity, able to be punished should you do things that people do not like – punished in the realm of the network itself. They are working on the means of retaining control of released data, where if you publish a document, Tor may see to it that it is stored all over the network, but you as the publisher, still specify an expiration date by which it might delete itself once it is up on the network. In other words, these tools that take the private and make it into a new public do not inevitably mean that the new public has an utter lack of accountability, or is utter chaos.

More important, processes already being used might be appropriated to help balance newfound individual privilege. Think about Google and other search engines. They offer little new content of their own. Rather, they crawl the web produced by everyone else, index it, and then rank search results in major part by consulting the natural rankings that the public at large assigns as people actively link (or fail to link) from one site to another.

This taps into people's judgement, and I believe there are ways in which we can tap into people's judgements for public spirited and public interested purposes. In the case of evading filtering, if that is what we want to do, or at least identifying it, we can set it up so that people as netizens can ask that their computers alert each other or a trusted 'mother ship' if they cannot get to a certain website. If they cannot, that mothership can ask other nearby machines to try as well, and with a process of triangulation, a digital shot spotter can say "this block is thanks to your own misconfiguration, please just fix your computer" or "it is thanks to your parents putting on filtering software" or "it is your internet service provider (ISP) slowing things down perhaps for mercenary purposes", or "it is your government not wanting you to see certain sensitive, political content".

We can create a collective dashboard of gauges where the more of us who step forward to do it, the more accurate it becomes. It is a whole new way of understanding how to collect and distribute information in real time, one that takes what happens down to the nanosecond, a technology then not merely used by supermarkets to individually tune the price of a loaf of bread, but by netizens to understand and measure the network itself. Our emerging network is consummately unmeasured, in part because its architecture does not easily lend itself to central points of flow.

A lesson learned from global warming is the value of developing a science of measuring and understanding important phenomena, one that at least keeps pace with our abilities to affect those phenomena. Understanding where the internet is going in the future ought to be paralleled with measuring the growth and direction of the internet. Just as we can do it with detecting filtering world wide on a real-time map through the individual actions of people volunteering their connections, so too can we use it to judge the quality of code that people might be about to run, and where network bottlenecks exist and who might be causing them.

Dangers of the new public

I now wish to refine this so-far optimistic story. People who feel newly empowered to move their private selves into external spaces and then to collectively create something new, can do wrong as well as right.

I've written that the number of internet security incidents has been on the rise in large part because of the *generative* characteristics of the internet: anybody can write code anywhere in the world, and over the internet get it onto a computer by simply clicking "sure, yes, I would like to download and run this file". Before you know it, you are running a new piece of software that might as easily be a virus or spyware as something useful and empowering.

With security incidents and spam on the rise – I am waiting for the moment when more than 100 per cent of email is spam – we have seen the rise of collective responses, such as the Mail Abuse Prevention System. An internet pioneer named Paul Vixie decided he had had enough of spam in the mid 1990s.

He started keeping a list of those internet addresses that he believed were involved in originating spam. The first thing he did with the list was make sure the people on it could not email him. The second thing he did was to make his list available in real time so anyone who wanted to could free ride off of his effort to distinguish between spammers and non-spammers. Hotmail decided to do just that, so if Paul Vixie did not like you, you could not send mail to anybody at hotmail.com.

Paul was not the only social entrepreneur; others existed as well, with varying tools for deciding what was spam and who was sending it – and varying tolerance for appeals from those incorrectly flagged. At one point, Paul thought that another such system, ORBS – one that sent test emails sent through others' email dropboxes to figure out who maintained so-called 'open relays' – was a spammer. He blacklisted them and they blacklisted him back, spurring an all-out digital war.

Today we see lots of individual decisions being aggregated, or single decisions being aggregated and sent back out to the individual for implementation. For example, Microsoft's new anti-spyware program displays a welcome screen during installation inviting you to "meet your computer's new bodyguards". These bodyguards tell you what you can and cannot run on your PC if you want to be safe, as far as Microsoft is concerned. Giving such power to third parties may be seen as both necessary and worrisome – it takes a network endpoint famously configurable by its owner and transforms it into a network middlepoint subject to only nominal control by its owner.

When so deployed, these private programs serve important functions that might otherwise be undertaken by public authorities. Microsoft's bodyguard metaphor is apt, and most of us rely on the police for analogous protection. The responsibilities when the

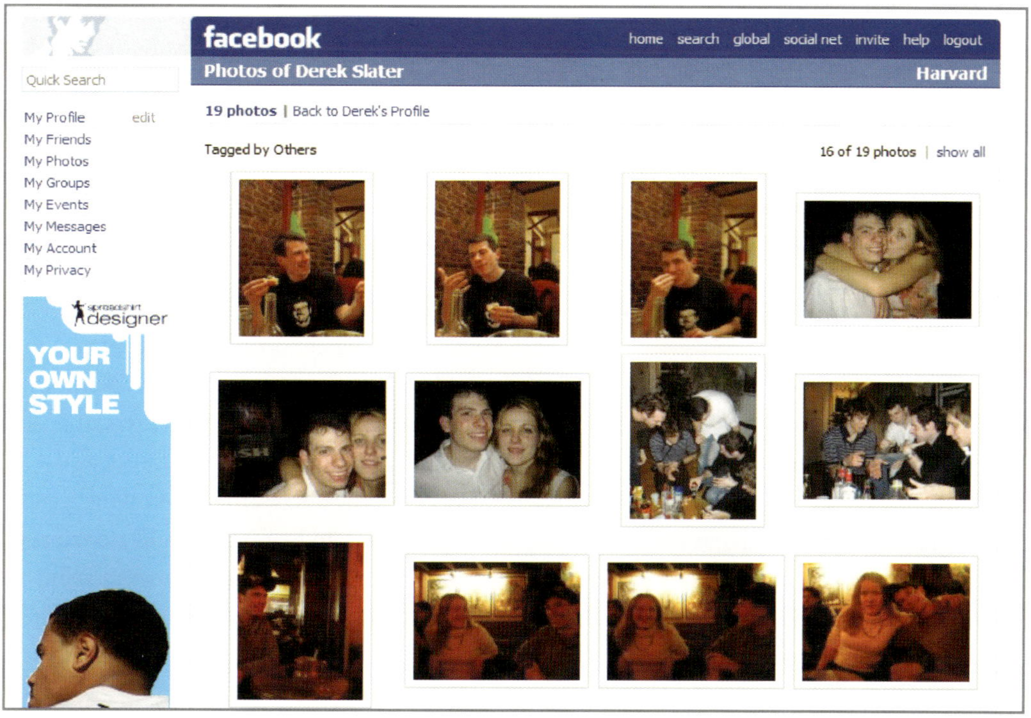

Figure 3

private becomes the public were addressed in the United States in the 1940s, when the town of Chickasaw, Alabama was owned lock, stock and barrel by the Gulf Shipbuilding Corporation.

A Jehovah's Witness was prosecuted for trespass for distributing literature on the town's streets because they were private property. In a regular town, the First Amendment would have protected her activities. The Supreme Court of the United States took up the situation in *Marsh v. Alabama*, and held that the private property was to be treated as public property, and her conviction was reversed.

I and others have speculated before that *Marsh* offers some wisdom in cyberspace, where certain chokepoints can arise from private parties. *Marsh* advises that sometimes government can defend the individual against a disproportionately powerful private party. This view can put public governments in a position of encouraging and defending the free flow of bits and bytes, rather than seeking to constrain them for particular regulatory purposes.

It would be a complex theoretical leap to apply the *Marsh* substitution of public for private for Paul Vixie's anti-spam service or Microsoft's bodyguards – asking each to give certain minimum due process to those they deem bad or malicious, and to be transparent about the judgements they make. It is even harder to apply to the collective power I am talking about today, where there is not a Paul Vixie or Microsoft channelling it, but rather, a collective peer-to-peer consciousness generating judgements and the data on which they are based. How do you tell a decentralised network that it needs to be mindful of due process?

Let me illustrate with a couple of examples. You may be familiar with facebook.com. In the United States, more than 85 per cent of university students are said to have an entry in it. Consider an entry for a former student of mine, Derek Slater (figure 3). The site includes a number of photos of Derek, and even photos that it turns out were taken by others and tagged as having Derek in them. One set of photos comes from another Facebook user's entry that says "Paris is better with Derek", and when you look up Derek, you see this user's photos as well because the tagging automatically linked them up.

Armies of people are out there right now taking photos and putting them onto Facebook and Flickr, another photo-sharing site. Soon there will be cameras that automatically send your photos to Flickr as you take them, stamping them with the time and date, and in the meantime mobile phones can do so. There will be a Global Positioning Satellite (GPS) system in the camera, too, so the system will know exactly where the photo was taken, something then made searchable to others. Plus, thanks to face recognition technologies developed by such companies as Riya and myheritage.com, once one of those photos is tagged as being Derek, others can be so tagged automatically. This means that if I were to take a picture of this room in a little while and put it up on Flickr, many of your names would be automatically filled in because somewhere, your name already appears tagged next to a photo of you.

This has troubling implications. Recall the Christian Gallery News Service, something that was started in the mid 1990s, which, among other things, took pictures of women in the United States seeking to get abortions at clinics and made sure to get them coming out of the car with the licence plate visible (figure 4).

With Riya and other image recognition technology, this means that suddenly your identity is searchable online the moment you step out of that car or make a protest that someone captures on film or video. As a result, these peer-to-peer technologies oddly make for a much more identifiable world rather than a more chaotic one.

These are natural inputs to a service, if one can call it that, known as Gawker Stalker at gawker.com, where people are encouraged to send in sightings of celebrities. As celebrities are spotted, Gawker encourages users to post the sightings within 15 minutes, so that if Jack Nicholson is at Starbucks drinking a cup of coffee, you can be tuned in and alerted and be there before he is done in order to stand awkwardly near him before he has a chance to leave.

Combine this with the kind of reputation systems developed to recommend new music at Amazon or worthy transaction partners at eBay, and I see a chance to enter a café in Paris with a Star Trek-style tricorder, a PDA, and say "first of all, is there anyone on my buddy list in this café, or within 100 metres? Second, are any of *their* ten closest friends within 100 metres? If so, and if they have also said they would like to meet, we should meet. Are there people who are graduates from Oxford?"

Oxford graduates, as I gather from attending Congregation, are a rather large group. How can we narrow it down? Enter reputation systems. On eBay, your rating is represented by a succession of symbols depending on others' assessment of you as a seller or a buyer. On Cyworld.com, where 18,000,000 people have accounts, largely thanks to South Korea, you have a virtual social network that you decorate and in which you display your identity more thoroughly than an American would on MySpace. You rent your room's virtual accoutrements, and should you stop paying,

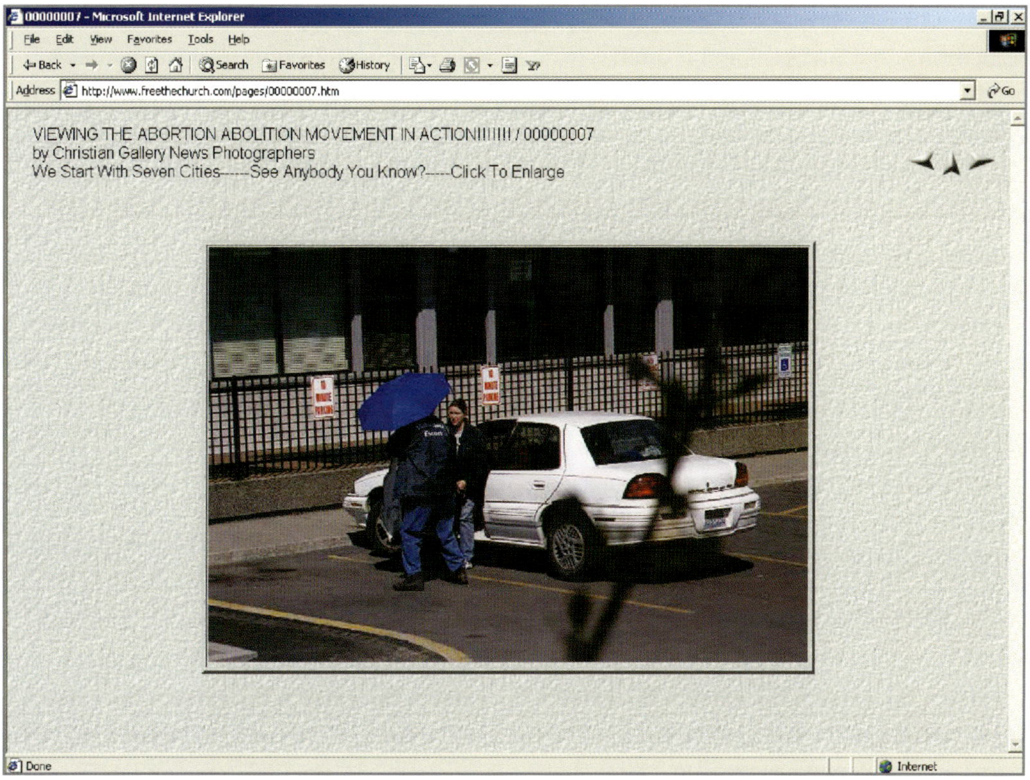

Figure 4

you lose all of your possessions. Cyworld includes rating systems that make it so that every day you can check in and through the actions of others anonymously rating you, obtain a rating of your sexiness, your fame, your friendliness, your karma and your kindness. As you interact with people, you try to maximise the kinds of behaviours that would have them in turn think well of you for these purposes.

This has led to someone writing the *I Was a Cyholic, a Cyworld Addict* article for OhmyNews, the online newspaper in South Korea with citizen-written articles. (Such writers include their national identity number for editors, so articles are not penned anonymously to OhmyNews.) If you ask me, moving from Cyworld to OhmyNews is going out of the frying pan and into the fire for an online addict. In any case, the *Cyholic* article describes someone who sees that her existence can be defined and measured precisely, if not accurately, by the reactions of others.

That trend is increasing as social networking takes off, and it too will cross over into the real world. Some who enter cafés with the 'tricorder' system will be deemed undesirable through a reputation system, perhaps on the basis of ratings generated from previous encounters with people the system believes to be similar to the people in the café. In this case, a chill will pass through the crowd, and people will avert their gaze. Enter someone projected to be likeable, and a summer breeze comes through – people will be delighted to find such a person sitting next to them.

More and more, we are going to be indicating our identity in daily transactions and interactions – there are even fingerprint readers on some laptops – and it can be aggregated together for collective judgments to be made. These judgments are not always right; sometimes you get accidental juxtapositions. Sometimes, even when you try to make good associations, it goes wrong, such as when Amazon makes recommendations based on other customers' purchases. Then we get a page for the *Official Lego Creator Activity Book* where we are told you can buy it with a perfect partner: *American Jihad: The Terrorist Living Among Us Today* (figure 5).You then rightly wonder, "What does Amazon know about me that I don't?"

As Google and other western firms enter the Chinese market, imagine redirecting some of collective judgements that Google harnesses to make sure that you are targeted with just the right advertisement when you run a certain search. What if that technology were to generate the maxim: "Subversives who liked reading this tract also liked reading this tract; this guy just read two of them, he is probably a subversive." It makes one wonder which is worse: emergent judgement mechanisms that are very good at what they do, or mechanisms that are easily gamed to be inaccurate?

Governing the new public

How do we deal with this new power of the masses to determine our own judgements about each other when they are not channelled through a public authority or through something at least as sueable as the *Daily Mirror*? I do not here want to propose laws for implementation by sovereigns, alone or by treaty. Rather, I look to three supranational institutions, each of which has grappled with the line between public and private.

Institution number one: the internet Engineering Task Force (IETF). The group that was founded to build the internet's protocols and to maintain its principles was porous – anyone could join the IETF. Members were simply asked to keep the network simple so that anybody could set up shop on it and be able to add applications to it. When they argue about how to make the network work better, they agree to argue on merit, not democracy and voting. Whoever has the best argument should win. How do they know? "Rough consensus and running code."

The IETF also made two key assumptions that have been sorely tested in the past few years: that people are reasonable and that people are nice. Those kinds of assumptions are built into the very fabric of the internet such that when your Ethernet card sends a signal, it might collide with the signal of your neighbour's Ethernet card. It is understood that both cards will wait a decent and random interval before sending their signal again, in order to make sure that there is not a collision the next time. You could get a card to say, "I'm not waiting, I'm just going to keep hammering home my signal", which would lead to better throughput for you, but bad community ethics at the Ethernet level. I have yet to encounter a card that violates this norm.

Other norms are a little harder to maintain, such as that of not asserting that you are Miriam Abacha and have millions of dollars to share from your Nigerian ex-husband when you send an email that appears as if it were coming from that very person. (I pity the real Miriam Abacha, should she ever want to obtain and use a hotmail account.) When it's so easy for a few bad apples to flout community norms, how can a system dependent on such norms function?

At the protocol level – such as that of the Ethernet card – there were many doubters.

Figure 5

That is why the IETF's mascot is a bee, because it is said, probably apocryphally, that scientists never figured out how bees fly; their wingspan to fur ratio is too small. (I am pleased to tell you that in January 2006 scientists finally figured out how bees fly. I will spare you the details, but it turns out that they flap their wings very quickly.) As a network designed for only 'nice people', the internet has held up remarkably well under the strain of a bunch of bad ones.

Institution number two: Wikipedia. Again, Wikipedia is not utter chaos. There are administrators – drawn from the ranks of users by other administrators – and they seek to impose order as a last resort, after simple persuasion fails. When people are found to be vandalising an entry, a Wikipedian sends a fairly nice note to the person saying: "unconstructive edits are considered vandalism; please stop and consider improving rather than damaging the hard work of others". Is this naivety, or does it stand a chance to blandish people out of their cynicism and into a new way of thinking? Wikipedia so far functions surprisingly well – the bee flies – though it has not yet experienced the likely onslaught of public relations executives looking to make the entries for their respective companies (or political candidates) look great.

Another set of institutions, the Internet Corporation for Assigned Names and Numbers (ICANN), the International Telecommunications Union (ITU), the World Summit of the Information Society (WSIS), represent what I believe are the wrong ways to think about internet governance. The best features of these entities are that they keep the busybodies in a room with themselves, not bothering the rest of us. Why is that? Because, when I think about an internet governance research agenda, a hypothesis about what makes the internet work comes to mind that goes in the opposite direction from the ICANNs, the ITUs, the WSISs and the Working Group on Internet Governance (WGIG)s. These organisations presume a set of 'stakeholders' who will gather under a tent – sometimes real, sometimes virtual – and talk or trade horses until formal agreement is reached.

Such discussions can be helpful, but these organisations succeed the most where the parameters of what they are to do are fixed – for example, allocating responsibility for management of a given 'top level domain' like .eu or .museum. These are, in the context of the internet as a whole, small decisions, and even they can become deadlocked, with progress (or perhaps put better, finality) only when a handful of powerful governments intercede or make demands.

I would say that the right question for our nascent, really not yet existent field of 'internet governance', is the following: what are the digital environments that inspire people to act humanely, and that maximise the power of those who do? This is not the typical lawyer's question. It probably ought to be, but it is not – I understand that. Let us be more analytic about it; it is not just closing eyes, hiding under a blanket and hoping that people can get along. It is saying that, sometimes, the smaller the group, the better.

A town hall works as a town; it does not work when it gets too big. It is hard to hold a referendum that is meaningful among too many people across too many walks of life. If the lessons of digital organisations like the IETF and Wikipedia are to be heeded, it calls for a model of apprenticeship where people actually understand the political culture of the instrumentality they seek to master from people who are already there.

Wikipedia strives mightily to encourage this kind of apprenticeship as new people enter the community and start creating entries. It has the availability of exit: if you do not like what you are experiencing in one realm, you can go to another. It suggests carving up the internet into discrete cantons; if you do not like Wikipedia, start your own, rather than thinking that one-size-fits-all governance once and forever is the way to go. (This clearly doesn't solve every problem; if one is defamed on Wikipedia, it's a weak remedy to simply start one's own website with the true story.)

Wikipedia also asks that people have a stake in what they are doing; that they are involved in something that matters. If you present somebody all at once with "what is your view on Ethernet cards?" it is not obvious what your stake is, and the amount of time it would take to tell you would means that by the end of it, you would be ready to vote for a dog catcher more intelligently than you would want to figure out internet governance.

Making the slices of decisions matter can help. Recall Rachel Corrie's Wikipedia entry. The community of people built up around that entry in Wikipedia is not the same as the community of people arguing about an entirely different topic on Wikipedia. Wikipedia is not actually a single encyclopaedia, but is a set of miniature communities organised around an idea, which is a particular article. Sometimes you see two articles that really ought to be the same one, reflecting two different communities that then negotiate on whether and how to merge the two together. This brilliant carving of the world's issues into differentially small slices helps ensure a productive tension between meta-rules (the three-reversion policy) and small but diverse communities that can come to build trust, to work together effectively without any explicit rules at all.

Wikipedia also embraces the freedom to do wrong. The thing that makes Wikipedia work over the mimicry of sites that pre-approve all 'suggestions' is that you have an opportunity, if you want, to do the wrong thing. You can vandalise the page. Wikipedia will clean it up later; within five minutes it will likely be gone. But you have the opportunity to do it, and every time you go in there and you do *not* do it, you are affirming something to yourself about yourself and to the community that is different from what happens when everything is vetted.

There is a third institution that I want to close on, and that is the university. I believe the use of the internet within the university has so far largely been uninspired, maybe even a failure. When I see laptops in classrooms and view the screens, I see students playing Solitaire or Tetris. When they go onto the network, I see them playing Networked Hearts. Even worse, although I have not seen it yet at Oxford, people are making and losing money by playing poker online during class.

It is not only the students who are misusing the technology. It is the faculty as well. Consider essaygrader.com. It promises to automatically grade student essays. The Essaygrader website boasts that students are happy with it because it gives them instantaneous feedback. Here is a sample of the feedback: "You appear to have included only six concepts from this chapter instead of the ten required. The concepts included that are from this chapter are: crime, labelling, primary deviance, time, and white-collar crime. Put in the others."

It turns out, of course, doing the kind of automated semantic analysis behind services like Essaygrader may mean that if you were to rearrange all the words in the essay so that they were alphabetical instead of in their sentences, you would get the same grade. To me, this says that we are at risk of mechanising the wrong things.

It is why I think we saw fathom.com, started by my university and others, flounder. It was started in a dot com kind of way and it is now an archive; its materials are available for free, but it had not yet figured out that the internet enables, indeed calls on us, to do more than simply digitise what we already do. It calls upon us to develop new tools for learning, ones that enable students to take the same kind of effort that they put into arranging their musical playlists, and give them the tools so that they can arrange intellectual playlists, and say "these are the books and readings in the order I think are great to really understand this period of history", then share it with others and let them make their own versions.

The system should know when two groups of students somewhere in the world are reading the same text, and should put the same effort that Google puts into knowing that you ought to see an advertisement for dog food into saying "these two classes would benefit from being connected right now because they are reading the same thing. Let us get them talking to each other". This is the kind of functionality that the internet invites us to build.

It is about having a stake in what we learn. I am marvelling at the fact that today we ask to write essays and turn them in to one person who reads them, and that is it, when there are whole Wikipedias out there in need of criticism, argument and expanding knowledge that matters to the world. What if we unleashed our students upon them so that it mattered?

Closing remarks

I think it is a wonderful thing that our library at Oxford and others elsewhere are scanning the works of dead people to make them available to the rest of us forever, and in a searchable format so that they no longer have to be chained to a wall. That is great. Our next challenge is how to make sure that the works we produce anew are ones that stand on the shoulders of the kinds of technologies that get us working, talking and collaborating in a way that simply was not possible before the internet came around.

Looking backwards, we see something like the Internet Archive, oddly enough, not started by a university but by someone named Brewster Kahle, an internet entrepreneur who ran a very successful search engine. The Archive seeks to ensure that everything that appears on the web stays there forever so that tomorrow's historians can look back and see it.

That is looking back. Looking forward, we have a chance to build monuments to humanity that are not the pyramids of old, created under slave labour made as a tribute to one vain person that we look at as a curiosity now. Let us invert that and ask, "What can we build together as a group?"

Yes, there will be bad people and inaccuracies among us, but our role as netizens is to actually join the fray, not to wait for anybody to do it for us. Our role as academics is to see how we can connect to that world at large and invite it in. In doing so, we take our own ethos – our own 'private' – and add it to a new kind of public, one whose opportunities and dangers will reflect our own warring impulses: generosity and greed, joy and fear, connection and isolation. To embrace it is to place a bet that the better half can prevail.

Why TV regulation will become telecom regulation

Eli M. Noam

Professor of Finance and Economics, Columbia University

Society's control over media has been around, in one way or another, since our stone age ancestors danced around the fire. When print media emerged in the 15th century, they were censored and licensed, though eventually they emerged with substantial freedoms. More recently, television media were tightly controlled through a variety of means, such as the requirement to obtain a scarce licence that came with ample conditions.

The trajectory of print towards substantial deregulation raised expectations that, in time, television would follow a similar path, when the scarcity of spectrum would cease to be a bottleneck and when any and all could become providers of TV. This withering away would happen when, as expected, television migrated to a delivery over the internet.

A second, more 'techie' school of thought believed in the integrative power of digital technology, although with a somewhat similar ending. The convergence of electronic media would lead to converging regulation: to a symmetrical regulation of technological 'layers' such as transport or applications, rather than along the traditional lines of different media industries. What would emerge would be some combination of the regulatory approaches, most likely with a fairly minimalist scope.

Unfortunately, both expectations are flawed. I will argue that, as most media will be provided over the internet as their major delivery platform, internet regulation will become the major form of media regulation; and that this internet regulation will come to resemble many elements of traditional telecom regulation. Hence, telecom regulation will become, for better or worse, a foundation of media regulation, instead of the hoped-for *laissez-faire* system.

The roots of TV regulation

The first issue to address is the notion that it was spectrum scarcity that gave governments a tool for control of television. Arguably, the opposite was the case. TV spectrum was scarce because governments chose to make it so, by allocating frequencies only grudgingly. One reason was the fear of the power of private broadcasting over politics and culture. State-controlled radio had played a major role in the internal and external propaganda efforts before and during the Second World War and in the emerging Cold War. The extension of such power to private parties was seen as quite dangerous. It also competed with the powers of governments over the means of information, and with the traditional media such as newspapers. Finally, even private broadcasters liked the notion of profitable exclusivity that emerged through a state-created scarcity.

But if scarcity was not at the heart of television regulation, what was? Each society has its concerns, problems, issues, traditions and priorities. Americans worry about sex more than the French, who in turn are concerned about their linguistic purity.

Swedes fret about violence. Germans, burdened by their past, are sensitive about racist speech. Iceland, given its size, worries about ownership concentration. Canada worries about its national identity; China about party control; Saudi Arabia about religious orthodoxy. Malaysians are concerned with ethnic diversity. The main purpose of media regulation is to advance such goals.

None of these objectives will vanish just because television signals travel over digital pipes rather than analogue airwaves. Some of these concerns might decline, but others will increase, and new ones will emerge. It seems unlikely that societies will simply give up on their societal priorities just because the video information now takes a different path, or is encoded in a different way. Instead, they will simply adjust the tools to the new environment.

At this point, internet stalwarts will object that one cannot regulate the internet, and therefore neither can one regulate TV over the internet. But this, too, is wishful thinking.

Many people believe that one cannot regulate the internet, even if one wanted to. After all, don't high school kids run electronic circles around flat-footed government and corporate enforcers? And can't you locate media servers in offshore locations, far from the regulatory powers of a particular country?

But that only proves that it is difficult to go after the electronic part of their communication. If one cannot reach the bits themselves and their source, one can still go after the physical elements of delivery: the networks. Those cannot hide, and in a two-way medium cannot easily operate across borders without permission.

One of the principles of regulatory enforcement is that it is easiest to regulate the least mobile and least elastic elements, such as land and physical goods. A second principle is that it's easier to regulate the element with the fewest providers. Not only can such regulation deal directly with problems that result from market concentration in the provision of the element, but it also provides an indirect regulatory tool to reach other related segments of an industry more easily.

Both of these principles favour, for internet-based activities, the choice of the delivery networks as the nexus for regulation. First, these networks are quite limited in number. Though the notion of the internet as being wide-open and competitive prevails, the fact is that a number of its basic instrumentalities are quite concentrated, foremost among them the 'last-mile' delivery networks. In most countries only one such infrastructure is prevalent: that of the incumbent telecom company. Some countries add the cable TV infrastructure. A few last-mile alternatives may emerge, such as mobile-based TV, satellites, fixed wireless, or powerline communications. But when the dust settles, these alternatives will serve market niches, and often be owned by the incumbent companies anyway.

Why is the delivery of bits typically such a concentrated industry? While the details vary from country to country, the primary reasons are economies of scale and sunk cost. Many elements of the internet are competitive but face the market power of other segments of the internet – in particular that of terminating networks once an end-user has selected a carrier. This results in battles, both in the marketplace and increasingly also in the policy arena.

A prime example is 'net neutrality', an issue that has achieved much prominence in America. Its meaning is a bit fuzzy, there being at least eight ways to define it, but the central concern is clear: the power held by the last-mile delivery networks – mostly traditional telecom or cable companies – to select, price or differentiate among the internet information streams that pass through their pipes.

Fearing such gatekeeping power, internet content and applications providers, as well as traditional media companies, have banded together. They seek protection from the network companies' power over access prices, quality, price discrimination and the likelihood that they will favour their own content subsidiaries.

Internet issues, telecom solutions

How things have changed. Barely a decade ago, the internet's pioneers declared in their 1994 Charter for Internet Liberties: "Government, leave us alone, we did not call you, we don't need you." Today, the internet community is seeking, under the heading of net neutrality, the policy that dares not speak its name: telecom common-carrier regulation for broadband connections. The body of regulation that deals with access, price discrimination, gatekeeping and vertical squeeze are just the problems that internet providers fear.

And this is not surprising. Telecom regulation, though derided as 'legacy', has evolved for a reason. We must first jettison the view that it is backward. If regulate we must, then telecom regulation is really quite a sophisticated tool relative to regulation of other industries. Take the concept of forward-looking incremental cost pricing for unbundled network elements. Neither aviation, pharmacological drugs, environmental controls, rail transportation nor electric utilities have anything that comes close in terms of economic sophistication and institutional complexity.

This is not to say that it is a 'better' regulation, just a more complicated one, dealing with numerous factors and conducted on an economic level of significant expertise. While broadcast regulation could be learned by an attentive student in a busy afternoon, and while aviation regulation is heavy with details but feather-light on concepts, telecom is quite complicated. Even experts can be lost in explaining the relation of reciprocal compensation to access charges.

A major reason for the complexity and sophistication has been the large number of goals that telecom regulation tries to accomplish. In America, these range from general coverage and affordability (universal service); to openness to users (common carriage); to control of market power (price and return regulation); to integration of networks (interconnection); to international collaboration (accounting rates); to encouragement of competitors (wholesale retail pricing); to consumer protection (quality regulation); to the protection against interference of transmissions (spectrum licensing); to innovation (information services); to vertical protections (divestitures and fully separated subsidiaries); to national security (CALEA); to personal safety (911); to consumer choice (number portability); to federalism (state and federal jurisdiction); to rural-metropolitan equity (high cost fund); to social equity (lifeline); to promotion of the internet (e-rate) – and quite a few more. The result has been a highly complex set of rules, which try to balance the multiple objectives and accommodate the various political forces behind them.

It has been easy to belittle the resultant complexity as old-fashioned legacy telecom regulation. And it is of course true that any accumulated set of rules will have its share of inconsistencies and holy cows. And of course, as in any balanced system, everyone despises those aspects of the regulations in which they have to concede something. But on the whole, telecom regulation is complicated and advanced in balancing multiple objectives, as regulatory schemes go. But they are, of course, regulations, not the free market. A *laissez faire* approach is a principled alternative; a selective libertarianism is not.

Internet libertarians belittled telecom regulation as part and parcel of old-fashioned telephone networks, and a reflection of dinosaur phone companies and co-opted regulators. Instead, they expected, the new sprit of digitalised, decentralised, packetised, nimble, innovative, entrepreneurial new media companies would transform the system, and obviate the entire anachronistic regulatory system.

But just a few years later, the tune is quite different. Now, the internet's advocates seek a large set of supportive governmental policies. And in particular, they seek governmental protections from the powers of the telecom and cable companies' powers over pricing, quality, discrimination, access, content, favoritism for own subsidiaries, and so on. They seek an opening of incumbent networks through unbundled network elements and re-bundled platforms, line-sharing of frequencies over local networks, etc. And who should institute these protections? The same derided, old-fashioned telecom regulators.

There is nothing wrong with these goals. Of course, they are not exactly libertarian. Yet they are not hypocritical, either. They just reflect the discovery of the reality of the market power inherent in last-mile delivery networks, and its implication for regulatory rules and institutions. For a while, hope ran high that competition would open wide the delivery network environment. One decade and one burst bubble later, the reality looks different. Last-mile delivery has either remained dominated by an incumbent telecom company or by a small oligopoly. The reasons are not those of conspiracy or of co-optation of regulators, but of economies of scale and sunk cost. In competition, prices are pushed onwards to marginal costs, which are too low for most firms to survive; ergo, consolidation takes place.

What might a system that protects content providers over the internet look like? It would assure entrants' ability to interconnect, resell, access, collocate and be compensated. There may be good public policy reasons to adopt such policies, but they cannot be described as deregulatory. Hence, for all practical purposes, the libertarian flag has changed hands: today, the incumbents wave it over their market power, while the internet community seeks an increased role for government – while still mouthing the after-dinner speech slogans of minimal government.

This does not apply just for the interaction of different media companies – those of content/applications providers versus the networks. Societal media policy goals enter as well. Content diversity. Source diversity. Balance. Privacy. Morality. Trade. National culture. Ownership diffusion. Protection from defamation. Consumer protection. Revenue generation. Truth in advertising. Innovation. Interoperability. Coverage across geography and income. Redistribution. Protection of competition. Protection from spam. Cyber-security. Redistribution.

To achieve these societal goals and others will probably require regulating the infrastructure providers by direct actions, as well as make them police their users and make them act as tax collectors. An example is the e-rate in America, in which telecom carriers collect subsidy funding for internet connectivity for schools, libraries and hospitals. Another example is that of liability for internet transmissions, which the Chinese government, as well as American music companies, tried to impose on the delivery networks.

As mentioned, a basic principle is that regulation will go after the most limited and accessible segment. And that normally seems to be the last-mile transmission element. Under traditional broadcasting, that element was the spectrum. Access to it was subject to a variety of conditions. For cable and satellite, orbital slots, frequencies and construction franchises provided the regulatory nexus to establish a variety of conditions. And under internet-based video media, the delivery transmission path is the regulatory nexus, too.

It is not difficult to conceive of means to do so. Conditions can be set for carriage or interconnection. Hardware can be required to have screening chips and software. Prices and quality levels can be set.

And there is more, unfortunately.

First, there is the network as a national *cordon sanitaire*. Internet TV cannot be expected to be only a time- or place-shifted version of regular TV or film. Inevitably, it will be used for more controversial uses. Imagine the reaction when we face a serious influx of on-demand Bangladeshi tele-medicine, Monaco tele-gambling or Thai child tele-pornographers. Because most of the originators of the content cannot be easily reached directly by a national government, it is likely that the responsibility to block it before it reaches citizens will become the responsibility of network operators in each country.

Second, the internet provides a tool for the customisation of everything, including regulation. Contrary to the conventional notion that 'a bit is a bit', internet bits come in identifiable packets with address and sender included. And once information is identifiable, it becomes differentiable – and hence regulatable. One can regulate packetised and identified information in a granular fashion. Preferential treatment, pricing and must-carry become possible with identification. Internet regulation actually becomes a more powerful tool for the state over what existed in the past.

The significance of all this lies in the fact that most next-generation media applications will run over the internet, loosely defined, and its fixed-line and wireless pathways: music, video, film, interactive games, online news media. For a time, internet distribution will be supplemental, but in time, with broadband use spreading and becoming ultra-broadband, it is likely to become the major way to deliver TV and other content. And for the emerging shared media creation or interactive uses, the internet is already the principal medium.

And therefore, as the internet becomes the main platform for most media uses, the regulatory rules for the internet become the rules for much of the media system as a whole. And with the regulatory rules for the internet moving toward those of telecom, because the issues that led to telecom regulation – discrimination, market power, national security, consumer protection – are precisely the same issues being raised by the internet. It then follows that the telecom rules become the rules affecting all media. Therefore, television regulation will become telecom regulation. And this is not a happy conclusion.

Serving the public good in the digital age: implications for UK media regulation

Carolyn Fairbairn

Partner, Head of UK Media Practice, McKinsey & Company

"The charm of history and its enigmatic lessons consist in the fact that, from age to age, nothing changes yet everything is completely different."

Aldous Huxley

Regulators are currently grappling with the latest technology revolution in media. The internet, broadband, digital television and mobile devices are offering consumers new chances for self expression, mobility and control, causing entry barriers to fall and creating a collision between the worlds of media and telecommunications. What does this revolution mean for people as consumers and citizens? Does the regulatory rulebook need to be rewritten? And is the time now right to replace the traditionally interventionist model of UK media regulation with more free-market principles?

In this essay, we set out our assessment of what the digital revolution might mean for individuals, the media industry and future regulation. Our conclusion is that, despite massive changes, the two most compelling reasons for regulation of the traditional broadcast market are likely to continue into the new media era. First, the public benefits of high-quality news, drama, comedy and sport will be as great and as difficult for the free-market to capture as ever; and, second, the tendency towards highly concentrated control of media distribution in the hands of a few operators will still exist. Now is not the time to throw away the broadcast rulebook.

However, each new decision to regulate needs a serious examination of its costs to avoid undermining innovation, investment and risk-taking. Rather than adopt a 'one-size-fits-all' model, policymakers should take a case-by-case approach to each new media platform, based on the answers to three questions:

- First, is this new media service or platform likely to become a 'mainstream' means of media consumption for many people?
- Second, is there a risk of spectrum scarcity and/or highly concentrated platform ownership that could limit the range and quality of available content?
- Third, are the benefits of regulation greater than its impact on the commercial marketplace?

If the answer to these questions is 'yes', it means that at least some of the original reasons for regulating broadcast media also apply to the new platform, and failure to intervene would be likely to result in losing public value. Regulators should therefore consider using the levers of established broadcast regulation to ensure that the UK's media industry continues to serve the public good. However, if the answer is 'no', free-market solutions are likely to offer a better route forward.

Later in this essay we give examples of how this approach might be applied to different new media platforms: digital television, the internet and mobile broadcast. But first, we look at the choices facing today's regulators, and at what the digital revolution means for audiences and the UK's media industry.

1. Regulatory choices

Ofcom was established in 2002 to address the converging worlds of telecoms and broadcasting. With some prescience, its creators anticipated the convergence between the two that is now underway. But in bringing together these two distinct industries, it also brought together two distinct models of regulation: the heavily interventionist broadcast model and the more free-market telecommunications model.

In deciding how to approach the emerging media industry, Ofcom now faces a choice between these two models. Do the dramatic market changes mean that it can now be regulated more like the telecoms market – with a focus on promoting competition, auctions for scarce spectrum, price but no content regulation – or is it better to carry forward the established principles of broadcast regulation – with intervention to secure investment in, and easy access to, high-quality content? Or is the answer a mix of the two? It is a vital and difficult decision, and one that will lead to very different outcomes for the UK's media landscape.

UK broadcast regulation: a public service model

By any measure, the UK's broadcast regulatory model involves intense public intervention. It is based on the principle that a rich range of universally available programmes, made free at the point of use, is vital to the health of society and could not be guaranteed by the free-market alone. Its system of public service broadcasting has been shaped by four main policy levers:

- **Publicly-owned broadcasters:** the BBC is the most significant public policy lever in UK broadcasting. It is the largest public service broadcaster in the world, receiving £3.1 billion a year (in 2005) in public funding and accounting for over 40 per cent of all UK viewing and listening.[1] Channel 4, though commercially funded, is also publicly-owned and does not have to make a profit.
- **Licensing or 'gifting' of spectrum** in return for specific public broadcasting obligations, such as achieving reach, diversity and quality targets. ITV and Five are both licence holders of analogue and digital terrestrial spectrum licences.
- **A range of access requirements** placed on digital television operators to ensure public service content is easily found. For example, all cable operators must carry public service channels on their networks.
- **Direct content regulation of taste, decency and impartiality** on broadcast platforms.

The UK has arrived at this unique regulatory landscape through a curious mix of economics, politics and historical happenstance. The BBC was founded in 1922 to drive the sale of

1. BBC Annual Report and Accounts 2005/2006

radios, but over time has developed an underpinning economic rationale. Economics is the most developed language for the analysis of broadcast regulation, and a comprehensive analysis of the six main market failures in broadcasting was included by Gavyn Davies (who later became the BBC's chairman) in his 1999 study of the BBC's licence fee. They are:

i) **broadcasting produces externalities**, meaning that it can bring benefits to people even if they do not themselves consume it, for example through its contribution to a better-informed society;

ii) **economies of scale exist in broadcasting**, leading to high fixed costs and low marginal costs, which makes it difficult for new firms to enter and tends to lead to a concentrated industry structure;

iii) **spectrum is scarce**, limiting competition and potentially leading to natural monopolies and high barriers to entry;

iv) **high quality broadcasting is a merit good**, bringing benefits to people that they only recognise afterwards. Free markets do not supply enough merit goods;

v) **broadcasting is a public good**, meaning that it can be supplied to many people at the same cost as a few; and

vi) **consumers of broadcasting are not fully informed**, so they don't know what they are buying until they have experienced it.

Together, these six market failures mean that, left to itself, the UK's broadcast market would not serve the public interest as well as it could or should. Over time, efforts have been made to render this dry economic language more accessible, while still relying on the same logic of poorly functioning free markets.

Most recently, the BBC has proposed the creation of 'public value' as its sole purpose. In its public case for renewal of its Royal Charter, *Building Public Value*, the BBC aims to describe the potential of high-quality drama, comedy, news, current affairs and sport to unite people and to reflect their differences; to create opportunities and extend horizons; and generally to improve the quality of life for people, both as individual consumers and as citizens.

So has the UK's unique model of broadcast regulation been successful? Comparisons of programme quality are notoriously difficult to make but overall the evidence is positive. By international standards, UK television carries one of the highest proportions of home-grown programmes and has the lowest level of imported programmes outside the US. It also spends more per head on original television programming than any other country in the world, apart from the US (see figure 1). As a result, the UK's broadcasting system strongly reflects the culture, tastes and values of its society. Effectively, this model of regulation represents a trade-off of personal freedom for a 'better' broadcasting system.

UK telecommunications regulation: a price and access control model

The telecommunications regulatory model is very different. The telecoms industry has been regulated largely by price and access control. Unlike in broadcasting, regulators have been agnostic about the content that is carried over telephony networks, and limited their concerns to perceived abuses of market power: for example, locking out competitors from monopoly networks (resulting, for example, in BT being required to give other operators access to local telephony networks), or to raise prices unfairly.

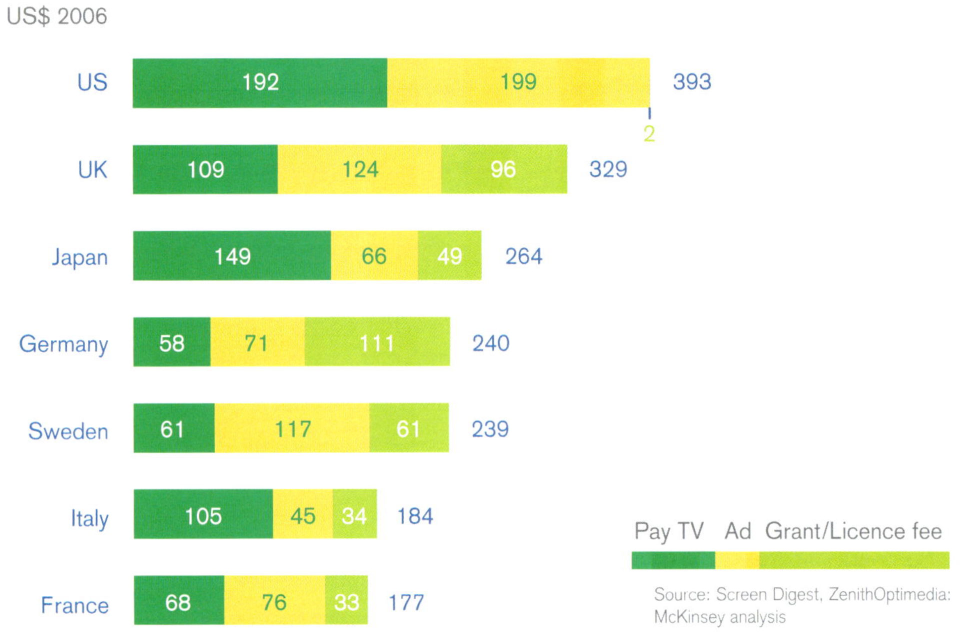

Figure 1: Spending per head on television by country

Again, most people would agree that it has been a successful regulatory model. The UK has a competitive telephony market for both fixed line and mobile, while broadband penetration has recently accelerated and has now almost caught up with the US.

One of the biggest differences between the two models is the way in which scarce spectrum is treated. In the telecoms model, it is almost always auctioned. In the broadcast model, it is typically allocated by non-market means to give regulators control over the kinds of content that are carried. These non-market methods can include straight 'gifting' of spectrum to publicly-owned broadcasters, or by licensing of the spectrum to commercial broadcasters or other operators based on a range of public service factors such as reach, range and quality of content.

This difference was highlighted in two very different spectrum allocations in the late 1990s, held by the two main regulators of the media and telecoms industries at that time. In 1996, the Independent Television Commission (ITC) chose to gift scarce spectrum to broadcasters at no cost, to enable them to launch digital terrestrial television services. Four years later, in 2000, the Radio-communications Agency chose instead to hold a free-market auction for spectrum to enable telecoms companies to offer entertainment services over 3G. This spectrum was sold for the heady price of £22.5 billion.

The choice ahead

We are now at a fork in the road. Ofcom could choose to wind down the broadcast regulatory model in favour of the telecoms model. Its early public comments on the appeal of auctioning new spectrum (including the UHF spectrum that will be liberated when analogue television is switched off) suggest it is attracted to transferring at least some of the principles of telecoms regulation over to media. For example, in 2004, Stephen Carter, Ofcom's

Chief Executive, said: "Where possible, we should leave it to the market to decide the best use of spectrum. It's not our job to decide how a piece of spectrum should be used."[2]

Ofcom is certainly right to be looking hard at media deregulation. It is right that regulators should only intervene where there is a strong case to do so. In today's risky environment, it will be particularly important to avoid dampening private investment and innovation. However, the right regulatory balance needs to be determined with great care. Striking that balance will depend crucially on how media markets are changing and whether these changes diminish the rationale for intervention. This is the subject of the next section.

2. The changing media marketplace

In the past five to ten years, a range of technologies has emerged that will have a considerable impact on the evolution of modern media. They include new distribution technologies such as broadband and mobile television, and new consumer devices including MP3 players and personal video recorders. While each innovation will change how people use media, collectively they have the power to transform it.

Power to audiences

For consumers, the first stage of the digital revolution has largely been about expanded choice: many new channels on digital television and radio, and easy access to information on the internet. The second stage will be significantly more far-reaching as most people gain access to full video services via the internet and on the move. Three trends will matter most:

- **Fragmenting media consumption** For most of the last 50 years, UK audiences consumed a predictable diet of radio, television and newspapers as their staple sources of daily news and entertainment. Until the turn of the century, the time spent on each was roughly constant at 3-4 hours per day per adult for both television and radio, and roughly 50 minutes per day of newspaper and magazine reading. But in the last decade there have been four major waves of digital innovation (sometimes known as platforms), each offering consumers new ways to watch, read and listen to media content. They are digital television, the internet, digital radio and, most recently, broadband. A fifth – mobile media – is just beginning. Each seems to be following a similar pattern (see figure 2), achieving 50 per cent take-up between six and ten years after launch. These new platforms are taking a growing share of people's time and spending power.

- **Audiences taking control** The second feature of the consumer revolution is that audiences can increasingly find the song, programme or clip they want, when they want it. Video and audio are going 'on-demand', and audiences are no longer guided in their content choices by broadcasters. Between 35 per cent and 50 per cent of households are expected to have personal video recorders by 2012. Accessing a vast library of video and audio from the internet will become commonplace for many people. At the same time, television and the internet will become as portable as radio: news and entertainment content will be easily available on mobile phones and other portable devices.

2. Quoted in *The Times*, 16th December, 2004

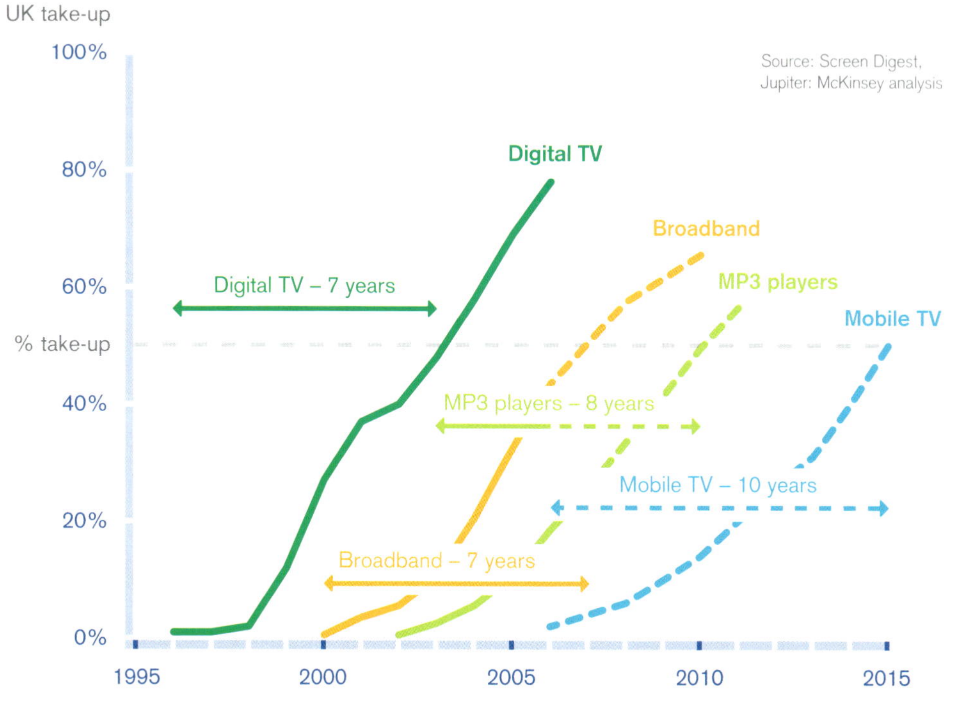

Figure 2: Actual and projected take-up of new media innovations, 1996-2015

Grass-roots contributors. Media used to be a one-way affair. One of the most powerful aspects of the digital revolution is the power it gives people to contribute their own thoughts and talents to others. The market for content produced by individuals and small, often amateur, groups is growing astonishingly quickly. More than 30 million blogs – personal opinions posted on the internet by individuals – now exist. Sites such as MySpace, YouTube and Bebo enable anyone with an internet connection to reach anyone prepared to watch, read or listen (see figure 3).

The benefits of these changes are potentially exciting for us – both as consumers and as citizens. Significant increases in choice, flexibility, convenience and mobility will enhance enjoyment and value for many people. Evidence lies in the money they are choosing to spend on it: UK consumer spending on video services (excluding equipment) is one of the fastest-growing categories of consumer expenditure, ahead of cars, holidays and clothes. It has grown by 40 per cent in the last ten years and is projected to accelerate over the next ten years as well (figure 4). Evidence from the US suggests that much of the future increase in spending will be on new video services, such as online video.

The changes will also benefit people as citizens. For example, the growing interactivity of media through the internet and broadband has the potential to create a more powerful educational tool than television or radio. Active democratic debate is being stimulated by a wide range of websites and interactive television programmes that encourage participation – the chance to vote, find out more about a local MP, or express views to democratically elected representatives. And the

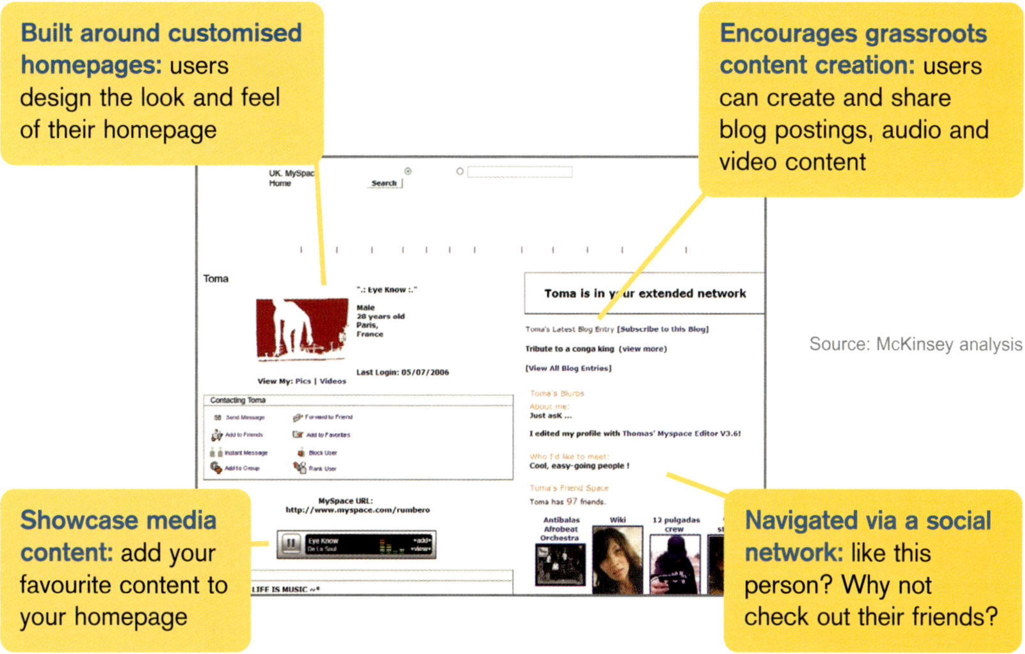

Source: McKinsey analysis

Figure 3: Features of Myspace.com

internet is creating opportunities for people to connect in new ways, creating new communities around hobbies, passions and interests.

A changing industry

At the same time, the structure of the media industry is being transformed, bringing a wide range of new competitors, many of whom are taking share from long-standing incumbents. Four main changes are shaping the emerging industry picture:

- **Explosion of new entrants** Over the last ten years, falling entry barriers have created opportunities for a wide range of new media participants. For example, the number of channels available on television has grown from five to more than 300 (for satellite subscribers) over this period, taking share of viewing hours away from traditional channels such as BBC1 and ITV1.
- **Rise of the aggregator** One category of new entrant is becoming particularly powerful: the aggregators and organisers of content. Google and Yahoo have become the goliaths of the new media world, offering consumers largely unedited access to a massive range of content and, in the process, they're challenging the role of traditional 'editors' of content such as the BBC, ITV and newspapers.
- **Convergence of telecommunications and media** One of the most significant technological advances is the new ability of telephony networks to carry music and video as well as merely text and voice. As pure telecommunications revenues come under threat due to intense price pressure, telecoms companies around the world are entering the media industry in a search for new revenue sources. For example, at the time of writing, BT is about to enter the

digital television market with a consumer service that combines free digital television with on-demand video and personal video recording, while mobile operators such as Vodafone, 3 and O2 are developing increasingly rich mobile video and music services.

- **Uncertain advertising revenues** Advertising has been one of the staple sources of funding for media content because it has offered companies a reliable means of reaching mass audiences. However, as audiences fragment, it is becoming less secure, particularly for traditional mass media such as newspapers and network television channels. Advertisers in the UK are projected to increase their spending on the internet from 5 per cent in 2004 to 12 per cent in 2008 (see figure 5). Over time, more targeted forms of advertising will develop, taking advantage of the growing personalisation of media and the fact that, for the first time, some media companies know in detail who and where their audiences are.

Ongoing limitations of the free market

The scale of change and uncertainty in the media industry is not in doubt, nor is its potential to bring a wealth of new services to audiences. However, the pertinent question for future regulation is not the scale of the change, but whether it will reduce, or perhaps even eliminate, the market failures that provided the historic rationale for broadcast regulation?

The answer is mixed. Some market failures will be substantially reduced. For example,

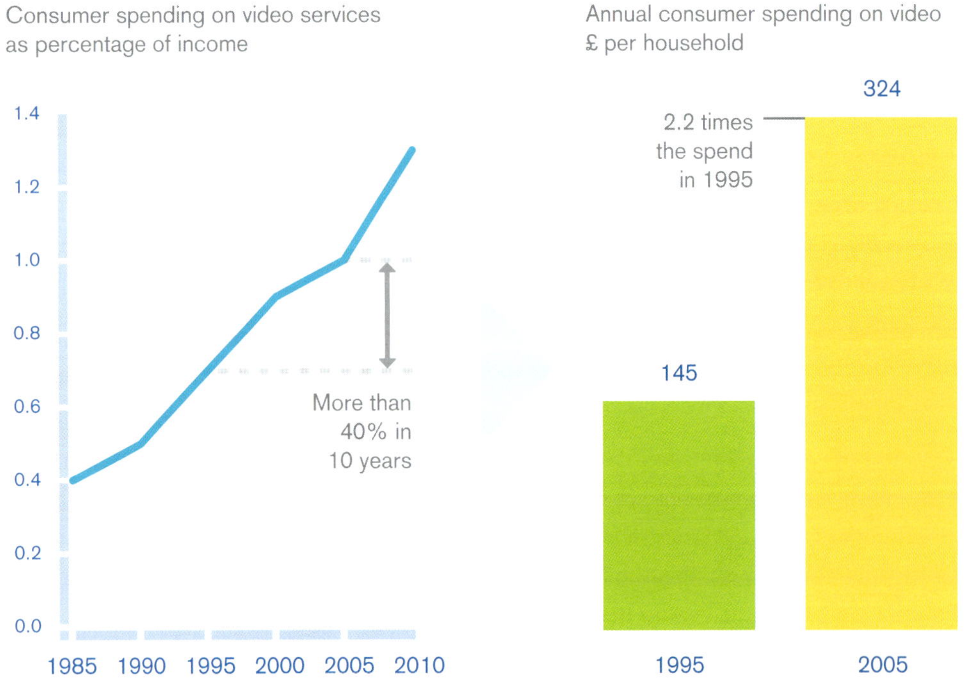

Source: Screen Digest, BBC annual reports: Office of National Statistics, McKinsey analysis

Figure 4: Consumer spending on television and video-related services

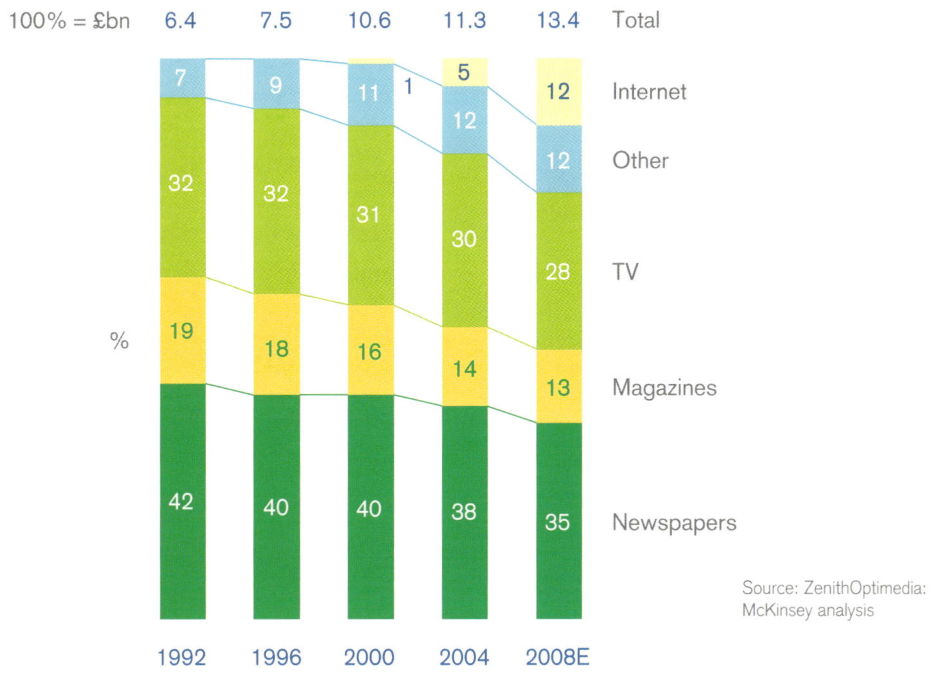

Figure 5: Share of UK display advertising revenues by channel

barriers to entry for content providers have fallen dramatically, and information problems will diminish as audiences are able to catch up with a missed programme on-demand. Also, it is now possible to charge people for the content they consume using encryption technology and secure payment systems. These factors reduce the need for regulation in the new media world. However, three problems will remain:

- First, the spectrum needed to deliver content to audiences will continue to be scarce. Spectrum shortages will be less extreme than in the analogue era, when there was room for only five broadcast television channels, but scarcity issues will not go away. The growth in video applications requiring more spectrum is accelerating. There are already signs that high-definition television (HDTV), mobile broadcasting and interactive services will compete with new digital channels as bandwidth becomes available. Demand is likely to outstrip supply for the foreseeable future. As a consequence, opportunities will exist for spectrum owners to create distribution monopolies and restrict the range of content available – one of the main reasons for media regulation in the analogue world.

- Second, the economics of distribution on some platforms may further contribute to the creation of monopolies, or at least operators with significant market power. Where the fixed costs of building a network are very high compared with the revenues, 'natural' monopolies are likely – that is, the most efficient number of operators may turn out to be only one or perhaps two players. Digital pay television has this characteristic, while other emerging platforms, such as mobile broadcasting, may share it. Monopolies in the provision of access to content are also possible, for example through concentrated ownership of access technologies such as electronic programme guides or subscriber management systems. Again, the risk for consumers is higher prices and a narrower range of content that may not include public service programming.

- Third, there is no evidence that consumers will suddenly start to take account of the broader social impact of their media decisions. Even a perfectly functioning media market has no way of pricing these externalities, and would continue to under-produce the kinds of programmes that create them. Left to itself, the UK's media industry would still produce and make available less – and more highly priced – good-quality news, drama, comedy and educational programmes than is in the public interest.

Overall, it is possible that less, not more, regulation will be needed in the media industry of the future, but the case is not yet proven. What we can safely say is that it is too early to abandon the broadcast rulebook altogether. But how should regulators decide when and how to apply it? In the next section, we set out some ideas.

3. A proposed approach

The complex nature of the evolving media industry means that regulators will need to explore each new platform in turn, rather than adopt a 'one-size-fits-all' model. Policymakers should take a case-by-case approach and extend established broadcast policy into the new world selectively, based on the answers to three questions:

- **First, is this new media platform likely to become a 'mainstream' means of media consumption?** If so, traditional social value and 'externality' arguments remain relevant for the new platform, and regulation is likely to be necessary to prevent a loss of public value as audiences switch consumption from television and radio to the new platform.
- **Second, is there a risk of spectrum scarcity and/or highly concentrated platform ownership that could limit the range and quality of available content?** If so, gatekeepers could emerge with an incentive and ability to limit the range and quality of content, or raise prices above the level that is best for consumers.
- **Third, are the benefits of regulation greater than its market impact?** Regulatory intervention should not disproportionately damage innovation and investment. Market impact assessments have become common practice for new BBC services and could be extended to all regulatory interventions.

If the answer to all three questions is 'yes', it means that the original reasons for regulating broadcast media are also likely to apply to the new platform, and failing to intervene could result in a loss of public value. Regulators should therefore consider using the established levers of broadcast regulation to ensure that the UK's media industry continues to serve the public good.

However, if the answer to the three questions is 'no', then a more free-market regulatory model should be considered. Computer games such as X-Box and Playstation are in this category – they are only lightly regulated because the externalities created by the content are relatively low; they do not substitute for mainstream news and entertainment consumption; and the market that delivers them is highly competitive.

Perhaps not surprisingly, the most interesting cases are those that give a mixed set of answers. Take, for example, a platform that is likely only to complement existing platforms rather than be a substitute for them – so that its public value is not clear or at least not proven – but that is likely to be highly concentrated. The expectation is that it should be regulated like telecoms, to ensure fair access for all service providers, but should not need traditional broadcast regulation.

But what about a platform that is clearly likely to be a mainstream means of media consumption, but which is delivered by a competitive industry structure? The answer is that there may still be a case for regulation, on pure public value grounds. If the externalities of a platform are high – perhaps because of its contribution to democratic debate, education, or social inclusiveness – then, as before, the market will not produce or distribute as much of this sort of content as it should, despite being competitive in all other ways. Market signals about what consumers are willing to pay still do not capture its wider benefits to society. Clearly, the benefits of regulation then need to be weighed against the market impact on innovation, competition and private investment. This balancing of public value and commercial cost is at the heart of the so-called 'net neutrality' debate about how the internet should be regulated (see case study below).

To show how this approach could work in practice, we have looked at three case examples: first, digital television (DTV), for which the regulatory regime is largely complete; second, the internet, where regulation is still evolving; and third, mobile broadcast, the least-developed of the three.

Case example 1: digital television

Digital television has been the first test case of regulation in the digital age. It provides a useful case study because its regulatory framework is well-developed and we are able to see what it has meant for consumers and the wider market.

The regulatory framework for digital television, largely built by the ITC during the 1990s, has been modelled strongly on broadcast principles. All four broadcast policy levers were used: the BBC was permitted to use licence fee funding to develop digital television channels such as News 24 and BBC3; spectrum was gifted to public service broadcasters to ensure guaranteed distribution; content regulation was extended from analogue to digital television; and a new set of access rules was established to ensure public service channels were easily available. In addition, when the commercial ITV Digital service on digital terrestrial television (DTT) collapsed in 2002, the ITC licensed a consortium involving the BBC to launch a free-to-view digital television service, known as Freeview.

This regulatory outcome is consistent with our proposed three principles.

- Digital television was predicted from an early stage to become mainstream, and indeed is now available in nearly 80 per cent[3] of UK homes and accounts for 32 per cent of viewing.[4]
- The risk of both spectrum constraints and new distribution monopolies was high, for two reasons. First, the economics of digital television favour only a very small number of operators who could therefore use their market power to restrict access and raise prices. Second, available spectrum for the

3. Screen Digest estimates
4. Ofcom Communications Market 2006 report

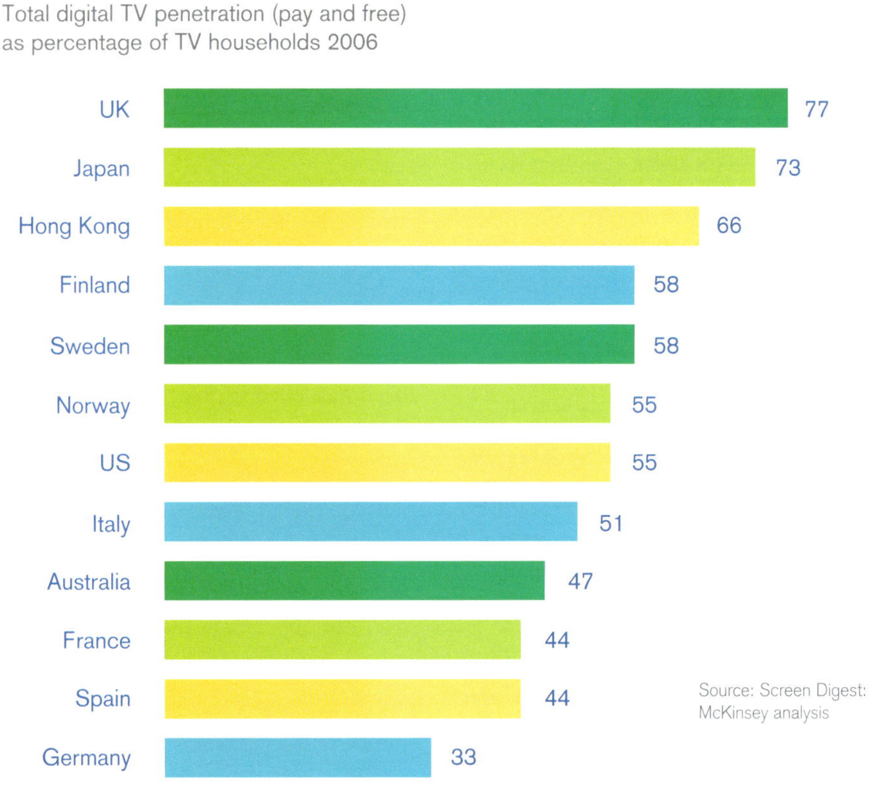

Figure 6: Digital television penetration by country

third digital television platform, DTT, was limited to 30 channels.

Finally, and with the benefit of hindsight, it looks as though the benefits of regulation have outweighed its costs by some distance. The UK leads the world in take-up of digital television (figure 6), and the range of available channels on both free and pay services is strong and growing.

Overall, the public interest seems to have been well served by extending broadcast principles into digital television.

Case example 2: simple and rich broadband internet

The internet is a more complex case, not least because it keeps changing. The 'simple' internet, offering mostly text-based websites, has one set of characteristics. But the fast-growing rich broadband internet, which can offer DVD-quality video-on-demand, is another thing entirely. As a result, the application of the three principles would give different answers for each.

The case for extending broadcast regulation into the simple internet world is weak. The simple internet has not been capable of becoming a mainstream means of media consumption (except possibly for news) because it could not carry video. Nor are there obvious capacity constraints or distribution monopolies – access to the internet for content providers has so far been fully open. And any attempt to regulate the internet could have dampened the creativity and investment that has been vital to make it work.

However, the picture for rich broadband internet is starting to look rather different.

- Broadband internet does have the potential to become a mainstream source of media consumption for news and entertainment content, and indeed is just that for countless young people. It will look increasingly like television, with many websites already rich in video content.
- Though the answer to question 2 is still 'no' – the market for broadband internet is likely to remain competitive and well-functioning – its economic characteristics are changing. This is because, to the surprise of some, the internet is starting to hit capacity constraints, and some internet service providers (ISPs) and network operators want to restrict access to, or extract charges from, content providers who put heavy capacity demands on their networks. Their demands are opposed by those who believe that the whole point of the internet is to stay 'neutral' between different types of content. The argument of the 'net neutralisers' is rooted in public value and the positive externalities associated with a free and open internet. They want regulation to guarantee it stays that way.
- But many ISPs claim that the costs of regulation would be unmanageable. Unless they are able to discriminate between different types of content in order to fund future investment in improving their service, the result would be that innovation and growth of the internet, traditionally so dramatic, would stall.

The net neutrality issue is a complex and important debate that is just beginning in the UK. Regulators will have to come to a view on the balance between the public value of the evolving internet and the negative costs of trying to regulate it – effectively a weighing-up of the answers to questions 1 and 3.

Case example 3: mobile video broadcasting

Mobile broadcasting is the youngest of the three platforms; indeed, in many countries, it doesn't yet exist. However, by 2010, most developed countries including the UK will have at least one mobile broadcast network enabling 'live' television channels to be broadcast to mobile phones and other portable devices, and that will complement mobile services over 3G networks.

So far, different countries are choosing different regulatory models. Regulators in Japan, South Korea and Finland are applying the broadcast model. They have imposed 'must carry' requirements for public service content and have gifted or licensed spectrum, rather than auctioned it. Meanwhile, regulators in the US and the Netherlands have opted for the telecoms model, including spectrum auctions.

UK regulators will soon have to choose which model to apply in this country, and it will be an important decision because of the precedent it could set. If regulators choose to regulate mobile broadcasting on a telecoms model, it will be the first UK broadcast platform to depart from established UK broadcast policy. So what insight might our approach offer? Again, taking each of the three questions in turn:

- Although the future take-up of mobile television is difficult to predict, research by McKinsey suggests that it could become a popular and mainstream service by 2015. Despite sceptics questioning the consumer appeal of viewing on a small screen, 59 per cent of UK consumers polled in a recent survey wanted mobile TV very soon, and a further 15 per cent wanted it immediately. Early launches in South Korea and Japan are showing high growth, especially among young people. Evidence from UK trials suggests that the potential for average daily viewing is in the region of 20-30 minutes, per day per person.

- It is likely that the distribution of mobile broadcast channels will be concentrated, at least in the early days, because the fixed costs of building a mobile broadcast network will probably be high compared with revenues, meaning that only one or two competing networks are likely to be viable. In addition, spectrum allocated for mobile usage may be limited. As a result, gatekeepers could emerge with the ability to restrict access and raise prices.
- The costs of regulation are difficult to estimate at this stage, but would need to take into account the deterrent effect that regulation might have on the willingness of potential major players, such as infrastructure companies and mobile operators, to invest in the platform.

The likelihood that mobile broadcasting will account for a significant share of future news and entertainment consumption, but will be delivered to consumers by a small number of distributors, suggests that a broadcast regulatory model should not be ruled out at this stage. Regulators should seek to understand as much as possible about the answers to the three questions posed in this essay – usage, industry structure and the market impact of possible regulatory measures – before taking this significant decision.

Conclusion

What regulators choose to do in the next five years will have a powerful impact on the evolution of UK media. They face a difficult and important choice between two very different models of regulation: the telecoms model, based on spectrum auctions, price and access control, and content neutrality, and the broadcast model which is based on preferential spectrum licensing, public funding of broadcasters, and content control.

An examination of likely market evolution suggests that, despite the dramatic changes in media technology, two of the most compelling reasons for regulation of the traditional broadcast markets are likely to continue into the new media era. First, the public benefits or 'externalities' of high-quality news, drama, comedy and sport are likely be as great and as difficult for the free market to capture as ever; and, second, the tendency toward highly concentrated control of media distribution in the hands of a few operators will still exist. For these reasons, regulators should examine each platform individually to assess the case for and against the extension of established broadcast regulation into the new markets.

The guiding principles for this decision should be based around three criteria: the likely future importance of the platform in delivering news and entertainment content to consumers; the likely amount of market power exerted by the main distributors; and what the market impact of regulation might be. This approach, rather than a single policy position for all platforms, should enable regulators to ensure they build a regulatory framework for this fast-changing new media world that demonstrably serves the public interest.

The consumer agenda on regulation

Ed Mayo

Chief Executive of the National Consumer Council

Philip Cullum

Deputy Chief Executive of the National Consumer Council

Regulation has become an established profession, no less important in a world of complex organisation than its forebears of medicine, law and the mediaeval guilds. Indeed, one distinguished US regulatory academic describes it as a 'craft'[1]. And yet, like all professions, what counts for approval and success is shaped not just by a neutral interpretation of the public good but also by the interests of those with the relevant expertise and knowledge, who participate. The challenge in the debate on regulation, as with every other profession, is to reconnect professional practice with public concerns without loss to the expertise that required such specialist knowledge and skills in the first place.

If the current policy debate on regulation is anything to go by, this will require very considerable cultural change. The public dimension is all too often an afterthought. Yet the truth is that active consumers do far more to regulate markets than any government and regulatory institution. Their sensitivity to price and quality is a powerful regulator of firms in itself, promoting efficiency and driving improvement. But the way in which this happens tends to be assumed or invisible to many regulators. This is symptomatic of a deep, contemporary flaw whereby regulators increasingly act in the name of consumers, but all too often fail to understand them.

Beyond the myths

Over the last two years, we have challenged those engaged in the regulatory debate to move beyond the urban myth that regulation is a cost to business. In a competitive market, this is self-evidently not true. The true costs of regulation are charged either to consumers, as business costs are passed on, or to taxpayers, as the costs of regulation grow. What matters in regulation, we have argued, is that we should start from the consumer and public interest, rather than treat it as an afterthought. The costs of getting this wrong, in terms of economic performance and the wellbeing of society, are considerable.

To explore how to get this right, this chapter looks at:

- how consumers approach risk and regulation;
- a framework for understanding the consumer interest in regulation;
- a case study of consumers and regulation – weights and measures; and
- a menu for action.

1. Malcolm K. Sparrow, *The Regulatory Craft: Controlling Risks, Solving Problems and Managing Compliance*, Brookings Institution Press 2000

How consumers approach risk and regulation

If you only read newspaper headlines you could be forgiven for thinking we are a nation obsessed by risk. The view that risk dominates our lives finds support in the opinion of commentators such as Frank Furedi, who sees widespread public panic over issues ranging from GM crops and mobile phones to global warming and foot-and-mouth – in his words, an "epidemic of fear".[2] Rick Haythornthwaite, Chairman of the Better Regulation Commission, claims that *caveat emptor* is heavily anaesthetised in many markets, that we have seen an erosion of personal responsibility on the part of consumers, and that there are moves towards regulating every last risk.[3]

So what is going on? Is the nation indeed gripped by fear? Is anxiety paralysing our lives? At the National Consumer Council (NCC), we have conducted research into how people deal with risks and, frankly, we don't recognise the picture that has been painted.

What we found was that:

- consumers are most concerned about the issues of risk that most directly affect their lives and families;
- higher-income groups and those aged 30-60 tend to think and engage with risk issues more proactively than other groups;
- lower-income groups, while used to dealing with risk, tend to be more reactive in their approach to it;
- family, friends and relatives are the most important influences when it comes to assessing risk; and
- regulators and government do not command a great deal of trust. This is down, in part, to high-profile examples of regulators not handling risk well, from transport accidents and mis-selling of pensions through to the classic case of BSE in food. People are particularly concerned with what they see as a failure to present information in a clear and consistent manner.

Risk-based regulation, rightly, is the order of the day. But who ultimately decides what risk matters? One assertion that we come across in most 'expert' discussions on risk and regulation is that the group least able to make this judgement is the ordinary public.

In fact, our research with consumers challenges and overturns a second urban myth: the professional assumption that the public will get things wrong. Of course, consumers are rarely experts in matters of service, technology or economic valuation. But on the whole, they are very adept at weighing up risks and benefits for themselves, assessing information sources and coming to their own conclusions. In fact, compared to some models of regulation on risk, where ministers and committees are more concerned to protect their reputations by being zealously over-cautious, the public may even outperform the experts. The House of Lords Economic Affairs Committee [4] recently concluded that while there is little evidence to support the view that Britain has become an increasingly risk-averse society, "public sector reward and assessment systems may… encourage excessive risk-aversion".

2. Frank Furedi, *Epidemic of Fear*, www.spiked-online.com

3. Rick Haythornthwaite (2006) *The Regulation of Risk – Setting the Boundaries*, Lecture for the Centre for the Study of Regulated Industries, University of Bath School of Management

4. House of Lords Select Committee on Economic Affairs, Government Policy on the Management of Risk, 7 June 2006

What about attitudes to regulators themselves?

In truth, very few people talk about regulators with their neighbour over the garden fence – despite recent advertising campaigns by the Advertising Standards Authority and Ofcom. Ofsted, the schools inspectorate, is probably one of the very few exceptions. But when it comes to the safety of the food you buy or the money you have on deposit at the bank, most consumers will want the reassurance of having someone there to make sure that things do not go badly wrong.

Even so, we have seen little evidence of regulators investigating what value consumers seek from what they do. Regulators, like businesses, can disappoint equally by failing to deliver or indeed promising too much. What is the Healthcare Commission offering the public when it allows its logo to be widely and prominently used in advertisements for private sector health and plastic surgery clinics? Conversely, what benefit do consumers get from the rattled-off regulatory caveats on the end of radio adverts for financial investments?

What we have tended to find in qualitative research is that, even where consumers are not conscious of regulation and enforcement, when they consider it, they think that some protection is important to prevent business taking advantage of its power. They want the confidence that they are going to get, and be able to make, reasonable choices. They want to know roughly what to do, or where to go, if something goes badly wrong. And they understand that they pay for regulation, so they tend to prefer the minimum amount needed to achieve agreed objectives.

But while regulators help to underpin visible benefits for consumers, they may often be close to invisible in doing so. In reality, many regulators' well-intentioned attempts at transparency – whether through open meetings or consultation documents – reach only the usual suspects. This can still be valuable, but it is not the same thing as involving real consumers.

Above all, regulators and other decision-makers need to understand the diversity of consumer attitudes, experiences and needs, and put this at the heart of their strategy. This means having real insight into complexities such as:

- how consumers behave in real life, away from theoretically 'perfect' markets – in particular, how they use information and make decisions;
- different attitudes towards choice. With growing disposable income, more consumers now prize convenience – valuing products and services in terms of value for time, not simply value for money;
- the extent to which minority ethnic groups are themselves far from homogenous, in particular across generations. Of these, some Muslim communities, in particular Bangladeshi and Pakistani, fare considerably less well relative to others in terms of educational outcomes and social exclusion;
- the UK's ageing population. The number of people over 65 (the 'grey pound') is projected to overtake the number under 16 by the year 2014. The most dramatic change is in the proportion of people over the age of 85: 1.1 per cent of all men and 2.7 per cent of all women;
- widening regional imbalances. Regional GDP figures vary more widely than in other European countries, and these are mirrored by other inequalities such as endemic long-term health issues;

- increased prosperity but also high rates of relative poverty (particularly in households with children). Women are more likely than men to live in poverty – in particular, lone mothers and retired women who live alone;
- the goods and services that consumers need. These include not only the everyday essentials to ensure health and well-being (such as energy for heating the home or nutritious food for the family) but also the things that allow their active and meaningful participation in society and the economy.

The consumer interest in regulation

In such a fast-changing context, it is important to be clear about why better regulation is needed. But time and again, unnecessary regulation is presented as a cost for business. The British Chambers of Commerce even publishes what it calls its business 'Burdens Barometer'. There's a real danger that politicians and regulators ignore the true meaning of better regulation, and instead engage in an unseemly race to cut as much red tape as they can.

The Better Regulation portal[5] calls for "proposals for regulatory simplification from business, the voluntary and community sectors, public sector front line staff and any of their representatives". Not consumers.

The DTI's website says "cutting red tape by simplifying regulation is about reducing burdens on business... This is your chance to influence the DTI's cutting red tape plans and help achieve real results for business".

At NCC, our concern is to achieve real results for consumers, and we think that's what regulators and government should focus on too. We're in favour of well-functioning markets. Our starting point is that competition works, with well-informed consumers exercising real choice. As we will see below, it is sometimes necessary to have rules protecting individual consumers who may be vulnerable in a particular market. But at the same time, policymakers and regulators can do much to promote consumer interests in the marketplace and enable consumers to help themselves – getting it right, not just putting it right.

Its not in consumers' interests to have unnecessary regulation – after all, they pick up the tab.

We also know from years of experience that incumbent businesses often use regulation to prevent or restrict competition, for example by setting unnecessarily high entry standards. Business may talk tough when it comes to regulation, but all too many companies – and indeed entire sectors – are willing to embrace (and sometimes lobby for) rules and restrictions when they think it will help further their own position.

At the time of writing, the Law Society has persuaded the parliamentary scrutiny committee examining the draft Legal Services Bill to recommend that will-writing services are regulated – even though consumer groups point out that there is no systematic evidence of detriment to justify it. If enacted, this would mean higher barriers to entry, less competition and greater profits for lawyers.

Regulation can inhibit business creativity and innovation. This can be a particular issue for small businesses, so often a driver of new ideas in our economy. In fact, it has been estimated that regulation has five times as much impact on them.[6]

5. www.betterregulation.gov.uk

6. Federation of Small Businesses, *EU dialogue is failing small firms, warns FSB*, 14 April 2005

A survey of members of the Federation of Small Businesses[7] found that the longer someone has been in business, the more likely they are to be dissatisfied with the rules affecting companies. This is the deadening effect of regulation. Unnecessary bureaucracy like this also increases costs, and as a result consumers and taxpayers pay more.

Indeed, the reality is that consumers have benefited hugely from much deregulation over the years. In markets as diverse as opticians' services, conveyancing and air travel, consumers now have greater choice and lower prices thanks to the opening-up of markets – moves supported at the time by NCC. In each case, scare stories put around in advance were unfounded, and quality standards have been maintained.

Many business people routinely denounce all regulation in the strongest terms. The CBI talks regularly about "a red tape tide", while the British Retail Consortium[8] says that "excessive regulation and red tape are damaging business". The British Chambers of Commerce[9] goes even further, declaring that "monitoring and measuring the cost to business of compliance with regulation and red tape is one of [our] top priorities".

According to one PWC survey[10], the world's top chief executives think over-regulation is the biggest potential threat to business growth – more important than competition from China and India, issues about human performance and skills, or the state of the wider economic environment. This seems unlikely to be correct.

The reality is that businesses benefit from much regulation, and they support it when it suits them – often justifying this as the need for a 'level playing field', when in fact they want quite the opposite. And while most in business say they do not like detailed regulation, many (not least those working in companies' compliance departments) like being given direct instructions by a regulator. This is apparently known in some circles as 'Nike regulation', after the sports firm's advertising slogan 'Just do it!'

John Sunderland, Executive Chairman of Cadbury Schweppes, is a fierce critic of much food regulation, denouncing 'traffic light' nutritional labelling as 'extreme'. But as President of ISBA, 'the voice of British advertisers', he has called for increased regulation of the BBC and for a proposed merger of two TV companies to be investigated by the Competition Commission.[11] When they talk about the detail, business leaders often have pretty nuanced views about regulation, recognising the pluses and minuses – yet such subtlety tends to be lost when the anti-regulation rhetoric starts.

There is a pressing need for a more measured, intelligent dialogue about regulation. We believe there would be much greater momentum towards better regulation if there were a more precise and upfront articulation of regulatory objectives, which would often (but not always) be about protecting or promoting the interests of consumers. This would also allow a much clearer assessment of what is working and what isn't.

7. FSB, *Lifting the Barriers to Growth in UK Small Businesses*, 2004
8. http://www.brc.org.uk/details04.asp?id=633&kCat=&kData=1
9. BCC, *Cutting Through Red Tape*
10. *The Times*, 30 January 2005
11. Speech to ISBA annual conference, 12 March 2003

Where regulation does fail, it is often because it lacks focus or has become disconnected from its original purpose. The application of the money-laundering regulations causes huge problems for many consumers who want to open a bank account, while doing little to deter criminals. The restrictions on the number of taxi licences in many towns simply protect incumbent operators – and mean you can't find a taxi. Failing pieces of regulation such as these restrict competition, limit choice and raise prices. Far from protecting consumers, such rules and restrictions harm them – so they should be swept away.

When regulation works

This doesn't mean that all regulation is unnecessary – far from it. But we have seen shifts in the focus for regulation, with different regulatory approaches, as the economy changes.

The consumer movement in the 1970s, when NCC was founded, was primarily focused on products; now it's the service sector that dominates the UK economy, accounting for 71 per cent of gross value added in 2001.

Communication technologies are accelerating this focus on service innovation, by changing the way that people (and things) interact. Three-quarters of adults own or use a mobile phone and over half have a home computer. But consumer concerns around services still revolve around issues of information and power – of not knowing what the quality of a plumber or builder will be; of being locked into contracts and costs; of not being able to gauge a fair price at the outset; of being open to mis-selling where the complexity of services mean you need to rely on the judgement of others.

We consider that there are two key types of regulation. In both instances, critics are simply wrong to characterise regulation as a problem.

The first is **lifeline regulation**, which safeguards consumers. The CORGI rules about gas safety save lives – around 47,000 businesses employing 98,000 gas fitting operatives are now registered, ensuring that UK households benefit from a well-trained workforce. Set up 35 years ago, CORGI commands wide support, and no-one today seriously argues that it has been anything other than a good thing for consumers and business alike. In a similar vein, lifeline regulation governs access to affordable energy and water for vulnerable consumers.

The second is **market-making regulation**, which enables proper competition and choice. The Building Societies Association says[12] that there is too much regulation and "an inadequate recognition of the long-term benefits of competition in delivering consumer value". But let's not ignore all those instances where regulatory intervention is needed to make markets work.

Common technical standards can help establish competitive product markets; and ensuring that advertisers are not allowed to make misleading claims helps to promote choice. The current Competition Commission inquiry into doorstep lending, prompted by NCC's supercomplaint, should sort out a market that is failing low-income customers. Standards on food labelling can help consumers make informed decisions and promote new ranges of healthy food.

Regulatory action to make markets work can be about cracking down on anti-competitive practices. But it can also mean more positive steps such as sweeping away barriers to

12. BSA press release, *High risk of over regulation*, 5 May 2005

switching between companies and promoting informed choice. It can also be about enabling consumers to share their experiences of different companies, so that those who deliver good products and high service standards benefit, while those who don't lose out.

As one measure of the trend to more consumer power, NCC researched the consumer experience of switching suppliers in core markets between 2000 and 2005.[13] The markets covered sectors as diverse as home insurance, current account banking, gas, electricity, fixed telephony and mortgages. The results, in the NCC Active Consumer Index, revealed a remarkable change in the economic geography of the UK, with a 52 per cent increase in consumer switching over the five years (figure1).

We'd like to see greater use of what NCC terms 'reputational sanctions'. We know that a relatively small number of poor-performing businesses – whether they're rogues or just incompetent – accounts for a huge amount of consumer detriment in different markets.

Figures from the Financial Ombudsman Service show that, in a single year, more than 22,000 businesses had no cases levelled against them. But just 277 companies accounted for 103,000 complaints, which means that one per cent of the companies accounted for 91 per cent of the complaints. In legal services, just seven per cent of solicitors' firms accounted for one-third of complaints to the Law Society over a 12-month period.

By publishing this kind of information about company performance; by encouraging consumers to share their experiences; and by working with journalists to train a spotlight on companies that are letting consumers down – we can provide a powerful incentive for businesses to change their ways.

Sometimes the best option can be self-regulation, allowing an industry to take responsibility for any necessary rules. The premium-rate telephone regulator ICSTIS – which covers information and entertainment services – is one example. Co-regulation has built the consumer trust needed to make this £1.6 billion market work – double the size of the market in the USA which is now belatedly adopting the UK regulatory model.

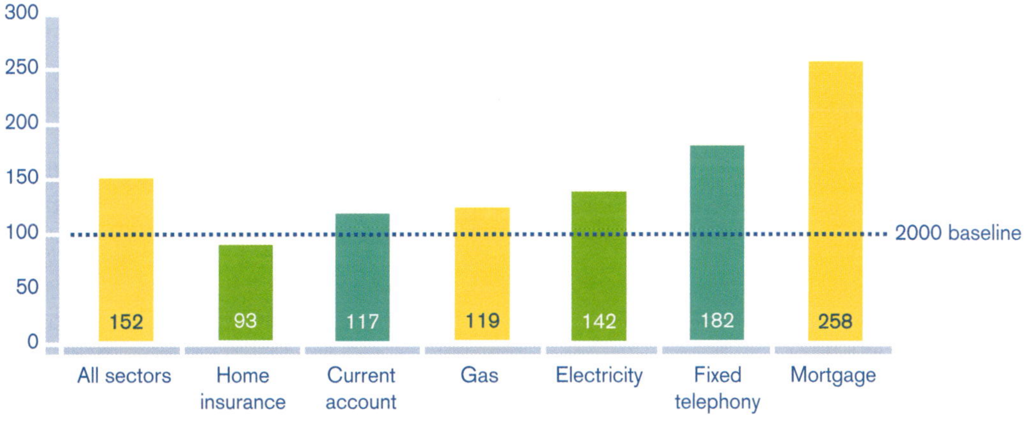

Figure 1: NCC Active Consumer Index

13. For full details, see *Switched on to Switching? A survey of consumer behaviour and attitudes, 2000-2005* and *The NCC Active Consumer Index: the rise of consumer power in the UK economy*, NCC, December 2005

However, in a failing market, self-regulation is a privilege, not a right. Our research suggests that it works less well in fragmented markets where there is a wide and diverse range of providers. Look no further than the car servicing and repair sector, a prime case study where there is overwhelming evidence of massive consumer detriment[14] in the wake of 11 failed attempts over the last 30 years to establish effective self-regulation.

Regulation as if consumers mattered

NCC has been working with the Department of Trade & Industry on a pioneering project to integrate consumer views by exploring a 'value chain' approach to the design of regulation.

The field that we have explored is weights and measures regulation. This is a classic area of consumer-focused regulation, but one in which the rules have been written and decisions about enforcement made without any obvious moves to ask the intended beneficiaries what they need.

When we conducted some qualitative research in this field, and explained the rules on units of measurement and prescribed quantities (which among other things govern the weights of a loaf of bread and a packet of teabags), many consumers simply laughed. They told us these rules are utterly alien to their day-to-day lives. One person searched in vain for some kind of rationale: "Can I just make an observation on the inconsistency of what you have just mentioned? Tea in set multiples of 250 grams which is a quarter of a kilo but then earlier you were talking about the standard weights for bread being 400 and 800 grams. Why make those particular weights? Why not 500 grams? Why not half a kilo? Why not 750 for the other?" Another said "It is just irrelevant".

As well as challenging the assumptions that have governed regulation in this field for decades, the research showed how consumers are able to engage with complex issues about regulation, risk and prioritisation – provided regulators approach this in the right way.

We can build on some positive precedents. Ofcom's predecessor, the Independent Television Commission was one of the UK's early adopters of citizens' juries, and the Food Standards Agency has a good record with this technique too. The National Institute for Health and Clinical Excellence (NICE)[15] has established a Citizens' Council – a deliberative forum of 30 people, a third of whom are replaced annually, facilitated by an independent organisation. The forum can cross-examine witnesses and make recommendations; while these may not be binding, NICE has to explain whenever its decision differs from the Council's view.

An Open University evaluation confirms NICE's view that consumer engagement in regulatory decision-making adds value – not by undermining professional expertise but by adding to it. NICE chief executive Andrew Dillon explains:

"It might be reasonable to assume that the public would to be prepared to accept that scientific value judgements should be made by those with the competence to do so providing, always, that it is done transparently. Such experts do not, however, have any special legitimacy in making the necessary social value judgements. Those judgements need to

14. This is estimated to be as much as £4 billion a year in a £6 billion market. See for example Steve Brooker, *At a Crosswords: Getting the UK Car Servicing and Repair Sector Back on Track*, National Consumer Council June 2005; and *Three quarters of garages leave faults after servicing cars*. Trades Standards Institute, 21 June 2005.

15. NCC annual lecture 2005, Andrew Dillon, *Involving the Public and Patients in the Work of NICE*, 23 November 2005

reflect the values held by those who pay for and use the services concerned."

Similarly the Human Fertilisation and Embryology Authority has involved the public in decisions about highly technical and morally complex issues, such as whether parents should be permitted to select embryos to provide gene therapy for existing children.

Future regulation could be a force for democracy in addition to wellbeing and economic advantage, by operating in ways that open out complex issues and trade-offs for public deliberation.

NCC's work on weights and measures is taking engagement a step further, by bringing together policymakers, regulators, enforcers and consumers in a collaborative dialogue, so that all the key players hear each other's views and can develop a common agenda about what's needed – and what isn't.

A menu for action

There needs to be a fundamental shift in the UK's regulatory culture. By far the best piece of government thinking on regulation comes not from our country but from the Republic of Ireland, where a 2004 White Paper[16] put the issues starkly:

"Regulation affects the quality of everyday life. This includes the quality of our food and water, the safety of our workplaces as well as the range of products and services available to us and the price we pay for them. Despite this level of influence, we have rarely paused to consider in a systematic way questions relating to the quality of our food and water, the safety of our workplaces as well as the range of products and services available to us and the price we pay for them. Despite this level of influence, we have rarely paused to consider in a systematic way questions relating to the quality of regulation. Questions such as: why we regulate, how we make regulations, what kind of regulation we want or what constitutes good regulation."

Here in the UK, the debate about regulation has seemed much less mature, too obsessed with the quantity – rather than purpose and quality – of regulation, and too often dominated by vested interests.

It's time to abandon the tired rhetoric about business burdens. Let's instead have a strong signal from government that the consumer interest should sit at the heart of much regulatory decision-making. Regulators need to start with a focus that sees the public not just as prime beneficiaries of regulation, but as co-producers of it. Consumer behaviour is complex, based often on habit, social norms and a range of decision-making shortcuts. But it can be understood and, in concert with action on the supply of goods and services, regulators are in a position to promote consumer power. We define consumer power as "where individuals have the confidence, capability and context to act purposefully, as interpreters of their own needs". After all, with consumer power, people not only benefit from competition, but they initiate and sustain it.

Regulation should also be seen in the context of more measured debate and analysis of the different policy and institutional tools that exist to address imperfect markets. NCC is currently developing a detailed methodology for action to address failing markets.

We'd like to see regulators doing far more to engage consumers in decision-making. All the

16. *Regulating Better: A Government White Paper setting out six principles of Better Regulation*, January 2004

evidence is that people are willing and able to get involved in helping make tough decisions; that they are comfortable about issues around risk and prioritization; and that public involvement leads to better outcomes in terms of regulation that commands widespread support and is fit for its purpose. Yet our 'regulator-knows best' culture persists.

As we noted above, some in industry like to be told by regulators precisely what to do – a phenomenon that has been described as business's 'dominatrix complex'. But regulators really need to aim at sustainable culture change in terms of how organisations relate to consumers, not just narrow compliance with detailed rules, so we support greater use of principle-based regulation. The Unfair Commercial Practices Directive is a welcome step towards this, as is the Financial Services Authority's 'Treating Customers Fairly' initiative. Both seek to encourage companies to take responsibility for their own actions.

Finally, once everyone is clear about when we should regulate and how best to do it, let's establish some kind of regular, independent assessment of regulatory performance. Local authorities have to account for what they've done through the Comprehensive Performance Assessment, and the National Audit Office already evaluates Regulatory Impact Assessments[17]. Why shouldn't regulators (including government departments) answer to consumers and business alike for how they've delivered on the better regulation agenda? The Prime Minister's new Panel on Regulatory Accountability[18], which aims to *"hold departments and their regulators to account for their regulatory performance"* is good news, but like all Cabinet committees it does not operate in public view.

This might not be the only way of increasing public scrutiny. While regulators have demonstrated increasing awareness and concern on issues of market competition, is there a case for the contestability of regulatory services themselves? Two recent examples demonstrate how this might work.

The Department for Constitutional Affairs has taken on powers to regulate the claims management sector, where there is evidence of malpractice with consumers being at their most vulnerable. They have contracted out the compliance and enforcement functions, after market testing, to one of the larger trading standards bodies in a local authority. This could prove to be both cost-effective and is an innovative example of decentralisation in the regulatory field, when the trend after Hampton[19] has been the other way.

In terms of dispute resolution, Ofgem has been looking at different parties to run an energy billing ombudsman – a contract that has been won by Otelo, a communications ombudsman service. It is possible that legal complaints might go a similar way. These offer a scenario in which some regulatory functions could be opened up on a competitive basis – not least to those people on the ground in trading standards service who are responsible for so much of the hard work of regulatory enforcement.

Consumers as an asset

There is, and will continue to be, a wide diversity of regulators, both in the UK and at EU and global level. Despite this, regulators tend to be entrenched within a particular discipline. Competition regulators tend, for example, to view consumers as passive beneficiaries of supply-side competition interventions,

17. National Audit Office, *Evaluation of Regulatory Impact Assessments 2005-06*, 28 June 2006
18. http://www.cabinetoffice.gov.uk/secretariats/committees/rbpra.asp
19. http://www.hm-treasury.gov.uk/budget/budget_05/other_documents/bud_bud05_hampton.cfm

or as active only through the narrow lens of utilitarian calculus. They tend to view consumer issues as about 'protection', characterised as an occasional constraint on market activity. Yet they still rarely integrate the findings of behavioural economics, such as the fact that consumer behaviour is not always utilitarian and that consumers are far from homogenous. A similar set of issues may apply in relation to regulators who operate within cultures of law, science or technology rather than economics. All share the rootstock of a regulatory tradition in which the public is seen in a 'deficit' mode, as lacking the knowledge and skills to engage in the design and evaluation of regulation.

Rather than see consumers as a deficit, we have pointed in this chapter to evidence that, in general, consumers are an asset to the outcomes prized by regulators. We have highlighted the consumer interest in regulation and pointed to a case study of innovation in closing the regulatory gap between professionals and the public.

Our vision for future regulation is one that sees the core of its relationship with ordinary people as one of assurance – starting from where people are, rather than assumptions of how they behave.

section 2

The changing nature of regulation in the public interest

The public interest challenges for the communications sector over the next 10 years: contestable public service funding

Chris Giles

Economics Editor, *Financial Times*

The future could look like this

It is the year 2015. New university graduates have never lived in a world without the internet. People turning 40, and close to the halfway stage of their working lives, have never experienced an office without email. Even the most technology-resistant pensioners now have multi-channel television since digital switch-over was completed last year, after the inevitable delays and teething troubles.

The Government is in a bit of a pickle, however.

Ministers know what they want. They are convinced that broadband communication is indispensable to modern life – not just to receive information from utility companies, government, schools, hospitals and private companies, but also to send information back, whether to report a leaking Victorian water main or for rural pensioners to authorise their medical details to be passed from a doctor to a specialist. The newly retired Beatles generation cannot imagine life without access to broadband communications (even if quite a few still choose instead to turn on, tune in and drop out).

Ministers also remain convinced of the power of video communication through the nightly television news; they want everyone to have access to the best information and entertainment available. The public, when asked in countless surveys over the previous decade, have agreed.

But ministers are unclear how to secure these public interest objectives.

Successful incremental liberalisation of the telecommunications market over the past 30 years has left many companies vying for customers in big cities, small towns and suburbs, using a range of technologies, suitable for different areas. But these companies, competing fiercely and making respectable returns, flatly refuse to subsidise services in rural areas or to provide below-cost tariffs for poorer people who still struggle to afford a broadband connection. Monopoly profits at British Telecommunications have all but gone, and it recently won a legal ruling at the European Court of Justice against the imposition of a universal service obligation that would require it to provide a copper line connection to all UK households.

In the broadcasting arena, only the BBC is left providing a regulated public broadcasting service. Its licence fee – still levied on the ownership of a television set, despite the considerable blurring between TVs and other visual display devices – pays for its audiovisual output across digital terrestrial, cable, satellite and downloadable services. Every private-sector competitor complains it has an unfair advantage, driving up their costs and reducing their commercial revenues.

This argument has run for decades and is no nearer resolution. But the BBC knows full well that it must remain connected with the British public if it is to keep its privileged position. So, to stay on the ball and to minimise the decline in its audience share and reach over the past decade, it has followed other broadcasters with trends in programming. In some areas it has developed new formats to keep audiences happy and stands accused, as never before, of sacrificing its public service remit on the back of a drive to maintain that share and reach.

In the commercial sector, it has been a couple of years since ITV and Five were relieved of their formal public service broadcasting objectives. The decision was taken when it became clear that, in a fragmented broadcasting market, advertising revenues could no longer fund the higher cost of programming that genuinely served the public interest in a way that was different from purely commercial broadcasting.

Channel 4 is now a successful niche player, producing programming that meets the desires of its relatively young audience profile. However, very few people watched the 'landmark public service programming' it highlighted in its 2014 annual report, when bidding to retain the remaining privileges that come with being classified in law as a public service broadcaster.

When Ofcom published its third public service broadcasting review earlier in 2015, it conceded that only the BBC now had the scope to provide a significant quantity of public service broadcasting. But it concluded that the lack of competition from rival broadcasters reduced the quality and quantity of BBC programming that could be easily described as fulfilling the purposes of public service broadcasting.

Across the communications sector, the realisation is that the public interest has not been well served by the regulatory settlements and compromises of the past decade.

My simple argument

Now, I cannot predict what will happen over the next decade. I know that the previous paragraphs will be wrong, and badly wrong in places. But the bad world I have outlined is a distinct risk. And one that sensible public policy should strive to avoid.

I will argue that thought needs to be given now to the introduction of contestable funding for public interest objectives, since the old cosy relationships and implicit funding of those objectives will struggle in a world of greater competition.

I am well aware that there may be little point in using the communications market to secure public interest objectives in the future: if people will not watch PSB programming there is no point in subsidising its production and if society is unaffected if some people choose not to adopt broadband, we should not waste resources ensuring that everyone has access. But I will not dwell on that possibility.

I will assume that there will remain some grounds for public interest interventions in communications markets and look at the best ways of securing them, using universal service obligations and public service broadcasting as examples.

No part of my case in favour of contestability – that providers should compete for the funding and the right to provide the services that elected governments believe enhance society – is new. None of it is radical or sophisticated. In fact, it is blindingly obvious.

Private companies contract with others on the basis of quality of provision and cost. They regularly sign long-term contracts with third parties but never guarantee to renew them without looking at the possibility of alternative suppliers.

In the public sector, companies bid to run public services, including running trains on the railways, building privately financed roads, operating the new NHS treatment centres, emptying our dustbins, putting parking tickets on our cars and designing new government computer systems.

I deliberately include controversial services in this list because, with the exception of the NHS treatment centres, the dissatisfaction with these public services has not led to a desire to end contestable funding; only to change the incentives faced by the contractors or, indeed, to change the contractors themselves – something you can do with contestable funding.

In the future, the operation of the proposed National Pension Savings Scheme will be put out to tender and contestability will play an even greater role in health and education services.

The reason the government is slowly moving in the direction of contestability was well made by Julian Le Grand, the LSE professor and former adviser to Number 10, in a recent article for the Centre for Market and Public Organisation: "Giving providers a monopoly has never been a good way to improve a service of whatever kind – and the 'old' health and education systems were no exception."

It is not just in the 'newer' bit of the Labour Party where this thinking is current. The Conservatives have taken contestability to heart in their philosophy for public service reform. In July, George Osborne, the shadow chancellor told the think tank Policy Exchange that his ambition was to introduce much more "contestability, so that poor performance is vulnerable to someone else coming in and doing better".

Contestable funding, therefore, shouldn't be controversial. As I shall outline, it isn't a big issue in telecommunications, in relation to universal service obligations in the future. But for some reason, it remains a very difficult concept for the current generation of broadcasting executives at the BBC to get their heads round.

Now that I have outlined the risks for the public interest in communications markets, and highlighted the general principle of contestability, I'll look briefly in the rest of this essay at how the changing world is undermining the tried and tested methods of securing certain public interest objectives in the communications sector, and highlight remarkable similarities between universal service obligations in the telecoms market and public service broadcasting in the TV market.

I'll then take a look at what the public might hope to gain from contestable funding.

Finally, using a universal service obligation and public service broadcasting as examples, I'll look at how governments, or regulators, could decide on the objectives they wish to secure and how to go about it.

Securing the public interest in communications markets

a) Telecommunications

Before privatisation in 1984, the telecommunications sector was often assumed to be a natural monopoly in which competition would be wasteful and would duplicate expensive capital investment. But in the following 20 years, new thinking about potential competition in many parts of the telecommunications network, as well as new technology and entirely new networks (such as cable and mobile phones), has lowered the barriers to entry and increased competition.

The main public interest served by the regulators, Oftel and then Ofcom, has been to increase productive efficiency (through price control and gradual introduction of competition), lowering the price of communication and increasing the opportunities of innovation to users of the telecommunications network.

The telecommunications market is complicated and, though competition has increased, barriers still exist that require continued regulation. Nevertheless, the scope for British Telecommunications plc to make monopoly profits is reducing every year as competition on its networks, and between telecommunications networks, increases.

In a world of less monopoly power, BT has less scope to cross-subsidise loss-making activities from profitable ones. That means that regulators have less scope to come to unwritten or written agreements with BT, forcing it to do stuff which serves the public interest in return for being allowed to make higher profits in other parts of the business. It is simple: if the monopoly rents are not there, you cannot demand that BT lowers its profits for the greater good of British households.

Since 1984 in the UK, the main bargain agreed between regulators and BT (and, in Hull, Kingston Communications) has been that it provides certain universal services as part of its licence to operate. The services include connection to the network at a uniform price across the UK, schemes to help low-income households, public call boxes in both profitable and loss-making areas and help for disabled users. The Government determines what these universal services are, although it is constrained by European legislation.

Once competition in the telecommunications market is strong, it becomes an empirical issue whether the universal service conditions are too onerous for BT to bear. Ofcom is currently undertaking another cost-benefit analysis to try to answer the question. But what is clear is that as competition marches on, and regulation of the market improves, there will come a point where BT can no longer be forced to make these services universal, or to introduce new ones.

But this is progress. Any redundancy of old universal service funding techniques comes as a direct result of benefits delivered in the form of lower prices and greater efficiency from a better-working telecommunications market. But when that day arrives, it still leaves government with a funding problem.

b) Public service broadcasting

The TV broadcasting market, much as it likes to think it is unique, nowadays shares many of its characteristics with telecommunications and print publishing.

It was not always like this. Unlike publishing, which commissions content and distributes product to readers in reasonably competitive markets, the TV market has traditionally been characterised by huge bottlenecks in distribution. Those who controlled distribution (the airwaves) controlled the market.

Ever since TV was invented, successive governments have believed the medium to be too powerful to allow a few private sector companies to broadcast. Initially, that's why just one organisation – the British Broadcasting Corporation, an independent but wholly owned public company – was allowed to produce programming. The limited scope for competition, given the scarcity of available airwaves, ensured that the BBC has ever since offered its idea of public service broadcasting to the public. In the past 50 years, one, then two, then three private sector competitors were introduced, each of which also had obligations to produce programming that served the public interest.

The lack of available spectrum led to excess profits from advertising. And these profits were partly taken by the private sector broadcasters, partly enjoyed by the taxpayer in the fees government charged for access to the airwaves and partly spent in the production of non-commercial public service programming.

But as digital technology has taken hold and the technological constraints have eased, consumers have benefited from greater choice, innovation and new services. Spectrum scarcity and the opportunity for excess profits has also fallen. Just as the case for implicit funding of universal services gets harder to press as excess profits reduce, so does the argument for funding public service broadcasting by private companies.

Again, this is progress. Consumers have benefited enormously from greater choice in the TV market, but the price of this progress has been a drying-up of resources to fund public service broadcasting.

But unlike universal services in telecommunications, broadcasting has a huge advantage. It already has a massive public fund – £3bn a year, raised by the TV licence fee – to spend on public service broadcasting. The only problem is that the fund goes entirely to one company – the BBC – which, on its own, cannot be trusted to spend the money in the public interest.

The Government has decided to give the BBC a ten-year renewal of its Royal Charter from 2007 and, once granted, the incentive will be to show it can reach an audience – whether its programming will be in the public interest remains to be seen.

The new BBC governance arrangements may prove to be better, or worse, than the last. But PSB is so difficult to define, and its objectives so slippery, that I cannot foresee the management having much difficulty in persuading the new BBC Trust that it is following its public service remit, even if it is not in many areas. In fact, to assume that the management would not try to play games requires a naivety that is rare in the modern world.

Table 1 outlines the similarities and differences between technological progress and the ability to maintain public interest objectives as technology improves and competition is advanced.

Characteristic	USO in telecoms	PSB in TV broadcasting
Then		
Previously uncompetitive markets	✓	✓
Excess profits for private companies	✓	✓
Private companies told to meet certain public interest objectives	✓	✓
Implicit financing of public interest conditions	✓	✓
Severe technical constraints pose barriers to market entry	✓	✓
Now		
Lower barriers to entry	✓	✓
Competing delivery technologies	✓	✓
Lower scope for excess profits	✓	✓
Benefits of increased choice	✓	✓
Reduced ability for cross-subsidisation	✓	✓
Existing public finance to procure public interest objectives	×	✓

Table 1: Similarities in securing public interest objectives in telecommunications markets and broadcasting markets

Contestability in markets

The idea behind a contest to receive public funding is straightforward. Competition for the right to produce a good or perform a service will lead to lower prices and higher quality. If more than one provider receives a contract, that competition can be sharpened even further by direct comparisons between two or more winners of similar contracts. Those awarding the contracts can see how well each provider performs in head-to-head comparisons with others.

It is generally accepted that this kind of competition for funding is almost certainly better than setting up a public sector operator to perform the task, or handing money over to an incumbent.

But there are reasons for some caution. There needs to be genuine competition for the contracts; the winners must have the ability to provide the service and meet public interest objectives; there is a danger about generating too much bureaucracy in the grant awarding process; after winning a contract, the terms must be clear and easy to enforce; and those awarding contracts must have sufficient information to make effective decisions.

It is clear from this list of concerns that contestability is not a magic wand that can be waved to ensure private contractors are efficient and align their incentives with those awarding the contracts. No one should be under any illusion that in an auction for public funds, issues of poor or asymmetric information, market power and game-playing will not go away. But governance and regulatory concerns exist in spades if you simply award a contract to a preferred bidder without a competitive process at all.

The argument is not that contestability is the perfect solution: it isn't. But for all its imperfections, it is certainly preferable to the alternatives.

The practical application of contestable public service funding in communications markets

In this final section I'll return to my two examples where some form of public funding might be needed in future, since the traditional ways of ensuring universal service in telecoms and a plurality of public service broadcasting cannot be guaranteed in a more competitive landscape. I will argue that some form of contestable public sector funding is the best solution to the problem.

a) Universal service obligation in telecommunications

Since privatisation, BT has been subject to a universal service obligation (USO). It has made sure that the company has charged the same rates to almost all of its domestic customers, regardless of the cost, and has offered low-cost options for poorer families.

The costs of operating a USO were paid out of high-margin business elsewhere, although Oftel regularly claimed that BT actually benefited from offering a universal service.

For the future, there are still valid arguments for thinking that some universal obligations in the telecoms market are a public interest worth preserving. The telephone has become so vital to modern living, it's reasonable to regard it as a necessity that every family should have.

But as competition intensifies, so does the pressure on that higher-margin business that funded the USO in the past. It follows, then, that it will no longer be reasonable for regulators or government to insist that the legacy monopoly provides the USO.

As Ofcom's strategic telecoms review noted: "As competition continues to erode high margins across a widening set of products, at some point providing the USO may become an 'unfair burden' to BT and to Kingston. At that point it will be appropriate to introduce alternative mechanisms for funding and allocating the USO."

How should the public authorities allocate the funding for a USO? Well, Ofcom was again very clear and precise. It said: "It is important to note that competition for USO is feasible." But since the regulator will not know which company could provide the USO most efficiently, an auction is the best way of allowing the most efficient service provider to meet the USO obligations.

There are complications, of course. Incumbents might have an advantage in any auction, and might have to be forced to co-operate with a different winning company. These can create regulatory headaches. But at no point did Ofcom, nor the weight of responses to the argument, suggest that establishing some competition for the market was wrong.

Who would pay for the USO? Well, ideally it should be the taxpayer directly; after all, the money is designed to offset a problem for society. Tax finance is also transparent and the public can have some say, in elections, whether decisions about the merit of the USO conform to their views of the correct way to spend their money.

As Ofcom noted, however, the taxation route is unlikely to gain support since governments do not like transparency as much as voters. This failure of government implies that the most likely form of finance would be an industry levy: individual households would still pay, but in higher telecoms bills rather than in higher tax bills.

b) TV Public Service Broadcasting

One thing that people in the broadcasting industry tend to agree on is that they are special. This is profoundly mistaken.

TV executives are certainly powerful: they have huge marketing departments and an ability to publicise their views in ways other captains of industry could only dream about. But in truth, public service broadcasting can never compete with health, education, emergency services or, arguably, telecommunications in the essential role it plays in society. That much is clear from all the research. Indeed the Government's White Paper on the BBC in the digital age makes the argument that a regressive licence fee is the correct funding mechanism, precisely because the public regards TV as a 'luxury' and there is no need for explicit redistribution to vulnerable groups.

For all the ordinariness of the television industry, I am happy to assume that there will remain some rationale for public funding for public service broadcasting into the medium term. It remains, however, extremely tricky to articulate precisely why consumers would not pay themselves for a product if they value it so highly.

The question is how to distribute the amount of money that an elected government deems should be spent on public service broadcasting. Ofcom estimated in 2004 that the annual amount of public funds devoted to PSB was about £3bn a year, split between the BBC (£2.5bn), ITV (£0.3bn), Channel 4 (£0.2bn) and Five (negligible). Most of the money came from the licence fee (which raises over £3bn a year and also funds other BBC services), but the rest came from direct taxation, paying over £400m a year to the BBC from the free licences for the over-75s, and a whole series of hidden subsidies to broadcasters, which reduce the revenues collected by the Treasury.

Ofcom's public service broadcasting review asserted that many of these hidden subsidies would become impossible to continue: competition would erode the advantages enjoyed by terrestrial broadcasters, leaving the BBC as the only beneficiary after digital switchover of public service broadcasting financing, still receiving the licence fee and direct support from the Treasury.

Ofcom thought that the result would be a lack of "competition for quality", and that this would not be a desirable outcome. The Department of Culture Media and Sport agreed. The BBC half-agreed. I agree.

The question for the future for public service broadcasting is, in many ways, much simpler than it is for universal telecommunications services. The public authorities must decide how much PSB they think should be funded, and how to distribute the cash already allocated for the purpose in a way that best serves the public interest.

The Government chose to delay addressing the issue until digital switchover. This was a missed opportunity. But others do not need to be so timid. The BBC certainly decided to let its (unsurprising) views be known: that it should receive all the public money for PSB.

Its arguments were: a) that the public funds would be spread more thinly across BBC services; b) other broadcasters, but not the BBC, might engage in 'game-playing' and seek funding for programming they would have shown anyway; and c) competition for funding might add bureaucracy and undermine creativity.

The arguments are exactly what you would expect from an organisation that is dedicated to self-preservation and the maintenance of a privileged position. The fact that the BBC cannot see the distinction between the public interest and its own is a very powerful argument in favour of contestable funding. In short, the BBC's case is absurd.

The new competitive landscape threatens to thin the spread of PSB funding, by eliminating subsidies from the BBC's competition. That is the legitimate concern for the public, not the BBC's self-centred view that it is the only true provider of public interest television. Game-playing is a problem wherever incentives between the funding body and the provider are not exactly aligned. That applies to all broadcasters and perhaps, specifically, to non-profit making broadcasters who cannot see a distinction between self-interest and the public interest. Also, a contest to receive funding for PSB could increase bureaucracy, or it could reduce it. That depends on its design.

So, consider this example for a bit of fun.

Let's decide what we want from a system of distributing public revenues to public service broadcasting providers. The first thing that gets almost universal support is "competition for quality". I will split this into two and interpret this as we want competition among producers of PSB programming, as well as competition among commissioners to get that competition for quality.

Third, we want accountability and oversight of public money. And fourth, we want brand awareness that drives people to watch their own publicly funded programming; otherwise, there is no point.

Now, let's think about seven possible models that public funding could take.

1. **An 'Arts Council of the Air'.** This would fund individual programmes on broadcasters' schedules. There would be no special treatment for the BBC. It could apply for funding like any other broadcaster. Otherwise, it would be free to seek any other funds through subscription or advertising.

2. **Fully contestable funding at the broadcaster level.** Broadcasters would compete for block funding for a specified period – say three years – specifying what sorts of PSB programming they would produce.

3. **Fully contestable funding at the broadcaster level for a specific franchise.** Broadcasters would compete to run a PSB-style channel. This is along the lines of Ofcom's proposals for a public service publisher.

4. **The BBC competing with ITV/C4/C5 for public money.** A closed bidding process in which public money would be distributed to the current analogue terrestrial incumbents according to their proposals and costs of meeting public interest challenges.

5. **The BBC alone as a commissioner without any production.** This would turn the BBC itself into the 'Arts Council of the Air' with a schedule to fill that met the public interest.

6. **Competition to run BBC services.** The government could auction the right to manage and operate BBC services. For example, it could auction the management of BBC2.

7. **The BBC alone, doing the same as today but in the new digital world.** This is the outcome if current policies are left in place.

The outcomes are shown in table 2. I have given a maximum score of 5 each time if a model is successful in meeting each of my four criteria, going down to 1 if it fails. I stress that this is not scientific. Each objective is given equal weight and my scores are little more than considered assertions – which, in that respect, makes it rather like many consultants' reports.

The minimum score for a model is 4 and the maximum score is 20.

As you can see from table 2, many of the scores are quite similar, but two models score badly. The 'Arts Council of the Air' scored poorly because there would be one commissioning body, and with it the danger that its commissions would have little impact on the public as they would not necessarily be shown on a widely watched outlet. The default option of allowing the BBC to be the only big player in public service broadcasting also performed poorly because there would be a lack of competition in production and commissioning. Without an overall competition for quality, oversight of public money would be very difficult to judge.

However, all the others score much better and I tend to be attracted either to fully contestable funding of all PSB, or to auctioning the right to manage and operate BBC services for specified periods.

Please note that in this discussion, I have not mentioned the source of the funds. This is because it is largely irrelevant. TV broadcasting should be thankful for the fact that, unlike telecommunications, it currently has a

	Competition in production	Competition in commissioning	Oversight of public money	Brand awareness	Total
'Arts Council of the Air'	5	1	2	2	10
Fully contestable funding at broadcaster level	5	5	3	4	17
Auction of a specific franchise	5	4	3	3	15
BBC vs ITV/C4/C5	2	3	2	5	12
BBC as the 'Arts Council of the Air'	5	1	2	4	12
Competition to run BBC services	5	3	3	4	15
BBC as now	2	1	1	4	8

Table 2 Outcomes of various funding models for PSB in the digital age

dedicated pot of cash, both through direct taxpayer support and from the licence fee – which, quite correctly, has been newly classified by the Office for National Statistics as a tax.

What matters is the way the money is spent in the public interest. And that requires some form of contestability to stop vested interests monopolising what is considered to be the public interest.

Summary

No one can be precise about the future. But given the trends towards greater competition in communications markets, we can all be pretty sure that it will be more difficult than before to secure public interest objectives without explicit funding. This is a welcome development: it forces us all to think whether we really want to spend the money and, if so, how we should allocate it.

While all markets are different, it is a universal truth that outcomes are likely to be better if protagonists have to compete in some form for their right to public funds. This is the principle behind contestable funding; it is a principle that is universally accepted for a universal service fund. That consensus should also be achieved in public service broadcasting, but existing vested interests, particularly those of the BBC, alongside timidity in public policy, will make the achievement of good outcomes harder to achieve.

What citizens need to know. Digital exclusion, information inequality and rights

Damian Tambini

Department of Media and Communications,
London School of Economics and Political Science

General duties of Ofcom

(1) It shall be the principal duty of Ofcom, in carrying out their functions ***(a) to further the interests of citizens in relation to communications matters; and (b) to further the interests of consumers in relevant markets,*** *where appropriate by promoting competition.*[1]

Introduction. Why are citizens' rights different from consumers'?

T.H. Marshall, the leading UK theorist of citizenship, had the view that rights beget rights: the emergence first of civil rights in the 17th century led to political rights through the 19th century and, finally, to social rights in the 20th. But why does Marshall claim that these categories of rights depend on one another? He draws on the example of the right to receive an education:

"The education of children has a direct bearing on Citizenship, and, when the state guarantees that all children shall be educated, it has the requirements and the nature of citizenship definitely in mind. It is trying to stimulate the growth of citizens in the making. Civil rights are designed for use by reasonable and intelligent persons, who have learned to read and write. Education is a necessary prerequisite of civil freedom." (T.H Marshall 1950 (1992): p16)

The rights both to a fair trial and to vote contain within them the demand for universal rights to education, for without education, exercising those rights is a sham. And by the same token, a basic level of material welfare can be seen to be a necessary precondition of exercise of civil and political rights. For Marshall, the basic civil rights to speech, movement and to a fair trial facilitate the emergence of political and social rights: "The right to freedom of speech has little real substance if, from lack of education, you have nothing to say that is worth saying, and no means of making yourself heard if you say it." Citizenship is based on an intertwining of the various types of rights and duties that enable members of complex diverse societies to function together.

Consumers' interests, on the other hand, can be reduced to individual choices in a marketplace. The appropriate balance between public provision through a rights-based system

1. (Communications Act 2003 1 (3)) (bold added)

or distribution through markets is a question that is not peculiar to communications and broadcasting; it is a debate played out across all the public services.

In the context of a book published by Ofcom – a regulator with the duty to promote the interests of citizens – the next move would appear to be obvious: to loudly proclaim the 21st century as the age of 'information rights'. And, Ofcom's role during that century would be to extend citizenship into the digital domain by providing remedies and interventions to support the provision of a basic tier of information services to all citizens. Just as the basic provision of both social welfare and education were necessary to make social and civil rights viable in the last two centuries, in this one no-one can fully exercise their rights as a citizen unless they are able to use communications media.

A neat move perhaps – but not so fast. Regulators' general duties are largely symbolic, and Ofcom has the small practical problem of balancing hundreds of other statutory duties, so will be unlikely to view itself as an agency for promoting citizenship. Surely the regulator should avoid quaint notions of rights and get on with the job of maintaining competition and protecting consumers: take care of markets, and citizenship can take care of itself.

It could be argued that information rights are in an inevitable decline. We have long had a policy framework in place designed to protect citizenship in media and information sectors. Telephony was provided by a publicly owned corporation until the 1980s, which was under 'universal service obligations' to ensure, through cross subsidies, that poorer and peripheral citizens were included in the network. Public broadcasting is also regulated to ensure that it is universally available. But competition has proved much more effective at extending access than regulation did, and in a situation of abundance there is no longer any need to ensure that a basic tier of services is available to everyone. Critics of regulation argue that rather than use regulation to ensure strong foundations for digital citizenship, we should rather happily wave through the long-overdue twilight of information citizenship and usher in the age of choice.

Those calling for the state to withdraw from all remaining forms of regulation of communication are in respectable company. We have a strong framework of freedom of speech in the UK and elsewhere in the Council of Europe. Given the abundance of communication possibilities in the digital age, the familiar argument runs, the state should finally let go of the few aspects of communication it does manage, rather than modernise its regulation.

In our attempts to grapple with these huge questions we tend, surely mistakenly, to focus policy debate on institutions, rather than people. We have an extremely well-developed set of arguments about the role and value of the BBC in relation to market failure, citizenship and public value.[2] We have less of a sense of the wider framework for communications as a whole, and the ways in which development of the market may undermine citizenship.

Ofcom and others have amassed an impressive battery of evidence about the public broadcasters as a group, including the commercial sector, in its recent review of public service broadcasting (PSB).[3] But a convergent world

2. The recent collection *Can the Market Deliver* (2005) is the latest in a long line of publications.

3. *Ofcom Review of Public Service Television* vols I, II, III, 2004

will need a more holistic approach to the citizenship role of communications that encompasses the entire range of communications services; after all, citizenship is not just about the BBC and not even just about broadcasting. It is about whether communications undermines or supports every citizen's attempt to make their voice heard in the democratic process of the country.

This essay is a first exploratory foray into what might constitute that broader holistic approach. I will argue that Ofcom's duty to protect citizens' interests implies a notion of rights that is unlikely to be measurable through the standard approaches to consumer welfare. You don't have to be an ardent supporter of T. H. Marshall's grand historical theory to accept the basic point that rights are crucial in holding complex societies together, and that they tend to be interdependent. A basic education and basic welfare rights are required to ensure that citizens are able to exercise civil and political rights; by the same token, it is already becoming difficult for those excluded from key information services to participate in society as equal citizens. I will also argue against a crude technological determinist view that focuses on connectivity. Citizenship requires access not only to networks, but to particular forms of content, and to a shared space for views to be exchanged.

And it must be acknowledged that the task is an urgent one. Marshall was writing at a time when the UK (or at least, its southern middle class) was very homogeneous. But the imperative for formal citizenship rights – including information rights – that are held in common becomes stronger as society becomes more complex and multicultural, and subject to social fragmentation and exclusion, and where there is a danger that citizens may decide to exit from the normal processes of democratic communication.

In these early years of the 21st century, the framework for regulating communications in the UK tends to be based on a market-failure model. There is a general preference for avoiding state interference and a gathering orthodoxy says it is only justifiable to involve the law or public regulation where non-intervention would result in a demonstrable decline in consumer welfare.

The old era of unquestioned public provision of telecommunications and broadcasting services is now gone: the burden of proof is now firmly on the proponents of public intervention to demonstrate that a failure to intervene would lead to market-failure and harm consumer welfare. The understanding of market-failure is therefore absolutely critical. Given the importance of communications to citizenship, reflected in Ofcom's general duty, many attempts have been made to incorporate citizenship into this framework, in particular through developing measures of public, social, or citizen value.[4] A full critique of the application of the market failure framework will not be possible in this short chapter. In the closing sections I outline some policy implications for the long term.

Media and citizenship: much more than just access to the net

For Andrew Graham, there is a clear link between communications media and citizenship:

"Citizens are entitled to core information about their society, much of which now comes from broadcasting. Citizens are also entitled to

4. Ofcom PSTVR, Spectrum Review

participate fully in society; democratic discussion, much of which takes place via television and radio, is an essential part of such participation... The delivery of these rights is not the purpose of the market. To be met, they require the existence of broadcasters – public service broadcasters (PSBs) – charged with different purposes from those of commercial suppliers. In much the same way that we do not expect our schools or universities to be run by McDonald's or Nike, we should not expect those parts of broadcasting that exist to inform, educate and possibly even entertain to be run in the interests of maximising profits and the returns to shareholders." (Andrew Graham, *The Guardian*, Monday, September 1, 2003[5])

If access to certain media services has become necessary to citizenship, as Andrew Graham asserts, how should that affect the regulatory framework for communications media? Just how much digital exclusion should be tolerated, and at what point does citizenship become untenable for those excluded? Can't we leave it to the market to provide the information services that citizens need in this age of media abundance? We need to understand a little more about the nature and dimensions of information exclusion before we can answer these questions.

Although the i2010 strategy launched by the UK presidency of the EU in 2005 proposed that key targets and policies will be agreed by 2008, there is already an acknowledgement of the increasingly central role of ICT in social inclusion. The EC working group on digital exclusion reported in 2005 that:

"Participation in contemporary society now requires a minimum (and ever-increasing) level of access to, and competence to use, ICT tools and services. For a growing number of people, ICT is increasingly necessary for work, keeping contacts with friends and relatives, social networking, obtaining information, transacting with businesses and administrations." (*Kaplan Report* July 2005, p17)

There have now been at least ten years of academic research and public debate on the 'digital divide'. If there is any consensus at all, it is that access to the internet and other platforms is only part of the problem. Policy has often focused on the issue of physical access, when other issues may in fact be more decisive in terms of their impact on citizenship outcomes.

Figure 1 shows the complexity of the current shift in communications and its implications for citizenship. Internet access may be affordable and available, but if the economics of internet distribution favour niche subscription services in contrast with the public goods of broadcasting, then relative inequalities of access to content may increase. In the diagram, some of the existing regulatory interventions illustrate how, thus far, the centre has been held by regulation. And the issue of media literacy and competency cuts across the other matters: all manner of moves to make sure spectrum is available for public services will amount to nothing if less technically competent consumers don't know how to use them.

And the reverse scenario could also occur: the basic communications resources and forums (such as those provided by public service communications organisations) could remain free at the point of use, but large inequalities in access to new platforms (mobile, broadband and high definition) could grow. The notion that there are core services that

5. Graham develops this point about rights at length in his essay *It's the Ecology, Stupid* in *Can the Market Deliver?* (op. cit).

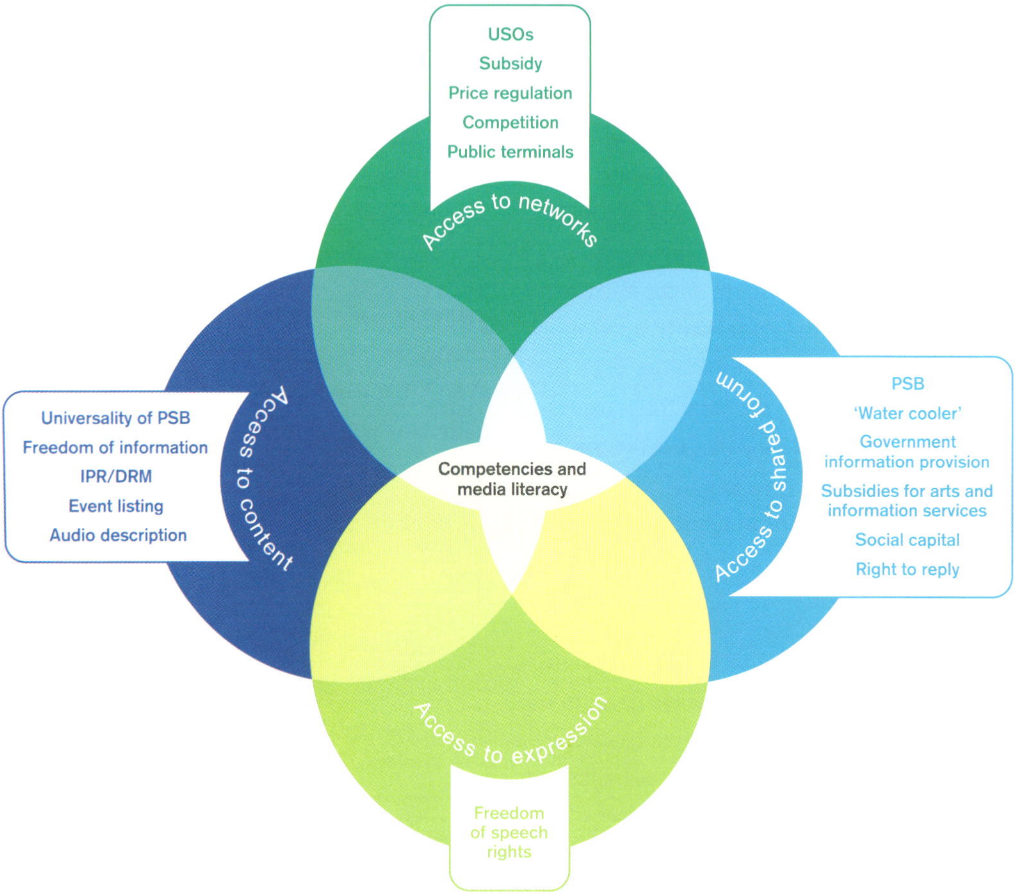

Figure 1: An holistic approach: four dimensions of information exclusion

should be made available under a 'must carry' regime may translate onto certain new platforms, but not all.

The media necessary for citizenship involve, at some level, all of these dimensions. If an unacceptable level of information-inequality or exclusion does occur, it could be as a consequence of each, or any combination, of these various aspects.

Citizenship requires a public sphere where rational critical, evidence-based debate on the decisions facing us can be carried out. The extent to which the public sphere survives – or indeed thrives – in this current digital transformation depends on a complex process, involving arcane areas of regulation: these include must-carry rules, intellectual property, and even defamation and rights to privacy, as these forms of regulation are redefining the boundaries and content of the public sphere of openly available content.[6] The technical standards and codes of conduct for release of public sector information, from crown copyright to geo-data, have come under increasing public scrutiny, illustrating the further politicisation of public policy for the public sphere. Citizens – and also businesses –

6. See Tambini, Damian. *Transformation of the Public Sphere: Law, Policy and the Boundaries of Publicness.* In: Harrison, J. and Wessels, B. (eds.) Mediating Europe. New Media, Mass Communication and the European Public Sphere, Oxford: Berghahn Books, forthcoming, 2007

want access to public information goods, and the tension between market exploitation of that information and provision to the public sphere will continue to grow.

In figure 1, *Access to networks* refers to the condition of being connected to a network, (e.g. a fixed or wireless telephone network, broadcast transmission network, broadband internet) through purchase of a consumer device, and/or as a subscriber. Public policy on spectrum allocation, subsidies, competition and price regulation all impact on overall levels and price points of access. So to do the package of measures referred to as Universal Service Obligations which are subject to the measures outlined in the 2002 Universal Service Directive and sections 65-72 of the Communications Act 2003.

However, merely having access to the network is not enough if key services are encrypted. *Access to content* occurs within additional legal, policy and economic frameworks. Some forms of information – such as government information – are subject to a particular framework that is designed to balance rights of access with other considerations such as data protection principles, privacy and secrecy. The business framework governing which content is unambiguously in the public domain is also undergoing fundamental shifts at present, as Digital Rights Management technology enables smarter approaches to exploit the maximum revenue in the digital value chain, but may impose a higher level of restriction, for example on users ability to share, copy or re-use it. As Ofcom has acknowledged in the case of broadcasting, it is becoming easier to exclude non-subscribers, and the 'public good' features of broadcasting are in decline.[7]

Citizens also require a voice and access to the relevant forum in which to make themselves heard. Clearly, constitutional protection of freedom of speech affords a certain level of protection, particularly from government restriction, but this does not entail a positive right to access to expression, i.e. the right to have one's speech heard, or to have access to key media for expression. The regulatory challenges in safeguarding this is a particular challenge in the case of groups with a specific disability, as the long history of obligations on broadcasters testifies. Redefining accessibility of interactive services and its regulation in new services will be a key challenge and one that would benefit from a notion of all users as citizens rather than merely users or consumers.

Access to the shared forum refers to the notion that citizenship practices take place in various spaces, but are increasingly played out in media environments that range from key news services and discussions to web forums, chat rooms and blogs. Not only do these offer citizens the possibility of accessing information and having their voices heard, but they may also offer social capital (Brookes 2004) and shared experience. This latter is particularly hard to capture using a market-failure perspective as there are collective action problems that increase with media fragmentation. As Brookes points out, markets regularly – and increasingly – fail to provide this scarce public good.

Market failure, communications and citizenship

Within a standard market-failure outlook, intervention is justified where markets fail. The assumption is that individuals are served best by the invisible hand of the market, and

7.Ofcom, *Public Service Television Review*

their utility will improve as the forces of competition drive prices down and quality up. Monopolies and market failure are the central problems for regulators in this perspective, and the proof of market power on one hand, and market failure on the other, constitute the key triggers for intervention.

There seems to be little reason to question this thinking. The enormous success of market mechanisms and competition in rolling out choice and new services to consumers is undeniable. But what about citizens? If our starting point is not the interests of each individual consumer but the overall public benefit in terms of maintaining a level of citizenship integration, we begin to see the limitations of an approach that is based on proving market failure as a trigger for regulation.

The presence of externalities is one reason why markets might fail to achieve optimum welfare outcomes, and the regulatory debate tends to focus on categorising communications benefits to citizens as positive externalities. While this may be feasible on a purely theoretical level, I would argue that when market-failure analysis of these citizenship outcomes fails, the problem lies with the method. It is extremely difficult to measure these outcomes and, as I've mentioned above, it is even more challenging to track consequences of these multi-dimensional processes across access to networks, content, speech and common public spaces for communication. Market failure measurement as a means to justify intervention (such as public service broadcasting) can be a useful intellectual thought experiment, but can we seriously claim to measure all the positive and negative externalities of different forms of drama, news and speech?

Market competition is often much more effective than any notion of rights or basic provision could ever be, in extending access to networks. Few would deny that two decades of liberalised telecommunications in Europe have led to intense innovation and huge consumer benefit. However, some questions remain that imply notions of access rights. The EC Universal Service Directive sets out a framework to continue to develop a dynamic notion of which services remain necessary 'for social inclusion'.[8] In particular, there are repeated concerns that a small minority will suffer exclusion from certain networks.

When the analogue television signal was switched off in the Berlin Brandenburg region in 2004-2005, the switch-off was delayed because a small number of viewers would be deprived of their rights to receive public service broadcasting. Public subsidies were also provided for receivers. In the UK, the Government has committed to provide 'targeted help' to consumers left behind as the signals are switched off between 2006 and 2012. How would universal provision of services provided by government and by the BBC operate in a two-tier internet, where the market enables charging customers for faster services? And as the role of broadband delivery (of government services, for example) changes,

8. In considering whether the scope of universal service obligations be change or redefined, the Commission is to take into consideration the following elements:

• are specific services available to and used by a majority of consumers and does the lack of availability or non-use by a minority of consumers result in social exclusion?; and

• does the availability and use of specific services convey a general net benefit to all consumers such that public intervention is warranted in circumstances where the specific services are not provided to the public under normal commercial circumstances? (Universal Service Directive 2002/22/EC, annex V)

you might argue that public policy will need to grapple with the notion of broadband as a prerequisite to participate; where access to the network is a condition of exercising, say, your civil, political and social rights.

What would constitute an unacceptable level of inequality or exclusion from the point of view of citizenship? There is an unavoidable element of judgement here. In the 2005-6 EC consultation on the review of the EC universal service directive, the EC proposed that broadband penetration had not yet reached the level where it should trigger the package of measures designed to ensure that everyone should have access at an affordable price. The DTI in the UK concurred in its response, but both agreed there was a need for constant review, since levels of access range from 5%-35% penetration in Europe.[9]

Constitutional protection of freedom of speech assumes that fundamental speech freedoms are too important to be left to the private domain. Since J.S. Mill, and increasingly in the US and, more recently, within the system of the European Convention on Human Rights, jurisprudence reflects the idea that democratic citizenship is one of the more important justifications given by judges granting protection from interference in the rights of speech. (Barendt 1985)

Access to content is more complex. The *Today* programme on BBC Radio 4 has, in a sense, become part of the democratic constitution of the UK. Crucially, it is one of very few news programmes with the ability to 'empty chair' a minister who refuses to appear. Yet a significant body of political opinion argues that the BBC should become a subscription service.[10] If this was the case, it is unlikely that any other broadcaster could achieve the same position. How would this form of 'market failure' be valued?

And a longer-term perspective also has to take into account changing perspectives on standards and technology. It is increasingly the case that the protection of intellectual property is a matter not for general principles and authors' rights, but for private contracts and smart software that will enable further exploitation of ever-narrower distribution windows. However, there is also growing support that this should be limited, and that the public interest in innovation and access to knowledge would be served by a rebalancing in favour of users' rights.[11]

While we do live in an era of manifold content, we can imagine a scenario where access to content becomes subject to citizenship-based claims and disputes. For example, you could argue that content that is funded and provided through public taxation – including government information – should be available to all citizens. Further, the existing regulatory framework for broadcasting contains many elements which have the specific objective of ensuring continuing availability. Not only do licensed broadcasters have coverage obligations, but platforms are subject to a regime whereby they 'must carry' certain television channels, and these channels also benefit from prominence in electronic programme guides. (EC Directive 2002/22/EC art 31)

9. See Dti *Response to Commission Communication (9592/05) on the Review of the Universal Services* (2002/22/EC) Available on the DG Infosoc/European Commission website (accessed June 2006)

10. Elstein recommendations. This review was commissioned by the Conservative Party but does not reflect Conservative Party policy.

11. (RSA Charter, Rufus Pollack Paper)

When the democratic health of the country is taken into account, we must acknowledge that widening and persistent inequalities of access – particularly when they overlay other social divisions – could lead to the kind of exclusion that ultimately causes democratic failure and, conceivably, even the use of violence as a political statement. It is imaginable that, at the very least, some audiences would drift away from core public services to niche services which, in turn, would become the new 'Today' programmes. In the very long-term, the most excluded would simply be unable to afford subscription services essential for the provision of information and debate that is critical to citizenship.

The notion of a shared forum

Two of the key trends across media industries in the last decade have been the declining reliance on blockbusters (Aris and Bughin) and the growing ability to exploit value in the 'long tail' – i.e. niche and specialist audiences.[12] This has led to a concern for the role of media products as central collective experiences, and what Greg Dyke inelegantly called the 'social glue' role of public service broadcasting. Just as markets may fail to make sure that uneconomic consumers of broadband are connected, they may fail to ensure that a good that consumers do value: social spill-overs and network effects, will be provided.[13]

"Network externalities are the particular type of externality involved here – the benefits increase with the number of people who have watched the programmes and shared the experience. This distinguishes the argument from conventional discussion of externalities and broadcasting. The other distinction is that the externalities concern concepts such as 'community' and 'citizenship' that are usually deemed to be beyond the reach of economics when discussing broadcasting." (Martin Brookes, 2004, *Watching Alone*, p26)

"Social capital results from broadcasting through externalities. These are a source of market-failure in economics. A particular type of externalities creates social capital – network externalities. My enjoyment of a programme increases the more my friends or colleagues have also watched it and can talk about it – there are positive external effects from people watching the same programme. A purely commercial broadcaster without public service obligations will seek to maximise profits. In general, this results in maximising the size of audience, especially among demographic groups (e.g. young, upmarket viewers) which are particularly attractive to advertisers. This is not the same as maximising spill-over benefits and shared experiences. Consequently, such a broadcaster would produce too few of the right programmes, resulting in an inefficiently small amount of social capital. A free market in broadcasting would not produce enough social capital. This conclusion is not a value judgement but follows from standard economic reasoning about externalities." (Martin Brookes, p6)

"Producing programmes with high viewer ratings does not necessarily mean producing programmes that a non-public service broadcaster would produce. A public service broadcaster might produce programmes which target audiences from all ethnic groups, all social groups, programmes which encourage debate or reflection, that highlight social

12. Chris Anderson Wired, October 2004. http://www.wired.com/wired/archive/12.10/tail.html

13. See also Sunstein, Cass. Republic.Com. 2001

issues, etc. Such programming should be designed to attract significant audiences and build social capital." (Martin Brooks, p40)

It is worth noting at the outset that the 'social capital' arguments do sit somewhat uncomfortably with the traditional arguments for source, content and exposure diversity of media. From the point of view of democratic pluralism, it is often argued that diversity of media outlets and content, rather than 'common experiences', should be projected. In this perspective, common texts are problematic rather than beneficial.

If a very broad notion of externalities is adopted, then it may be reasonable in theoretical terms to include citizenship benefits as being externalities. But the standard measures of the utility of such goods may not capture the general benefits associated. Andrew Graham outlined the huge problems facing those who try to measure definitively the value of broadcasting for citizenship. In particular he questions current methods of research on public service broadcasting, which ask questions about how individuals evaluate broadcasting's impact on other people as well as themselves.

"The individuals replying [to BBC Public Value surveys] to have captured **fully** the value to society of PSB, it would be necessary for them to be both perfectly informed and perfectly altruistic. The assumption about information is required for them (a) to understand fully all the external effects and (b) to forecast the merit good effects. The assumption about altruism is required in order that, having understood and forecast accurately, they would then be able to place themselves completely in the position of 'the other'." (Graham 2004, p92)

Conclusion

"The current circumstances, remaking public broadcasting and struggling to construct a viable public domain within the emerging moving image industries constitutes a titanic task. But it is a necessary one if there is to be any hope of defending and extending the rights of complex citizenship."[14] (Graham Murdock)

Citizenship constitutes a package of rights and responsibilities that are held equally by all members of society. Because they are the same for everyone, they are a powerful force for integration, particularly in those societies that are apparently becoming more diverse and fractious. Electronic communications are becoming ever-more central to exercising citizenship rights, just as a basic level of education and welfare have been.

"Citizenship is not just another sort of market failure, it derives from a different set of values. As long as it can be shown that the media plays a core role in the functioning of a society, and access to it is a necessary condition for a well-functioning democratic society, then the level, content and changing basis for public service broadcasting cannot be encapsulated solely in the market-failure paradigm." (Dieter Helm 2004, p5)

Deiter Helm's observation about citizenship should be extended across converging communications and not be restricted to a debate about PSB. There is a balance to be struck when it comes to the overall level of inequality in citizen access to communications,

14. Graham Murdock, *Rights and Representations in Television and Common Knowledge*, p16 in Gripsrud, J. (ed). *Television and Common Knowledge*. Routledge 1999

and it is a delicate and crucial balance for democratic development.

If all citizens access the same content, networks and public forums, we live in a totalitarian state. But at the other extreme, there is a point at which inequalities in access to information and voice render democratic citizenship untenable. In the current context, fragmentation is a greater danger than totalitarianism in the UK. If some groups of citizens are extremely well-informed and have access to the means to co-ordinate and mobilise around their shared interests, and communicate in a timely and strategic way, and others simply do not, then democracy is a step closer to being a sham.

Ofcom has a duty to further the interests of consumers. It also has a duty to further the interests of citizens. You can only give meaningful protection to this latter group if rights – civil, political and social – are taken into account. These rights cannot be given a cash value and measured as 'externalities' because rights are incommensurable with consumer value. An unacceptable level of information exclusion is one that undermines individuals' abilities to perform the rights and duties that society grants them as citizens. A dangerous level is where exclusion combines several of the four dimensions outlined above, and is overlaid with other forms of exclusion: for example, unemployment, lack of education and linguistic ability.

What, if anything, should be done in the face of rising fragmentation and information inequality? One very real possibility is that regulators and legislators should do nothing. It is entirely possible that the market provides. Until now, however, there has been a very well-developed framework of regulations that protects notions of basic provision and rights to information services. The discussion here has attempted to outline some of the complexity of the current shift in communications and citizenship.

Principles and guidelines

The services necessary to citizenship change over time, and are changing rapidly today. Until very recently, the basket of communication services considered necessary for citizenship would have included only a telephone, a postal address and access to free broadcasting. Whether broadband should be included in that package is largely a question of when, rather than if. And if new media standards (for example mobile, high-definition television) become dominant and necessary for access to public services, it may be that mobile, high definition or broadband services become part of the necessary citizenship package. Regulatory monitoring and research needs to respond to this by developing rights-based notions of consumer access research, to supplement the existing services. Without it, the provision of public services in a multi-platform environment will be problematic, and pockets of multi-layered exclusion may develop.

Secondly, and related, a relative rather than an absolute notion of inequality will be appropriate, rather than an issue of achieving a baseline safety net of services. Citizenship is fundamentally a dimension of equality between members of the same group. Democratic institutions depend for their legitimacy on the assumption that the rules and conditions of participation are equal for all. Apathy and cynicism will result if citizens get the impression that the field is not level, or if a small group of players don't have access to the means to play the democratic game. It is a problem for democracy if a limited group of the 'super connected' are seen to control things.

Third, from the point of view of social inclusion and the health of our democracy, priority must be given to structured inequalities that are multi-layered and affect particular marginalised groups. In the context of representative democracy, intractable problems result when particular interest groups and minority opinions are systematically under-expressed, under-represented and under-deliberated. The consequences of democratic disenfranchisement of key groups, and their subsequent involvement in anti-constitutional means of expression, are all too painful to observe. Monitoring of research on media should have a particular focus on key excluded groups such as disabled people and religious and ethnic minorities.[15]

So there are key implications for research and monitoring carried out by regulators and others. There is a clear need for guiding principles other than market failure. These can be provided by the notion of communications for citizenship. The policy consequences of all this are less straightforward. The pragmatic approach outlined in the 2003 Communications Act and the EC directives of 2002 set out a framework for constantly revising the scope of universal service. However, as I mentioned earlier, it is only by combining data on various aspects of exclusion and inequality, and assessing them in relation to other aspects of social exclusion, that judgements can be made on whether exclusion has reached unacceptable levels for particular groups.

Regulators or governments that profess to 'know what's best' for citizens are quickly dismissed on principle: at best, they are viewed as paternalistic; at worst, they represent a step on the slippery slope to censorship. But we must be clear about the dangers of the opposite view: that the market simply fails to deliver the level of shared communications and a baseline of equal access, and that the long-term social costs of this may turn out to be enormous.

To summarise: a market-failure approach, particularly one that neglects or under-estimates the role of citizen, social and public value, is likely to be indifferent to the key tipping points where digital exclusion begins to seriously undermine citizenship. Even notions of public, social value or externalities do not take an overall view of the level of overlapping inequalities and the extent to which they undermine citizenship *per se*. A notion of rights, and one that sees information citizenship as a relative poverty issue, is the best long-term framework to ensure that communications in the UK perform their integrative, democratic role.

This throws up many questions: should spectrum licensing policy support, or even permit, a new communications technology (let's say, a new standard for high-definition television) if only a select few would be willing and able to pay a very large premium for it? Or should there be a preference for technologies that are likely to be accessed by most or all citizens, even if they have less collective economic clout? To what extent should public broadcasters be funded when their audience share and reach is declining? And is it possible for public funding to be provided for non-universal services in

15. A useful model here is the report by Ian Hargreaves and James Thomas for the ITC: *New News, old news* (Independent Television Commission, 2002). This examined for example the differing perceptions of television news among various ethnic groups.

broadcasting? Clearly, there is an urgent need for reflection, and guiding principles.

T.H. Marshall had a quaint notion that somehow, in the progressive unfolding of citizenship rights, "all shall become gentlemen". This view was forged in the secure nationalism of the post-war British consensus. Ours is a different Britain: the 21st century is more affluent, but less equal. We have an abundance in connectivity and content, but there appears to be a growing gap in access, and the potential to eclipse the common reference points and spaces for debate that have helped hold society together through the post-war period. Digital exclusion, in all its complexity, is likely to increase, and a regulator with a duty to protect citizens (and not just consumers) must develop ways of monitoring its key dimensions. UK communications regulation must also develop a clear and open framework for justifying corrective intervention that goes beyond market failure. This will require an open debate about communications for citizenship, and about how Ofcom should honour its general duty to protect citizens' interests.

The continuing need to advance the public interest

David Puttnam

Member of the House of Lords
Chair, Hansard Society Commission on Parliament in the Public Eye

During the passage of the Communications Bill, which established Ofcom in 2003, I remember certain ministers becoming oddly exercised over the use of the term 'media ecology'. They claimed that concerns for the possible 'environmental degradation' of our media were largely the prejudices of a self-selected group of 'media-luvvies'.

A few years on, I think it's fair to say that this environmental analogy between climate change in the physical world and parallel events in the world of information, is strikingly clear – certainly when you look at it from a regulatory perspective. To perceive the marketplace as a force of nature in which effective and sensitive regulation can play little or no part is as misguided as it is downright dangerous. As with global warming, we have had more than enough international warnings – look no further than Berlusconi's Italy, or recent developments in Putin's Russia – to see what regulatory toothlessness and inattention can ultimately lead to.

In 2002, I was asked to chair a Joint Parliamentary Scrutiny Committee set up to examine the Government's draft Communications Bill. Over the months that followed, together with fellow members of the committee, I spent an enormous amount of time and energy examining every single clause of a large and very complex Bill.

It was a Bill designed to set the regulatory framework for the UK's electronic communications sector, coming at a time when digital technology was beginning to challenge some of our most cherished assumptions about how the media and telecoms sectors operate.

It was also a Bill designed to look forward and anticipate some of the likely changes to the media landscape that would result through the convergence of sectors such as telecommunications and broadcasting. Most important of all, it was a Bill that had to be 'fit for purpose' from the moment it received Royal Assent.

Four years on, and looking back on the work of our committee, I'm extremely proud of what we achieved – principally because the draft Bill, as originally framed by the Government, focused on technology at the expense of content and audiences.

It suffered from one of the classic symptoms of technological determinism: a belief that the new technologies could deliver a sufficiency of content that, in themselves, would serve as a guarantor of diversity and plurality. It has long been an article of faith with me that technology is a bridge, very seldom a destination. There is always a danger of confusing means and ends, and there is a lesson here for those charged with formulating and scrutinising future legislation, particularly where it concerns the communications industries.

Unquestionably, our most important conclusion was that, irrespective of technological development, primacy must always be accorded to the protection of that which underpins the principles of a vibrant and diverse social democracy.

Social democracy and the public interest test

This concept has to be embedded at the very heart of our approach to future regulation if we are remotely serious about promoting and protecting the public interest. If we fail in this, the rights of citizens and consumers could be quickly eroded, as a new generation of media giants takes advantage of every opportunity in their determination to 'optimise shareholder value'. In this respect, recent history offers some important pointers as to what we're likely to see.

That is probably why I take a great deal of personal pride in the fact that, at the heart of the Act, we were able to enshrine something that I believe will become ever-more important: a 'public interest test'.

This test requires Ofcom to investigate any matters of broad public interest arising from the merger of newspapers or broadcast media companies, should such an investigation be requested by the Secretary of State for Trade and Industry. The very existence of the test sends a strong signal that the UK regards the potential impact of mergers and takeovers in the broadcasting and newspaper industries as something that cannot be left solely to the essentially 'corporatist' criteria of competition policy.

Contrary to what some may have felt at the time, I've never had any instinctive antipathy towards the notion that companies exist to maximise shareholder value – but I do believe that blindly obeying this imperative can, and sometimes does, directly conflict with a number of the basic tenets of a plural and equitable society.

This is particularly true in the case of media companies; their ability to shape and define the way we see our world is at least as powerful – and sometimes more powerful – than the task of simply reflecting the realities of that same world back to us.

That's why I'm convinced that the public interest must lie at the heart of any thoughtful regulation of the media. It is one of the few weapons that can help enshrine and underwrite the basic rights of citizens in an evolving and participatory democracy such as ours.

With this in mind, it is worth restating exactly what the Communications Act requires Ofcom to do: to implement the media mergers public interest test, should the Secretary of State wish there to be further investigation of:

- a proposed merger involving newspaper enterprises;
- a proposed merger involving broadcasting enterprises; or
- a proposed merger between broadcasting and newspaper enterprises.

In the case of a newspaper merger, the Act defines the public interest considerations as:

- the need for accurate presentation of news in newspapers;

- the need for free expression of opinion in the newspapers involved in the merger; and
- the need for, as far as reasonable and practicable, a sufficient plurality of views expressed in newspapers as a whole in each market, for newspapers in the UK or part of the UK.

In the case of a broadcasting merger or a cross-media merger, the Act defines the public interest considerations as:

- the need for a sufficient plurality of persons with control of the media enterprises serving a given audience, in relation to every different audience in the UK or a particular area/locality of the UK;
- the need for the availability throughout the UK of a wide range of broadcasting which, taken as a whole, is both of high quality and calculated to appeal to a wide variety of tastes and interests; and
- the need for persons carrying on media enterprises and for those with control of such enterprises to have a genuine commitment to the attainment in relation to broadcasting of the standards objectives set out in Section 319 of the Communications Act 2003. (For example, standards governing matters of accuracy, impartiality, harm, offence, fairness and privacy in broadcasting.)

The public interest test challenges the notion that one can legislate at *all* times, giving equal weight to the imperatives of the citizen and the market. Many, if not most, of the challenges facing Ofcom present themselves as being capable of commercial or market-driven solutions.

But in introducing the public interest test, our intention was to show that not all of Ofcom's decisions can, or even *should*, be dealt with in that way. So while I feel we can be justly proud of what our committee achieved, the pace of change demands that we cannot be remotely complacent.

Grasping the nettle of the net

The communications world has changed a great deal in the space of just a few years. Blogs and community sites were really still in their infancy as the Act passed into law in July 2003. Skype was founded that year, and Myspace.com was launched the very month the Bill received Royal Assent. Three years on, both have users measured in tens of millions. And like many other brands on the net, their phenomenal growth has largely been driven by young people.

In an important speech made to the Royal Television Society in January 2006, Ofcom's then-Chief Executive, Stephen Carter, offered a salutary thought regarding the impact of these developments on public policy. He noted that while the word 'broadcasting' appeared 483 times in the 2003 Communications Bill, the words 'online' and 'internet' did not appear once, even though their impact on Ofcom's work is immense.

I don't believe that the absence of these words was a sin of omission, or an act of naivety, let alone an indication of bad faith. In fact, I suspect that *had* we insisted the words be threaded through the Act we would have been roundly accused of a sin of *commission*.

There's a real danger for legislators trying to anticipate the future of any rapidly evolving industry: either they end up constraining its development by throwing up unnecessary regulatory roadblocks or, having *mis*read the future, they simply legislate for a world that never comes to pass.

The Government was very clear that the purpose of the Act was not to extend the tentacles of regulation to the internet. I believe that, at the time, this was probably the right decision, precisely because of the unintended consequences of trying to anticipate what remains, even today, a very uncertain future. It's equally certain that legislation will be required in the not-too-distant future, extending Ofcom's remit specifically to include aspects of internet communication.

As the regulatory remit extends into new areas, it becomes doubly important that it is built around principles that retain a clear sense of cultural continuity with those that underlie our current approach. It's very likely that issues around market dominance and its potential abuse will continue to exist, or possibly even grow, in the digital era.

The primacy of the citizen

To the extent that the future is only likely to become more complex, our approach to regulation will need to be ever-more sophisticated, and our understanding of what is meant by 'public interest issues' ever-more acute. We cannot lower our guard in the belief that in a world of digital plenty, the market alone will provide. As with any *laissez-faire* approach to global warming, the potential for irreversible damage at the hands of the marketplace will be a constant threat.

In fact, I'd go further: in a *fully* digital era, the need to advance the public interest as a governing principle, underpinning the regulation of the *whole* of the communications industry, remains every bit as important as it was in an analogue era, when scarcity was the dominant feature.

I use the word 'advance' rather than 'protect' because I don't think of the public interest as something so fragile and enfeebled that it needs unnecessary cosseting. I start from the position that the public interest is just one of the compacts (albeit a fundamental one) that underpin the whole concept of citizenship and civil society.

As citizens we do have rights, some of which are, or ought to be, inalienable; but equally those rights bring with them a whole slew of responsibilities towards our fellow citizens. In this rapidly changing world, the public interest should increasingly become a central tenet of a free and evolving society; it is not something that can in any way become superfluous as a consequence of technological advance.

In this respect regulators must be primed to be vigilant, and never to drop their guard.

The media cannot be crudely reduced to the status of being mere commodities, any more than other vital aspects of a self-respecting democracy such as our systems of justice, health, education or, more recently, the security of supply of water, light and heat. It's for this reason I've consistently argued that the interests of the 'citizen' must at all times retain primacy over those of the 'consumer'; it's also why the Communications Act rightly, and importantly, separated the notion of those two critical entities.

In a digital era, I believe that the definition of the public interest in respect of regulation must continue to be informed by three overriding principles:

- first, that from the perspective of public policy, wherever the interest of the citizen finds itself in conflict with those of the consumer, the 'citizen interest' should retain primacy over that of the marketplace;
- second, that diversity and plurality are the key touchstones by which we measure the

extent to which the citizen interest is being sustained and advanced; and

- third, that we should seek to ensure that ***all*** citizens have the means, and the ability, to make informed judgements regarding the nature and value of the media they access.

We should not pretend that the ubiquity of digital technologies is, in and of itself, sufficient to ensure that the public interest will somehow prevail. This will remain the case even if the phenomenon of the 'long tail' – which is now well-established on the net – makes a far greater diversity of information and material available to consumers and to citizens compared with the analogue world.

For example, as the pre-eminence of companies such as Google and Yahoo demonstrates, scale is still a very significant advantage in the online world. The internet unquestionably allows innumerable websites and blogs to bloom, yet it's equally clear that size does not necessarily bring with it immense or even lasting advantages.

Of course, the issue is not simply about scale, any more than it was in the analogue world. Very large companies can, and often do, act to advance the public interest.

Yet equally, they can sometimes act in ways that serve to diminish choice and plurality. This then requires a regulator equipped with teeth, and a willingness to use them when the occasion demands. We shouldn't forget that there are clear limits to the effectiveness of self-regulation, and it would be a great mistake to ask too much of it.

It is equally true that smaller entities and communities feel that they face real obstacles in securing visibility and awareness. While the problem of access to platforms may have significantly diminished – the traditional barriers to entry are increasingly being removed – the difficulty of securing awareness (a form of market failure accurately defined as 'imperfect information') is likely to remain.

A huge number of websites, for example, get very few visitors simply because they are adrift in a sea of information dominated by just a few large brands.

The cultural tool of media literacy

If our understanding of the world is increasingly shaped by electronic media, then in order for the individual citizen to really make sense of this world, to navigate their way through it more effectively, they must be properly equipped: they need the ability to understand the means by which the media is able to shape, represent – and indeed when it wishes, even misrepresent – that world. These are the outlets that have the power to give meaning to our lives – sometimes a bewildering multiplicity of meanings.

Just try having Channel Four News, Google News and others send headlines to your mobile by SMS, and see their capacity for interpreting similar events in very different ways. Rather than simply and 'neutrally' reporting it, the media edit and represent the world as they see it. That is why the promotion of media literacy is so important, since it equips us with the creative, critical and cultural tools that help us understand the way in which the media help to 'construct' our understanding of the world around us.

As a former producer of both films and television programmes, I'm incredibly aware of the way in which they can influence our view of the world – for better and for worse. Movies and television, like all other forms of digital content, not only serve as entertainment; they are also an important form of expression.

For many of us they are, first and foremost, a way in which we access and share stories, ideas, attitudes and opinions. They are nothing less than the means by which we imagine our entire range of possibilities for the future, and nowhere is this more important than in the lives of young people.

Only by ensuring that we are all thoroughly 'media literate' can we be sure that we're sufficiently well-informed to exercise that whole swathe of judgements on which any healthy society is critically dependent. In fact, in a world increasingly defined by the moving image, media literacy is not just about our ability to understand and appreciate the way these images are put together – it is increasingly about underpinning the health of any fully functioning democracy.

More than that, it helps make possible the vitality and growth of our cultural and imaginative lives. Being media literate can only help stimulate our creativity – most especially by enabling us to engage with digital technology – whether making a 30-second film or creating impassioned websites that highlight those things that most concern us every day.

Thus media literacy, which the existing Communications Act already requires Ofcom to promote, can only become increasingly vital as a means by which we can advance the public interest.

IP: ownership and entitlement

However, to reap the full educational and cultural benefits of the digital dividend we also need to create public policies that provide for a new settlement between citizens and those who control intellectual property.

The digital world poses a number of challenges around the question of value – most specifically the value of intellectual property rights (IPR). In the past, much of the premium attached to IP was a natural consequence of its scarcity. How do we ensure that its value still inheres in a world in which the IP itself is infinitely reproducible, frequently at almost zero marginal cost?

This is a challenge that has largely been left to, on the one hand, content owners who are seeking a regime of stronger enforcement through legislation, because simple self-interest demands that they seek to protect their existing business models. On the other, there's the combination of technology companies and 'internet outlaws' who believe that everything should be free to anyone with the power to access it.

The battle between these opposing concepts of 'ownership' represents a challenge for all of us with a stake in the public interest – after all, unless citizens value IP it becomes impossible to maximise its contribution to the health and the prosperity of our society.

I do not for one moment believe that everything should be free. However, I'm equally far from convinced that putting all copyright material under lock and key is the answer either – most especially when public resources have helped to create that content in the first place.

The moment we start to conceptualise intellectual property in primarily defensive terms, to the point of adopting the language of the police state when discussing the value of copyright – that, for me, is the moment we fundamentally devalue the whole currency of media literacy. Some of the anti-piracy policies pursued even by different sectoral interests can look like the continuation of war by other means – hardly an effective strategy for winning the hearts and minds of consumers and citizens.

It is easy to see how such a restrictive view might be encouraged by tabloid hysteria regarding rival media content, whether disseminated via television or on the internet. It's easy to see how vested interests might only be concerned with protecting the revenues of existing business models, at the expense of more long-term thinking – because in reality, they always have. But had such vested interests triumphed in the past, the film industry would still be attempting to prevent the sale of movies to television. In fact, one of the very big issues for film is the length of the 'windows' through which the VCR, the internet and DVD find their way into homes. It is almost certain that those windows in their current form will not survive the transition to a fully digital age.

I am not any kind of anarchist. As a citizen, I believe in regulation, sometimes even tough regulation. As a copyright owner, I firmly believe in the adequate protection of authors' rights.

But I also believe that imparting an understanding of the need for valuing and respecting intellectual property is best achieved by creating a sense of empowerment, and legitimate entitlement among users.

What's certain is that those charged with advancing the public interest on a broad front will require foresight and a bold and insightful imagination. These are qualities we have not consistently displayed in the UK over the past few years. For almost two decades, I was a non-executive director of Anglia Television. During the whole of that time I can honestly say that we seldom, if ever, looked more than five years ahead.

I'd also have to concede that just about every move we made during those 20 years was primarily defensive. It was all about: how do we keep the franchise, how do we please or possibly even second-guess the ITC, how do we, year-on-year, incrementally increase shareholder value? There were very few leaps of imagination, but this is not a criticism of the Board; we were working within an environment in which it would have seemed almost foolish to view the future in any other way.

Not only was the world of electronic communications that now confronts us not planned for; with the possible exception of Rupert Murdoch, it wasn't even conceived.

In this respect at least, I take my hat off to Mr. Murdoch. He saw an opportunity, he moved, he manoeuvred, he was prepared to adapt, and did the one thing that the rest of us were frightened of: he believed in the possibility of a 'different' future.

So experience tells me that here is at least one extremely good reason for a strong, robust regulator such as Ofcom, to try to imagine and anticipate the future – its opportunities and its dangers – precisely so that society isn't continually outflanked by those who have historically shown themselves to be far better at responding to changing market conditions – sometimes at the expense of much of what is most firmly tied to the public interest. As we on the Joint Scrutiny Committee argued, 'regulation has to keep up with technology; *ex post facto* law is invariably bad law'.

Or as one of my personal heroes, the polymath Dr Albert Schweitzer, cautioned: "Never let your technology exceed your humanity." That seems to me to be an entirely admirable precept to which we should all adhere, if we are really serious about advancing the public interest in an environmentally dangerous digital era.

The revolution of audiovisual services: culture, economy or both?

Ruth Hieronymi
Member of the European Parliament

The development of audiovisual services

During the process of the global economy developing faster and faster, the special role of audiovisual services for safeguarding freedom of information, freedom of expression and cultural diversity has been recognised very late – hopefully, not too late.

The significance of audiovisual media and their special role emerged when the era of moving pictures started and became increasingly important in everyday culture. Already in the early days, audiovisual services could be bought and sold across national borders, as films could be dubbed and subtitled.

However, broadcasting was, and remains, primarily a *national* issue. It has traditionally been a matter for national laws, intrinsically linked with national cultural and public interest concerns. European television and radio stations are normally national in coverage and in their audiences, primarily for linguistic reasons. In the European Union, the competence over fundamental decisions in the domain of culture has always remained with the Member States. At present, Member States have the exclusive competence in licensing, spectrum planning, all aspects of media pluralism and the definition of the public service remit.

This, however, does not mean that the European Union does not have a role to play in the regulation of broadcasting. To the extent that broadcasting is a service, it comes within the ambit of the European Treaty and there is a role for the EU to remove restrictions in trans-national broadcasting, harmonise by minimum standards broadcasting legislation and ensure fair competition. As a consequence, a common legal framework was established by the EU Television without Frontiers, (TVWF) Directive[1] in 1989.

This facilitated the beginning of cross-border broadcasting. The advent of commercial television in the mid-80s and of digital TV in the mid-90s contributed to the growth and development of this sector.

The audiovisual sector today

The television sector has gone through a phase of continuous dynamic growth during the last years. Today, there is television in 98 per cent of all European households, something that confirms the social influence of television and its significance for opinion forming and the democratic process. At the beginning of 2004,

1. Directive 1989/552/EC dated 03 Oct 1989, amended by Directive 1997/36/EC dated 30 Jun 1997

there were approximately 220 regional television channels. In addition, more than 160 television programmes are broadcast on a cross-border basis and are thus available in several of the 25 EU Member States (compared to a mere 68 channels of this kind in the EU-15 in the year 2001). This development has been accompanied by a respective increase in the total turnovers of the TV stations and the advertising turnovers of private and public providers.

Regarding digital technologies and the emerging new audiovisual services, there are some figures indicating their dynamics in terms of demand. In 2005 23.2 per cent of all households in the EU had reception of digital TV. In 2003 there were 230 interactive digital TV services in the EU-15.

These figures do not only indicate a fast penetration of new audiovisual services and the use of digital technologies, but also their power to substitute traditional TV and the willingness of consumers to use these kind of services.

The increasing importance of digital technologies for providing audiovisual services is underlined by the availability of computers in households in the EU.

In 2005, 58% of all households had a computer and 48% had access to the internet in the EU-25, with a growing rate of access to the internet of 14% in the year 2004. This is even clearer for the EU-15: here 63% of the households have computers and 53% access to the internet, with an increase in access to the internet of 10% in 2004.

ICTs and the Lisbon strategy

During their special Lisbon summit in March 2000, the EU Heads of State and Government set themselves the high goal of "the EU becoming the most competitive and dynamic knowledge-based economy in the world".[2] In this context, the new information and communications technologies, as well as the new media services, have a key role. The share of the information and communications technologies in the productivity growth in Europe amounted to 45 per cent in the period between 2000 and 2005. The European Commission forecasted that in the EU in 2011, 22 million people would be employed in the multimedia and software sector alone.

The expectations for economic growth were very optimistic and not all have been realised as forecasted. Nevertheless, the media sector will without doubt be one of the most dynamic growth markets in the years to come. At the same time, the European Union continues to show a dramatic trade deficit compared with the United States in all audiovisual markets. For example, up to 80 per cent of the films, audiovisual services and e-learning products on the European market are of US origin. In this context, it is important to ensure that Europe remains an attractive region for economic investment, while ensuring that the rich diversity of European culture will continue to be protected. This needs to be achieved through a co-ordinated economic, media and telecommunications policy both at European as well as at national level.

Television with new frontiers

Up to now, the TVWF Directive has been the cornerstone of the EU media policy. When it came into effect in 1989, the predominant task of the TVWF Directive was to ensure that television services could be provided without

2. European Council of Lisbon, 23/24 March 2000, Presidency Conclusions

barriers within the EU internal market. This was possible through the operation of the country-of-origin principle, which results in the application of the rules of the country of establishment. Europe-wide consensus on the country of origin was only possible when common minimum standards for advertising rules, protection of minors and human dignity, and a requirement on minimum shares of European and independent productions were agreed in the Directive.

As regards television advertising, a number of crucial rules apply at European level. For example, an obligation to separate editorial contents from advertising, a prohibition of surreptitious advertising, as well as detailed quantitative requirements for advertising breaks. There are also strict qualitative restrictions on advertising directed towards children and young people. There are important safeguards for the protection of minors against contents that are harmful, such as pornography, violence and content that is discriminatory. Finally, with a minimum quota of 50 per cent for the transmission of European works and a prescribed minimum share of 10 per cent for works of independent producers, the Directive takes account of the double-sided character of the television sector: cultural and economic.

In 1997, the Directive was revised for the first time to take account of technological and market developments. The most significant innovation was Article 3.a of the Directive, which regulates the transmission of listed events considered to be of major importance to society. It provides the Member States with the opportunity to ensure live or deferred transmission for all viewers in free TV, even if the transmission rights were exclusively sold to a pay-TV operator by the rights holder. In most Member States, the listed events include the transmission of major sports competitions, for example the football World Cup, the Champions League or the Olympic Games.

An interim assessment at the end of 2005 confirmed the overall success of the EU TVWF Directive. All Member States now apply it, establishing a common legal framework for the audiovisual field. The regulations in the field of the protection of minors and human dignity have been implemented and complied with. The advertising regulations were, as expected, more controversial, but their adherence was finally agreed. The share of the works of independent producers as well as the prescribed 50 per cent share of European works is considerably exceeded in most Member States. In times of increasing media convergence, however, the question arises whether this legal framework is suited to take account of the revolutionary technological developments that we have witnessed in the audiovisual field over the last ten years.

The answer is no. With the development of digital technologies and the growth of the Internet, national borders are becoming increasingly irrelevant. Today, audiovisual services are available worldwide and crossing borders in a digital form – if and when the prices for the appropriate technological requirements can be paid – and have become one of the most powerful markets in the world.

Looking beyond Europe: the WTO regime for audiovisual services

Against this background, it is understandable that in the WTO rounds and in the GATS (General Agreement on Trade in Services) rounds concerning services, there has always been a struggle for agreement on if, when and which audiovisual services should be included in a worldwide trade regime.

A part of this WTO world trade system is the GATS, which transfers rights and obligations related to free trade to all service sectors and thus also to audiovisual services. The EU refrained from any liberalisation in the field of audiovisual media in the Uruguay Round (1994).

Against this background, the new negotiation for liberalisation, the so-called Doha Round, started in the year 2000. It is not surprising that the audiovisual sector has again proved controversial, with the United States, Japan, Brazil and South Korea demanding further liberalisation of services in the audio-visual sector.

There is ample proof that the marketplace alone cannot ensure cultural diversity. Thus, the public policies in Europe are premised on the conviction that the responsibility of policymakers to foster cultural diversity extends to the digital environment. This also means that governments must safeguard their flexibility to pursue national pro-active cultural and audiovisual policies at international level. But how should governments defend their audiovisual policies successfully against global developments that could put these policies in jeopardy?

If audiovisual services were liberalised under the GATS as some countries request, the most basic and fundamental audiovisual and cultural policies would be put at risk. This is because the main principles of the GATS are incompatible with many audiovisual policies, which typically strengthen the cultural identity and discriminate against foreign audiovisual service suppliers. This kind of discrimination is prohibited under GATS through the Most Favoured Nation principle and the National Treatment principle. Examples of national policies directly at odds with GATS include all financial support systems for local audiovisual productions; quota systems; must-carry-rules in favour of local content; and a number of media ownership rules. In addition, public funding of public services broadcasting may one day be found to be in violation of disciplines for subsidies under GATS.

In the European Union, there is an intense discussion between the Member States, the EU Commission and the Parliament on this question but there is general agreement on the premise that audiovisual services are not only economic but, more importantly, cultural goods and services. Accordingly, the European institutions have in the past argued for the exclusion of audiovisual services from these purely economic negotiations in the WTO and GATS rounds – especially the present Doha Round.

Furthermore, on 20 October 2005, the Unesco Plenary presented a worldwide Convention on the protection and promotion of the diversity of cultural forms of expression "being convinced that cultural activities, goods and services have both an economic and a cultural nature, because they convey identities, values and meanings, and must therefore not be treated as solely having commercial value". In the member states of Unesco, the extremely important ratification process: the precondition for the entry into force of the Convention, is taking place at present. The Unesco Convention for the first time establishes a worldwide legal instrument for the protection of cultural diversity that has to be respected by the community of states, just as it does worldwide trade law.

Legal regulations in the European Union

Despite this position towards WTO/GATS, the status of audiovisual services inside the EU has not yet been legally clarified up to now. Only for traditional television has the legal status been clear, since 1989 due to the TVWF Directive. The *Electronic Commerce* Directive of 2000 comprises the regulations for the exchange in the field of electronic services on demand, which are defined as 'services of the information society'. Broadcasting services have so far been excluded from the *Electronic Commerce* Directive because of their double nature as cultural and economic goods, and because they have not yet been on-demand.

The EU Services Directive adopted in 2006, reiterated the special nature of audiovisual services: "audiovisual services, whatever their mode of production, distribution and transmission, including radio broadcasting and the cinema" are expressly not included due to their significance for democracy and society.

The convergence of technologies, platforms and media services, as well as the development of on-demand TV services and new business models, means there is a significant legal grey area for these audiovisual media services, which are not identical with traditional TV. This has triggered discussions on what is the appropriate regulatory framework at the European level, and a second revision of the *Television Without Frontiers* Directive is now on its way.

At present, there are different regulations in different Member States for the new audiovisual services; something that creates legal uncertainty and which is weakening the competitiveness of the European industry in the audiovisual sector. In order to fully benefit from the potential of audiovisual services, a common European legal framework for the internal market is required.

The key question is whether there should be a sector-specific media law at European level (*lex specialis*) or whether TV services on demand should fall exclusively into the scope of the e-Commerce Directive.

The European Parliament has resolutely pleaded for the creation of a common legal framework for all audiovisual services that would address their double character as a cultural and economic good.

The European Commission and the Council have reacted to this request and have demonstrated with a large majority their support for maintaining the sector-specific legal framework for traditional and new audiovisual media. They have also rejected the classification of the new media services as a pure economic good.

In December 2005, the EU Commission therefore tabled a proposal for a sector-specific directive for audiovisual media services[3] to amalgamate traditional and new audiovisual media services with graduated regulatory rules, depending on the impact of these services on public opinion and the degree of choice and control exercised by the user.

This proposal of a Directive cannot be applied to all audiovisual services in general, however. It creates a *lex specialis*, a sector-specific media law for audiovisual services that are of special significance for the freedom of information, media pluralism and cultural diversity, and which are included in the definition of 'audiovisual media services'.

The draft proposals define audiovisual media services as services as defined by Articles 49 and 50 of the EC Treaty, with or without sound, under the editorial responsibility of a media service provider, and which are directed

3. COM(2005)646

towards the general public via electronic networks for the purposes of information, education and entertainment. The Commission has proposed to differentiate between linear services with a fixed programme schedule that cannot be changed by the user, and non-linear services that can be watched on-demand and changed interactively by the user.

In order to take account of the different character of non-linear services, only basic requirements on issues of protection of minors, human dignity and advertising are mandatory. Furthermore, for these services Member States are encouraged to use co- and self-regulation.

A future without media-specific law

I truly hope that there will be a majority support for this approach in the European Parliament and the European Council. Otherwise, what's the alternative?

An alternative approach that is supported by parts of the industry – in particular by the telecommunications, cable and internet providers – favours the application of sector-specific media regulation at European level only to traditional television with a fixed programme schedule.

In my view, the consequence would be that all those audiovisual media services that are not distributed according to a fixed programme schedule will not fall under a sector-specific media law, and will be left in a legal grey area where decisions as to whether the TVWF Directive or the *Electronic Commerce* Directive is applicable are to be made on a case-by-case basis. The position that all services provided on-demand fall under the eCommerce Directive, independent of their content, should scarcely be legally tenable. Such a unidirectional commercialisation of audiovisual contents is a fundamental contradiction to the current audiovisual European approach.

Thus, without a legally fixed status for audiovisual media services, neither would the competitiveness of European providers be strengthened nor pluralism of information and opinion be safeguarded. The necessity of establishing a level playing field, therefore, is in the interests of both the internet economy and cultural diversity in compliance with Article 151 EC Treaty.

At this stage of the debate, it can be said that the vast majority of the European Parliament and the Member States strongly support the proposal of the EU Commission concerning the *Audiovisual Media Service* Directive.

I would like to re-iterate that audiovisual services that are not audiovisual media services are not taken out of the application of competition and trade law at European and worldwide level. They continue to be in a legal grey area. We hope for a future regulation that takes account of the Unesco Convention on the protection of cultural goods prepared and decided in co-operation with the European Union.

Conclusion

It is absolutely necessary to keep a sector-specific media law at European level. The extension of the TVWF Directive to the new audiovisual media services is the only way for safeguarding EU-wide freedom of expression and freedom of information, and furthermore for maintaining cultural diversity. These values are considered to be most important to the European Union, both as a basis for democracy and as a key feature of the EU.

All in all, a sector-specific media law does not neglect the economic character of audiovisual media services, but allows the implementation of certain regulations to safeguard important and basic values connected to the cultural character of these services. Only under this common agreement on standards can the country-of-origin principle be accepted.

There should be a clear distinction between the audiovisual media service providers that deliver content to the European citizens and the providers of communications networks, used for the transmission of content. The former are to be regulated under the *Audiovisual Media Service* Directive, while the latter would be regulated under the umbrella of the eCommunications strategy of the European Commission.

This eCommunications strategy involves the management of radio spectrum and the revision of the 'regulatory framework for electronic communications'. For the management of radio spectrum, the EU Commission states that the regulatory approach should be updated since electronic communication services such as mobile telephony rely on radio technologies, and these radio frequencies do not know borders. Recently, new and more market-based approaches have been developed. Also in this field, an update in legislation is urgently needed.

As for the regulatory framework for electronic communications, it was established in 2002. Currently, its revisions process is ongoing. In June 2006, a communication was published by the European Commission,[4] *On the Review of the EU Regulatory Framework for Electronic Communications Networks and Services*. The Commission's proposal for a revision of the Framework Directives is expected for the end of the year. Important aims to be achieved are:

- to strengthen incentives for investment (which should be ensured through competition to provide infrastructure);
- to promote innovation through openness of the rules for new technologies; and
- to safeguard the European model of freedom of information and freedom of expression, especially through the policy option of structural separation. Structural separation means that telecom regulators could require a dominant operator to provide non-discriminatory access to all operators by separating infrastructure provisions from service provisions to a greater or lesser extent[5].

All these EU approaches on services of the information society and electronic communication have one thing in common: their regulation will be based on a liberal market approach, competition law and rules for safeguarding cultural diversity at the same time, because audiovisual goods are both cultural and economic goods.

4. COM(2006) 334

5. Speech/06/422 by Viviane Reding held on 27th June 2006 at the annual meeting of BITKOM

section 3

Economic regulation beyond 2010

Creating an environment for rapid innovation

Irwin Stelzer

Senior Fellow and Director of Economic Policy Studies Group, Hudson Institute

"Obviously the face of the earth... looks as it does... due to the unremitting efforts of people to improve, according to their lights, upon their productive and commercial methods, i.e. to the changes in technique of production, the conquest of new markets, the insertion of new commodities, and so on. This historic and irreversible change in the way of doing things we call 'innovation' and we define: innovations are changes in production functions which cannot be decomposed into infinitesimal steps. Add as many mail-coaches as you please, you will never get a railroad by so doing."[1]

The creation of an environment in which innovation can flourish requires that we get four major economic policies right:[2] macroeconomic policy; the rule of law; competition policy; and regulatory policy. Each one of these presents daunting problems for policymakers; to get all four right might be beyond the reach of merely mortal politicians and regulators. Worse still, it is difficult to measure the success of any policy designed to stimulate innovation: measurement difficulties and the fact that innovations, especially those of the most disruptive sort, can take decades to work their way into day-to-day business operations. Still, it is better to have a clear view of the policy goals that might maximise innovation, than to stagger from issue to issue, with only ad-hockery as a guide.

Macroeconomic policy

The relation between innovation and macroeconomic policy is not as straightforward as it would at first seem. It seems obvious that a stable economy, growing at a relatively predictable, non-inflationary rate, is one that maximises boardroom confidence, keeps capital costs at a minimum, and provides the general setting in which entrepreneurs feel comfortable investing in new ideas and technologies. Perhaps.

Disruption and crises can be an important goad to innovation. However, as a general matter it is safe to assume that a growing, crisis-free economy provides the optimal environment for innovation. Such an economy, in turn, requires the development of sensible monetary, economy-wide regulatory, and fiscal policies.

1. Joseph A. Schumpeter, *The Analysis of Economic Change*, Reprinted from Review of Economic Statistics, May 1935, in Richard V. Clemence, ed., Essays on Entrepreneurs, Innovations, Business Cycles, and the Evolution of Capitalism. New Jersey: Transaction Publishers, 1989, p.134.

2. It is beyond the scope of this paper to consider the non-economic factors that importantly contribute to the rate of innovation, most notably the degree of individualism a society endorses, as against deference to hierarchy, community and/or nation.

But there is more to producing an optimal macroeconomic environment than getting monetary policy just right. There remain the nontrivial matters of macroeconomic regulatory policy, and fiscal policy.

There is less controversy surrounding the choice of national regulatory policies that stimulate steady macroeconomic growth because we have had the benefit of a laboratory test: a comparison of the relatively unregulated US economy with the more closely regulated economies of the European Union. The core EU economies – France, Germany and Italy – have experienced declining labour productivity over the past 30 years, and in the most recent five years for which data are available have averaged productivity growth rates of 1.8%, 1.3% and 0.2%, respectively, compared with 2.8% in the US.[3] Per capita income in those countries now averages 70% of that in America, "the largest gap since the late 1960s, thanks to lower levels of employment, falling work hours and Europe's failure to harness technology to boost productivity".[4]

It would be foolish to claim that the EU's greater reliance on regulation of product and labour markets is the sole reason for this disparity, but equally foolish to ignore the role played by Europe's regulatory constraints on managerial and entrepreneurial activity, in discouraging innovation and the associated increase in productivity and living standards.

Which brings us to the role of fiscal policy in maximising the rate of innovation. Policymakers have learned less than those who concentrate on monetary policy. In part, the greater improvement in monetary policy since the 1930s, when the Federal Reserve Board's credit tightening at an inopportune time contributed to prolonging the recession, is due to the combination of greater experience and improved quantitative analytical tools.[5] But only in part.

In important part the greater progress of monetary policy, as compared with fiscal policy, is due to its depoliticalisation. Following the lead of the United States, most leading countries have adopted the decidedly (non-democratic) policy of turning monetary policy over to independent or quasi-independent central banks. Although there is no agreement among economists as to the effect of central bank independence on the rate of inflation,[6] there is little question that investors prefer to gamble their wealth in an environment free of politicians manipulating monetary policy to suit their short-term ends – which, more often than not, results in increased inflation. There's little doubt that inflation has a "significantly negative" effect on an economy's growth rate,[7] and hence on the propensity of its entrepreneurs to invest their sweat and capital in the risky process of innovation.

Fiscal policy, by contrast, remains the playground of politicians, and the focus

3. OECD, *Labour Productivity Growth*, April 20, 2006

4. Allister Heath, *Only Fools and Europhiles*, The Spectator, 1 July 2006, p.38

5. Although it is at times difficult to determine just how much we have learned since 1873 when, to cite just one example of his wisdom, Walter Bagehot advised that "it is the role of the central bank to lend... most freely in time of apprehension...." See his *Lombard Street: A Description of the Money Market*. Reprinted in 1999 by John Wiley & Sons, Inc., p.64.

6. Harvard University professor Robert J. Barro concludes on the basis of a study of 67 countries that the correlation between the extent of central bank independence and inflation rates is "negligible". *Determinants of Economic Growth: A Cross-Country Empirical Study*. Cambridge, Massachusetts and London: The MIT Press, 1997, pp.104-111.

7. Ibid., p.117

of fierce arguments among economists. Some are in an old tradition that government must balance its budgets, and that deficit financing is "stigmatized as frivolous and unworthy of respectable governments"[8] and should be avoided, if necessary by raising taxes. Others contend that the budget deficit does not provide "a decent measure of the federal fiscal condition"[9], that measured deficits do not present a clear and present danger to the US (or most) economies. Still others, the now-famous supply-siders, argue that budget-balancing is a laudable goal, best achieved by a combination of spending constraints and tax cuts that stimulate economic growth and thereby increase tax revenues.

Again, this is not an argument to be settled in this brief essay, beyond stating the obvious: wildly unbalanced budgets require growth-stifling, tight monetary policy as an offset if inflation is to be avoided; some tax cuts, especially those on marginal rates, do indeed produce revenues that in part at least offset the losses due to the lower rates.

For our purposes here, we can conclude that if the rate of innovation is to be maximised, a combination of monetary and fiscal policies that produce steady non-inflationary growth is essential. As one observer puts it, "Low inflation, government solvency, and slowly adjusting exchange rates are necessary for economic growth to match growth potential."[10] So, too, is minimising regulation of labour and product markets.

And growth fosters a rate of innovation unavailable to stagnating economies, which innovation feeds back into more rapid growth, as the members of the European Union have learned – or at least should have.

Rule of law

But growth is no more essential to innovation than is the rule of law. It is, according to Professor Stiglitz, one of the "essential elements of a healthy economy", and a system that provides "general confidence" that disputes surrounding property rights are resolved in "a fair and equitable manner".[11] Without the rule of law, a capitalist system cannot survive – contracts become meaningless, the state can intervene at any time to destroy private wealth (cf. Vladimir Putin and the sad tale of Yukos), and the rules governing business operations can change without notice, and certainly not in the "fair and equitable manner" that Professor Stiglitz counts as an essential characteristic of the rule of law. And without capitalism, we have found that efficient, durable economic growth is somewhere between impossible (the former Soviet Union, much of Africa) and difficult to obtain. As William Baumol, perhaps the closest student of the process of innovation has put it:

"The capitalist economy can usefully be viewed as a machine whose primary product is economic growth. Indeed, its effectiveness in this role is unparalleled."[12]

In short, no rule of law, no capitalism; no capitalism, no economic growth and no innovation.

8. Joseph A. Schumpeter, *History of Economic Analysis*, edited from manuscript by Elizabeth Boody Schumpeter. London: George Allen & Unwin, 1954, p.769

9. Robert L. Bartley, *The Seven Fat Years and How to Do It Again*. New York: The Free Press, 1992, reprinted in Neconservatism, Irwin Stelzer (ed.). London: Atlantic Books, 2005, p.183 (paperback edition)

10. William W. Lewis, *The Power of Productivity: Wealth, Poverty, and the Threat to Global Stability*. Chicago and London: The University of Chicago Press, 2004, p.230.

11. Joseph E. Stiglitz, *Globalization and its Discontents*. New York and London: W.W. Norton & Company, 2002, pp. 84 and 260-261

12. William J. Baumol, *The Free-Market Innovation Machine: Analyzing the Growth Miracle of Capitalism*. Princeton and Oxford: Princeton University Press, 2002, p.1

This is perhaps clearest when it comes to rights to intellectual property rapidly becoming a principal source of the wealth of nations[13]. Intellectual property is the product of the laboratories of giant corporations and the garages of college undergraduates. It is produced in the face of risks that the research will produce nothing of commercial value, and that some other inventor/innovator will be first to the finish line.

Where there is no firm rule of law, including most notably effective laws covering copyright, patents and trade secrets, add the risks that the rewards to innovation will be diluted by pirating, illegal copying or government action to appropriate the gains of entrepreneurs to itself. For the level of reward for innovation is a direct function of the innovators' ability to deprive others of its use. To the extent that an innovator cannot appropriate the benefits of his work, either by producing and marketing it himself or by charging licence fees to others, his incentive to devote time, energy and money to the innovation process is diminished. When free-riders can use the results without sharing the costs and the risks of research and development, investment in innovation will be at a sub-optimal level.

That, at least, is the story told by those who believe that Edison had it right when he said that innovation is "one percent inspiration, ninety-nine percent perspiration". As Joseph Farrell and Carl Shapiro point out in their perceptive contribution to the Raffaele Mattioli Lecture Series, if one accepts this view of one of the world's great innovators, it follows that "Perspiration will be more forthcoming if it is well paid.... On this view, innovative efforts, like many other investments, are driven primarily by the return they can generate, after adjusting for risk."[14]

It should be noted that this view of the incentive to develop innovations is not unanimously held by economists and policymakers. After all, man does not live by bread alone, and there are substantial non-monetary rewards accruing to an innovator. Whether the rewards are pecuniary or psychic, the right to profit from one's intellectual property is an important driving force – but one that cannot be allowed to become a tool for stifling lawful competition, for example by tying a patented product to an unpatented one, thereby reducing competition in the latter market. It is important to strike a balance "allowing the inventors of new technologies to reap the benefits of their innovations, while encouraging the timely diffusion of new technologies and limiting the damage from monopoly power".[15]

Competition policy

Which naturally brings us to the third policy that must be got right if the pace of innovation is to be maximised: competition policy. In both developing[16] and developed economies such as the UK, the experience of privatising monopolies without vigorous application of competition policy taught that

13. I borrow in the following passages from John H. Shenefield and Irwin M. Stelzer, *The Antitrust Laws: A Primer*. Washington, D.C.: The AEI Press, 2001 (fourth edition), chapter 10.

14. Joseph Farrell and Carl Shapiro, *Intellectual Property, Competition, and Information Technology*, in The Economics of Information Technology: An Introduction. Cambridge: Cambridge University Press, 2004, p.59

15. Roger W. Ferguson Jr. and William L. Wascher, *Distinguished Lecture on Economics in Government: Lessons from Past Productivity Booms*, The Journal of Economic Perspectives, Vol. 18 No.2, Spring 2004, p. 22

16. In connection with developing economies Stiglitz points out, "Privatization, unaccompanied by competition policies and oversight to ensure that monopoly powers are not abused, can lead to higher, not lower, prices for consumers." Stiglitz, op.cit., p.84.

detailed regulation is required to prevent monopolistic extortion – and is likely to make a less-good fist of it than enforcement of competition law.

A vigorous competition policy should maximise the rate of innovation and the rate of introduction of new technologies, in the process denying what Hicks identified as the greatest monopoly profit: a quiet life.[17] To those who say that even monopolists have incentives to innovate so as to maximise profits, the practical answer seems to be that they misunderstand the value of a quiet life. Innovation is hard work,[18] often resulting in failure that requires embarrassing explanation to colleagues, board members, shareholders and others. Absent the threat to survival that competition presents why not ensure a quiet life by deploying the weapons often available to monopolists to make certain that challengers have difficulty raising capital, or getting distribution outlets, or obtaining patent protection. Big, competitive firms might be necessary for some forms of innovation, but monopolies are not.

I recognise that the economic literature on this point is ambiguous. Mancur Olson argues that supernormal profits created by innovation increase in the presence of "difficulties of entry and imitation", increasing the rewards to "the innovations that mainly explain economic growth and progress!"[19] But even he worries that cartels, governments and lobbies can constrain new entry and permit the "lethargic or inefficient" to "shelter from a Darwinian struggle for survival..."[20], suggesting an important role for an active competition policy.

Other economists point out that competition in product markets might deny innovators the ability to appropriate to themselves the fruits of their inventions, thereby discouraging expenditure on research and development.[21] In response, John Vickers, in a paper considering the relation between competition and productive efficiency, notes: "Competition seems very well in practice, but it is not so clear how it works in theory."[22]

A commonsense appraisal of real-life experience – the comparison of productivity growth in countries with and without competitive market systems; Microsoft's efforts to use its dominance of the PC operating systems market to deny potential competitors access to related markets and to venture capital; the rapid introduction of new electric generating technologies when competition was opened up in the electricity market; the innovative services offered by the airline industry when competition replaced regulation; the flood of innovations when the monopolies of BT in Britain and AT&T in America were successfully challenged – all suggest that competition means a fiercer gale of creative destruction of old technologies than does the cosier world of cartels and monopoly.

17. J. R. Hicks, *Value and Capital.* Oxford: Clarendon Press, 1939. This section draws on a lecture I delivered at No. 11 Downing Street on 15 November 2000 under the auspices of The Smith Institute and reproduced in my Lectures on Regulatory and Competition Policy, pp.85-101, published in 2001 by The Institute of Economic Affairs.

18. Lewis, op.cit., pp.230-231

19. Mancur Olson, *The Rise and Decline of Nations.* New Haven and London: Yale University Press, 1982, p.61

20. Ibid., p.59

21. In this connection see Donald Hay and John Vickers (eds.), *The Economics of Market Dominance.* Oxford: Basil Blackwell, 1986, pp. 8-9

22. John Vickers, *Concepts of Competition,* Oxford Economic Papers 47 (1995), p.1

This is especially important in this day of the internet and related technologies. It is of course true that the claims for this technology are somewhat overblown, and that electricity, the telephone, the automobile and perhaps other inventions can claim to have had a greater impact. No matter: we are in the midst of what Schumpeter referred to as a 'cluster'. Some 65 years ago he wrote:

"Why should the carrying into effect of innovations...cluster at certain times...? One answer suggests itself immediately: as soon as the various kinds of social resistance to something that is fundamentally new and untried have been overcome, it is much easier not to do the same thing again...so that a first success will always produce a cluster... This is indeed the method of competitive capitalism..."[23]

If we are indeed entering into, or are in the early stages of, a 'cluster' of innovation, we should be especially alert to efforts by those threatened by change to deploy anticompetitive weapons in order to thwart the diffusion of new inventions and techniques. Baumol put it in characteristically trenchant fashion:

"...What differentiates the prototype capitalist economy most sharply from all other systems is free-market pressures that force firms into a continuing process of innovation, **because it becomes a matter of life and death for many of them**."[24] Baumol goes on to attribute "the unprecedented and unparalleled growth performance of the capitalist economies...primarily to competitive pressures...that force firms... to unrelenting investment in innovation and that...provide incentives for the rapid dissemination and exchange of improved technology throughout the economy."[25]

As with macroeconomic policy, so too with competition policy: it is easier to call for appropriate, innovation-oriented policy than to devise one. This is especially true in an age when it is tempting to believe, as many do, that competition policy developed in the days of smokestack industries is not applicable to our high-tech world. Even assuming that we can divide modern economies into high- and low-tech industries, (which, considering the use of new materials and robots to produce cars in the supposedly 'low-tech' auto industry, would be no easy chore), some competition policy truths remain self-evident, or should:

- price-fixing and cartel behaviour operate against consumers' interests, tend to result in extortionate prices and stifled innovation, and warrant criminal penalties;
- a firm with substantial market power, even power fairly won in the marketplace, cannot be allowed to leverage that power by tying other products to the one that it dominates;
- a firm with substantial market power cannot be allowed to use that power to bludgeon independent manufacturers not to deal with its competitors, or impose a pricing system that accomplishes that same result; and
- a firm with substantial market power over a product, access to which is crucial for firms that compete with it in other product

23. Joseph A. Schumpeter, *The Analysis of Economic Change*, loc.cit. pp.141-142
24. Op. cit., p.viii
25. Ibid., p.3

markets, cannot be allowed to deny access to its product. This latter point, I hasten to add, is more than a bit more controversial than the prior three.

With those principles as bedrock, competition policy, intelligently applied, is sufficiently flexible to determine whether any market dominance is likely to be merely transitory; to distinguish products competing in worldwide markets from those facing competition from less global sources; and to determine which business practices increase or at least do not diminish competition, and which unreasonably reduce the force of the competitive gale of creative destruction.

Regulation

Any discussion of the relation between micro economic regulation and innovation must start from the premise that no matter how clear the governing statutes, no matter how intelligent the administrators of those statutes, regulation is an inadequate substitute for competition in stimulating innovation. Alfred Kahn, the dean of America's regulators (and deregulators: he was instrumental in the deregulation of the airline industry) points out in his classic study of regulation:

"Competition is far more powerful than regulation in forcing businesses to explore the slope of their cost functions and elasticity of their demands, and to push down costs, if they are to prosper. In those situations in which competition is feasible, regulatory commissions clearly should welcome it rather than rush to restrict it."[26]

There are two sorts of regulation that can affect the rate of innovation. The first includes the host of rules designed to protect individuals from risks that bureaucrats and legislators see as threatening health and safety. Benjamin Franklin probably would have had difficulty getting a permit from the health and safety authorities to fly a kite tipped with a metal key into an electrical storm, as he did in 1752 to prove that lightning "is really a stream of electrified air, known today as plasma".[27] And Bill Hewlett and Dave Packard would undoubtedly have been unable to have the Palo Alto garage in which they invented the audio oscillator, and which is now considered "the Birthplace of Silicon Valley",[28] declared a safe work place.

More recently, either out of a desire to protect domestic producers from competition or consumers from health-damaging foods, Europe's regulators have fought to prevent the introduction of genetically modified foods. Meanwhile, American regulators have prevented patients with terminal illnesses from having access to innovative treatment methods, on the theory that it is more important for the regulator to avoid any blame in the event of some mishap than it is for dying patients to avoid pain.

The second set of regulations that can have an extraordinary impact on the pace of innovation are those emanating from the various sectoral agencies charged with overseeing industries such as financial services, energy supply and telecommunications. They include direct price and profit controls; rules on the terms of access to monopoly facilities (transmission

26. Alfred E. Kahn, *The Economics of Regulation: Principles and Institutions.* New York: John Wiley & Sons, Inc., 1971, Vol.2, p.112
27. *Electrified Ben*, The Franklin Institute Online, www.fi.edu/franklin/scientist/electric.html
28. www.hp.com/go/garage

facilities in the electric industry, the networks and 'last-mile' in telecoms); regulations concerning access to content (sports rights, movie rights) and the nature of content (suitability for children, product placement,[29] consistency with societal norms[30]); control of, or acquiescence in, the deployment of monopoly power and privilege in related markets (BBC cross-promotion): and a host of other terms on which business is done in regulated industries.

The role of regulators in stimulating or stifling innovation in each of the regulated industries would be worth separate chapters – the history of America's transport regulators in discouraging innovation would make a cautionary tale worth reading by today's regulators. One scholar of the history of regulation argues, in effect, that such separate, sectoral consideration is essential since general treatises on regulation are not particularly useful. In view of "the diverse and sometimes contradictory functions" of regulatory agencies, "all overarching theories and heroic generalisations about 'Regulation' (with a capital R) run an extremely high risk of being in error. No single theory from any academic discipline can predict precisely which industries will be regulated and which will not ... Like the application of regulation to industries, the behaviour of regulatory commissions, once they have been created, shows no clear single pattern."[31]

Support for the view that industry-by-industry studies can be more informative than generalisations about regulation can be found in R.H. Coase's discussion of how various activities should best be organised and financed. In his classic essay *The Lighthouse in Economics*,[32] Coase instructs that:

"Generalisations are not likely to be helpful unless they are derived from studies of how... activities are actually carried out within different institutional frameworks". Only after having done such studies, can we arrive at "generalisations that have a solid base".[33]

Aware of McCraw's warning, and with Coase's apparent blessing for "generalisations that have a solid base", we examine below some of the sector-specific regulatory regimes that affect the rate of innovations, and then venture a few tentative general conclusions about Regulation (capital R), at least as it relates to innovation.

Financial services

The importance of the decisions of regulators of financial services extends far beyond the industries over which they have direct control. As two officials of the US Federal Reserve System have pointed out, "A...major ingredient in promoting the productivity gains associated with technological innovation has been a complementary set of innovations in the financial sector that have changed the

29. Including or referring to a product or service in a programme in return for payment is banned by the European Commission's *Television Without Frontiers* Directive, a rule that not incidentally reduces the revenue of commercial broadcasters competing with tax-funded public service broadcasters.

30. "If you think pornography and/or obscenity is a serious problem, you have to be for censorship.... If you care for the quality of life ..., then you have to be for censorship." Irving Kristol, *Pornography, Obscenity, and the Case for Censorship*, The New York Times Magazine, March 28, 1971, reprinted in Stelzer, op.cit., p.177.

31. Thomas K. McGraw, *Prophets of Regulation*. Cambridge, Massachusetts and London: The Belknap Press of Harvard University Press, 1984, p.301

32. In *The Firm the Market and the Law*. Chicago and London: The University of Chicago Press, 1988, pp.187-213

33. Ibid., p.211

financial landscape in ways that were especially appropriate to the predominant form of business organisation in each period."[34] Thus the development after the US Civil War of deep markets for debt and preferred stock, and after World War I of markets for equities; the creation by Michael Milken of the market for lower-rated corporate debt ('junk bonds' to the WASP-establishment firms that found them a threatening source of finance for often-Jewish 'predators'); and the more recent emergence of methods for quantifying and managing risk – all represented innovations of importance extending far beyond the financial services industry. In each case, financial services regulators had to decide whether to facilitate or impede these changes, each affecting the cost at which American industries can raise the investment capital needed to permit firms to fund growth and investment in new technologies.

These regulators also have an obligation to offset the information asymmetry that at times gives innovators and marketers of financial products an undue advantage over buyers. The restless minds of investment bankers and share- and insurance-floggers are continuously creating new and complicated financial products, many designed to allocate risk to those best able to bear it. This challenges regulators to decide whether these products, with their potential to enhance efficiency, are fairly described to investors of varying degrees of sophistication and risk-tolerance. In that role, they can allow or prevent the sale of innovative products that offer attractive risk-adjusted returns; they can encourage or bar the emergence of firms offering new products (think hedge funds); they can decide on a degree of supervision that leaves the public at the mercy of the unscrupulous or, alternatively, one that is so costly that it drives up the cost of operation and hence the cost of capital unnecessarily.

The most difficult question currently facing regulators is how to deal with hedge funds, and the relation of those funds to the nations' banks. An attempt by the US Securities and Exchange Commission to require hedge funds to register was overturned in the courts. Whether that effort represents regulatory over-reach by nervous bureaucrats: or a response to pressure from interested parties that find the innovative activities of hedge funds threatening to their established way of doing business: or indeed prudential intervention that might reduce the risk of a systemic threat to the banking system – it remains an open question.[35] As does the economic question of the level of leveraging that can safely be tolerated in pursuit of financial innovation.

Financial regulation in America and Britain, whatever its defects, can claim two major accomplishments. First, despite a few failures in which scandals have shaken the faith of investors in the ability of regulators to prevent fraud and other abuses (US), and the mis-selling of financial instruments to buyers on the short end of the information asymmetry phenomenon (UK), regulators in the countries that are home to the world's two most important financial markets have maintained investors' faith in the integrity of those markets. We can be reasonably certain that a cost:benefit analysis (if such a quantification of the intangibles present is indeed possible) would show that the costs have been far exceeded by the benefits. Whether those costs have been at the minimum level attainable we probably will never know.

34. Ferguson and Wascher, loc.cit., p.14

35. See report in *Financial Times*, August 9, 2006

The second accomplishment in which financial regulators can take some pride is that they have not stifled innovation in the development of markets for financial instruments, or in the invention of new instruments, including new mortgage instruments; nor in the deployment of the technologies necessary to permit the efficient, massive, worldwide trading of securities and currencies; nor in the methods of managing risk that have reduced the risk of systemic failure of the financial system. Again, we are not faced with an uninterrupted set of regulatory achievements, witness how close the US came to systemic failure when Long Term Capital Management (LTCM) collapsed. Whether banking regulators are to be faulted for failing to foresee and prevent LTCM's debacle[36], and the systemic collapse that it almost produced, or praised for engineering the takeover of the LTCM portfolio and the interest rate cuts that, to cite a former Fed Governor, prevented "a wave of destruction [that] would burst through America's biggest financial institutions, posing a serious threat to the sustainability of the [economic] expansion"[37] – is still debated.

The lesson for students of regulation is this: restraint in stifling the introduction of new technologies – in this case, new forms of trading and risk management – is not without risk. Regulators must often balance the goal of allowing innovation against the goal of protecting the public from change that is not in the public interest. Economists generally hope regulators will err in the direction of allowing innovation. In the case of LTCM, the regulators' gamble seems to have paid off: the costs of failure were confined to the private-sector risk-takers who were hunting for extraordinary profits, while the public has benefited from the onward march of new instruments that are improving risk management.

Energy

Regulators of industries in the energy sector can adopt rules that pass on the gains of innovation to consumers so quickly that companies have no incentive to explore new ways of producing and delivering energy, or allow firms with monopoly power to benefit excessively from cost savings generated by investing customers' money. Whatever one thinks of the economic viability of nuclear power, there is no question that regulators' treatment of investment in US nuclear plants that eventually proved uneconomic contributed to the long period in which investors refused to make capital available for the construction of new plants. Nor is there any question that regulators' treatment of investment in equipment designed to ease the environmental impact of power production importantly, and probably adversely, affected the pace of innovation in such equipment. Finally, there can be little question that the deregulation of the electric and, probably, the natural gas industries, was not handled as well as they might have been.

To review all of the arguments on both sides of these questions would consume – indeed, has already consumed – volumes. Suffice it to say in this brief chapter that the electric power industry in Britain finds itself the beneficiary of the introduction of competitive markets, but with insufficient capacity to prevent serious price spikes; that in both the US and

36. "With regard to derivatives, the policy-making arm of the Fed took a laissez-faire approach – starting with Greenspan, who was enamoured with the seamless artistry of the new financial tools." Roger Lowenstein, *When Genius Failed: The Rise and Fall of Long-Term Capital Management.* New York: Random House, 2000, p.105

37. Laurence H. Meyer, *A Term at the Fed: An Insider's View.* New York:HarperBusiness, 2004, p.115

the UK, consumers are faced with choices between energy sources, and between consuming and conserving energy that are not guided by prices that reflect the true economic costs of consuming fuels of various sorts; and that regulators have yet to figure out how to regulate transmission pricing so as to produce construction of an optimal amount of capacity that is available on efficient terms to independent generators.

This is, at best, a mixed record, and perhaps a summation made excessively harsh by the author's too-close involvement in the energy regulatory process over many years.[38] After all, regulators can reasonably claim credit for the facts that: in most, although not all, instances the lights have stayed on, thanks in part to their vigilance in discharging the difficult chore of supervising monopolists not uniformly renowned for their efficiency and ability to adapt to change; innovations in the generation of electricity have had a nontrivial effect on efficiency; and important sectors of the energy industry have been deregulated.

Telecommunications/media

In the telecommunications and media industries, converging at a rapid rate, regulators have perhaps their most difficult role in devising strategies that optimise the rate of innovation. They have to cope with political pressures not present in other industries, except perhaps during crises such as energy-price spikes, and with a set of economic factors, including a rapid rate of technological change, that make regulation extraordinarily complicated.

Start with the political pressures. Legislators are sensitive to demands of entrenched interests seeking to bar entry by innovating challengers, and to constituents who often have conflicting demands: they want freedom of expression, but also protection of children from unwholesome programming; preservation of the key role of state-funded public broadcasting, but a desire for multiple channels and varied programming; access to sports programmes on a no-fee basis, but some method of funding improved quantity and quality of sports events; unbiased news reporting, but no government regulation to enforce such 'objectivity'.

Similarly, lawmakers want more competition, but they also want small and rural customers subsidised; they want to grant monopoly franchises to US cable companies so that they can wring various concessions from the beneficiaries of these grants,[39] but they don't want to confront complaints from customers about the monopoly charges and slothful service that inevitably accompany monopoly power; in Britain and in the EU, legislators don't want to offend powerful constituents such as BT, or the BBC and other state broadcasters, but they want users of telephone, broadband and television services to have options if they are unhappy with those providers.

The result is that legislators often seek to avoid voters' wrath by enacting vague laws that give regulators huge flexibility in – and responsibility for – determining how the telecommunications industries will develop. The decisions that regulators make while moving within the very broad parameters set by their legislator-masters importantly

38. During the course of that involvement the author counted as clients many electric and gas utility companies, and at this writing consults with a developer of wind farms.

39. In most cities cable companies pay franchise fees and allocate channels for 'public access programming'. The latter are used by politicians to televise council hearings and other programmes in which they feature. Incumbent cable companies generally insist that potential entrants, such as telecoms companies seeking to offer broadband services, bear the same burdens as they have been forced to carry.

determine the rate of technological progress in these industries.

Not that legislators simply pass laws regulating the media, and then move on to other matters. Because the media importantly influence the political agenda and a nation's 'culture' politicians do not hesitate to pressure regulators to adopt rules favouring their own positions and those of their constituents. In America, Michael Powell, former chairman of the Federal Communications Commission, was in effect ridden out of office by a congress that felt his attempts to reform the regulatory process were a bridge too far in the direction of a free market. In Britain, members of parliament and ministers often march to the BBC's drum, protecting its inflation-busting mandatory licence fee from close scrutiny and challenge by regulators, many of whom were drawn from the ranks of BBC staff. In the EU, eager regulators attempt to protect state broadcasters from private-sector competition by intervening in the process by which broadcast rights are acquired, and by limiting the advertising revenue of those broadcasters.[40]

So much for the politics of telecoms and media regulation. Now to the economics. It is fair to say that we have come a long way since regulators regarded themselves as protectors of the financial integrity of the companies they regulate. It is no easy thing for a regulator to welcome to the field of battle for consumer favour an entrant that might threaten the financial strength of the companies he is charged with supervising. As Kahn notes:

"Technological progress...confronts regulatory commissions with the dilemma of seeing their charges lose business and the regulated price structure threatened with undermining... This impact of new technology, and the dilemmas it creates for protectionist regulatory commissions, are nowhere more clearly illustrated than in the field of communications."[41]

Consider the complications America's regulators face in separating the innovators from the change-blockers. AT&T is developing video delivery systems that threaten the hold that cable and satellite providers have on the market, but is also alleged to be fighting a rearguard action to stem the advance of internet-based telephone services (voice over internet protocol, or VoIP) that have the potential of driving the price of calling to somewhere between negligible and nil. Consider, too, the difficult choice the regulators face when deciding whether the cable-telco duopoly is so powerful that its broadband activities must be regulated, or whether new investments in wireless technologies represent a sufficient challenge to warrant allowing market forces to operate, unimpeded. Finally, consider the difficulty Ofcom has in deciding whether the move by the BBC into new services that compete with commercial broadcasters represents a passing on to consumers of economies of scale and scope, or a warning to private entrepreneurs that a state-funded entity with no need to earn a return on capital stands ready to drive any interlopers from the field, as it attempted to do in the field of 24-hour news. Remember: when it comes to technological competition, "no firm can be certain of being the winner – no matter how much effort it puts in – *unless it can make it unprofitable for rivals to enter the race*".[42]

40. The reader is reminded that the author has in the past been a consultant to BSkyB and is currently a consultant to News International and to News Corp.

41. Kahn, Ibid., pp.29-30

42. Partha Dasgupta, The Theory of Technological Competition, in *New Developments in the Analysis of Market Structure*, Joseph E. Stiglitz and G. Frank Mathewson (eds.). Cambridge: The MIT Press, 1986, p.541. Emphasis added.

If regulators are not to get in the way of innovation in the rapidly changing world of telecommunications and media, they must keep that change in perspective. First, the emergence of 'new media' does not mean that the 'old media' are disappearing, nor the regulatory issues associated with them. The history of the industry is that television has not destroyed the film or radio industries, cable and satellite have not destroyed free-to-air broadcasting, and the rise of the internet has increased, rather than eliminated, the demand for 'content' – news and entertainment. So although regulation should not be used to stifle the emergence of new media, neither can it be biased in their favour, as may well be the case with state- and regulator-promoted digital television.

Second, regulators must recognise that the survival of old media does not mean that the world is unchanged. Competition between new and old is intense, and takes the form not only of price competition, but of the more important competition to innovate; to develop new delivery systems, to lower costs of providing service, to find new ways of promoting services, many of which consumers have never experienced. This has profound implications for how regulators, who prefer narrow market definitions that unreasonably magnify the market shares of such firms as BSkyB, must define markets when determining the presence or absence of 'dominance', and rule on the propriety of widely varying competitive tactics.

For example, it was not so long ago that competition for advertising was a battle between competing newspapers; now, electronic media are equal contenders for many forms of advertising, most notably the classified ads that historically have been the major source of profits for newspapers. Any newspaper's share of classified advertising must now be considered not only by comparing it with other newspapers in the relevant geographic area, but with the share garnered by competing electronic media as well.

Most important, regulators must recognise that "Technology changes. Economic laws do not... A few basic economic concepts go a long way toward explaining how today's industries are evolving."[43] For example, in the early days of telephony the Bell System was able to stifle the rise of local competitors by denying them interconnection privileges, or access; years later, Microsoft was able to do the same to Netscape, and BT to emerging competitors. The technologies changed, but facilitating access to those parts of a network that the incumbent's potential competitors "cannot viably replicate"[44] remains one of the key problems regulators face in trying to ease the path of enterprising innovators. It remains to be seen whether Ofcom's solution – requiring BT to set up an "operationally separate business unit" under the supervision of an Equality of Access Board, chaired by a BT non-executive director[45] – solves the problem all regulators face when dealing with a vertically integrated company with substantial market power in one horizontal segment of the industry.

43. Carl Shapiro and Hal R. Varian, *Information Rules: A Strategic Guide to the Network Economy.* Boston: Harvard Business School Press, 1999, pp.1-2

44. Office of Communications, Annual Report 2005/6, p.18. The definition of which parts can and cannot be viably replicated will, of course, provide work for regulators and interested parties.

45. Ibid., pp. 19-20. The requirements under which this unit, dubbed 'Openreach', will operate include separate bonus schemes and uniforms, and "Chinese Walls" between Openreach and BT Group plc.

In performing their pro-innovation function, regulators should keep in mind that sauce-for-the-monopoly-goose is not necessarily sauce-for-the-new-entrant-gander. Monopolist-incumbents should not automatically be allowed to employ competitive tactics that are perfectly appropriate for newcomers to use in order to blast their way into a highly concentrated market.[46]

Most of all, those burdened with regulating the telecoms/media/internet/broadband world "should eschew prescriptive regulation because ...the prospect of open, competitive markets is strong enough to justify a policy of non-interference",[47] or at least to create such a presumption, subject of course to vigorous and appropriate enforcement of competition policy. When such enforcement alone does not satisfy the requirements of the legislature, and regulatory action is required to supplement competition policy, "agile administrative adjudication"[48] should be the watchword so as to keep regulatory costs, a particular burden on new entrants, to a minimum.

Conclusion

The forces affecting the pace of innovation are too varied, and quantitative measurement of those forces too inexact, and the task of separating cause from effect too difficult, to permit firm conclusions concerning policies that encourage and, alternatively, inhibit innovation. But a few observations are possible.

First, the macroeconomic environment – monetary, fiscal (including tax), and regulatory policy – must be conducive to investment and risk-taking.

Second, the rule of law must prevail, with intellectual property protected, but with rules firmly in place to prevent its use to stifle competition.

Third, especially in periods of rapid technological change, and in industries in which such change is accelerating, less is more. A presumption in favour of allowing market forces to operate to choose the technology that best satisfies consumer needs, at prices consumers find attractive, should guide policymakers and regulators.

Technology can change more quickly than laws or rules can be amended; incumbents know how to use the regulatory process to disadvantage challengers, who rarely have either the nous or the funds to win the regulatory wars; regulators, often recruited from the companies they will be regulating, have a tendency to come to their task with the worldview of the incumbents and to prefer modest changes in regulation to the structural dissolution needed to end abuses by integrated companies such as investment banks/brokerages, and firms controlling network access.

That said, regulators cannot avoid the role thrust upon them by legislators with often contradictory and vaguely-defined roles, and by the inability of competition laws to cope with the entry barriers that incumbents are often skilled in erecting. In performing such functions, regulators must be aware of the fact that new entrants do not have sunk investments to protect – indeed they are, by their very nature, destroyers of the sunk investments of

46. Ibid., pp. 302-304. Se also Joel B. Dirlam and Alfred E. Kahn, *Fair Competition: The Law and Economics of Antitrust Policy.* Ithaca, New York: Cornell University Press, 1954

47. Jonathan Sallet, *Principles for Governmental Action in a Broadband World*, working paper No. 273, published in Washington, D.C. by the Economic Policy Institute, November 2005, p.31

48. Loc.cit.,p.32

incumbent service providers. They should therefore be permitted to deploy a range of business tactics that are properly denied to incumbents, who have a stake in denying consumers access to new technologies.

Finally, regulators should be aware of the tendency towards mission-creep. Regulation often begets regulation, as is evidenced by the desire of the European Commission to revise its *Television Without Frontiers* Directive "to extend the scope of statutory regulation to include non-broadcast audio-visual content on the internet", a move so far opposed by Ofcom.[49] This self-denying position fits well with its decision to allow markets rather than men to set calling rates now that competition is sufficiently effective. Other regulators might take these moves by Ofcom as a model when opportunities to expand their reach present themselves.

49. Loc.cit., p.57

The challenges of a digital world and the need for a new regulatory paradigm

Leonard Waverman [1 2]

London Business School

Introduction

The growth and change in the telecommunications sector since the waves of privatisations that started in the 70s is well documented.[3] Most developed economies have now successfully managed the transition from a monopoly provision model to a competitive market environment, with a multiplicity of players in various layers of the value chain. These developments have been achieved through a number of key regulatory instruments focused on allowing market entry and curbing the dominant position of incumbent operators.

The stated public policy objective was to develop competition in the telecommunications sector while maintaining universal service provisions to protect vulnerable customers. However, the regulatory paradigm and tools that have been instrumental in achieving public policy objectives such as this are becoming increasingly inadequate in the new, all-IP multi-access world we are entering. That world is one of multiple high-speed links to most locations.

This new multi-path IP world – what I call 'MIP' – has its own set of challenges for regulators and competition policy authorities. Attempting to apply the same old regulatory principles, designed for the PSTN world, to these new emerging challenges often has opposite effects to the ones intended.

There are major consequences for society of getting regulation wrong as we move from access issues on 100 year-old copper wire legacy networks to fostering investment in Next Generation MIP Networks. The productivity boom in the USA since 1996 has been attributed to investment in ITC (Information Technology and Communications hardware and software). Europe and the UK lag behind the USA in productivity and, indeed, that gap has been widening in latter years.[4]

The USA has recently moved to end access regulation and to emphasise facility-based

1. I also authored a paper on a similar topic for France Telecom. There is some overlap. My views expressed in both papers are my own and not necessarily those of France Telecom or Ofcom. These views are independent of the views of both France Telecom and Ofcom.

2. With the assistance of Benoit Reillier and Kalyan Dasgupta

3. Eli Noam, *Interconnecting the Network of Networks*, MIT Press, Cambridge MA, 2001

4. Bart Van Ark, Robert Inklar and Robert McGuckin (2002): *Changing Gear: Productivity, ICT and Services: Europe and the United States*, GGDC Working Papers GD-60

competition rather than 'me-too' service competition.[5] Hence, relying in the UK and Europe on any existing telecom regulation which is 'unfit for purpose' not only retards telecom facilities and service growth but has significant negative spill-overs to the economy – holding back growth and limiting productivity gains.

Today's regulatory instruments, which need to be reviewed, are:

- **the traditional view that 'access' is a copper wire essential facility that needs to be regulated.** This view may well be outdated for the reasons outlined below;
- **the economic tests used to identify, and prevent, potential abuses of dominance.** These may no longer apply to some of these increasingly complex, dynamic test and sometimes 'multi-sided' markets. As former facilities-based monopolies come into competition with each other, it erodes the 'market power' in any one supply channel. Customers of all telecom services are increasingly able to find alternative paths. Once, a copper wire pair to the home was in some sense an 'essential facility' as it could not be easily replicated by entrants. Today, most businesses and homes have alternative paths – cable, cellular phones, fixed wireless and, potentially, satellite and power line;
- **the presumption of 'market power'.** In the old paradigm, the infrastructure supplier had market power as they faced small buyers who had no offsetting buyer power. The growth of ISPs and other buyers and users of telecom services also mitigates against this traditional view of inherited market power;
- **retail regulation of prices and offers by so-called dominant incumbent owners of legacy copper networks.** The rapid introduction of 'triple' and 'quadruple' plays by converged operators is quickly redefining reference offers in the sector, and the traditional retail regulation relying on past definitions of markets and market power need to be revisited in this light;
- **the concept of regulatory 'certainty' in an uncertain dynamic investment world needs to be redefined.** Investors in new infrastructures – the so-called '21st Century Networks' (21CN) – seek regulatory certainty before committing to large investments that could be undermined by the application of old regulatory paradigms. Firms viewed as incumbents seek commitments from regulators that they will not be required to unbundle these new networks or be forced to provide wholesale services for each new retail service offered. Such certainty is, however, not easy to provide and regulators are struggling with these demands for regulatory forbearance and regulatory holidays;
- **new issues on the agenda.** As we turn our attention away from ensuring access to incumbent legacy infrastructure and focus instead on the conditions of competition in MIP markets, we need to address new issues: for example, the costs and benefits of policies such as non-discriminatory treatment by infrastructure providers to content and service providers.[6] The 'net neutrality' debate in the USA is a taste of the future, which will likely be one of mediating the change from today's one-

5. D.C. Circuit Court, 2 March 2002 in United States Telecom Association v FCC. US Supreme Court decided not to review this decision on 15 June 2004

6. See Christopher Yoo, (June 2006), *Network Neutrality and the Economics of Congestion*, 94 Georgetown Law Journal

size-fits-all definition of service to one where many grades of service exist;

- **regulation protects consumers.** We are in danger of continuing to rely on an institutional invention – independent regulation – that fitted the world of legacy copper and cable infrastructures that each provided single services. The new MIP world has multiple technologies, all providing similar services. In this world, one needs to rely on an effective instrument that all other dynamic markets rely on: competition policy.

My central thesis is that we need to examine all bases of existing regulation including its most central tenet – non-discrimination. We have come from a world of ensuring that former telecom monopolies could not discriminate – originally at the retail level among customers and, more recently, at the wholesale level among those who use its facilities.

This principle of non-discrimination was essential and central. But it emanated from a particular world which called for the opening-up of competition in voice services on the legacy infrastructure of a single firm. In those circumstances, reinforcing *ex ante* non-discrimination by regulatory fiat was beneficial to society. But the new world of MIPs is totally different: no longer do we have a particular service (voice, television) tied to a unique infrastructure (copper, cable/satellite). Hence in the new MIP world, we are no longer concerned with a particular service/infra structure pair. There are multiple infrastructures – and hundreds and thousands of services.

So what is 'discrimination'? Is it discriminatory for BT, for example, to offer Yahoo broadband or exclusive Premier League Sports? Discriminatory yes, but is it anti-competitive? Surely not. We must therefore be vigilant against the calls for regulatory enforced non-discrimination, such as the advocacy for 'net neutrality'. Instead, we must recognise that most so-called discrimination is actually differentiation and innovation. Mandatory imposed non-discrimination will likely reduce such differentiation and innovation and thus lower overall welfare. Instead of *ex ante* regulatory-imposed 'non-discrimination', we should rely on competition law to arbitrate on a case-by-case basis between discriminatory acts which are anti-competitive and those which are not.

From monopoly to competition: regulating telecom

Regulation was designed for an era where a single firm dominated an infrastructure that provided separate services from another firm's infrastructure. Thus telephony was essentially circuit-switched, person-to-person voice calls, provided over a unique copper wire pair infrastructure. In the past, cable provided broadcasting over its own separate infrastructure. Mobile phones provided mobility for voice calls, again over a largely distinct infrastructure.

So in the minds of regulators, regulating the PSTN network was equivalent to regulating voice services since this was the purpose for which the network had been designed. The second path to many homes – cable – was a one-way transmission of centrally-provided television signals. Telephony and broadcasting were thus totally separate offerings on both the demand and supply side. The key element of the past was thus a *one-to-one correspondence between infrastructure and services*. All this is changing but the shift from a network-centric model to a market centric one is still very much underway.[7]

This separation of infrastructure/service pairs resulted in market power over customers wanting these services. And, as a result, regulation of retail rates was deemed necessary for telecom, but oddly - in most countries - not for cable.

Customers of the fixed-voice telephony service were buying directly at a retail level from the integrated network provider that had rolled out the infrastructure. This was the model in most countries and resulted in strong retail price regulation - originally rate-of-return regulation and more recently (since 1984) price cap regulation (RPI-x%)[8] often with the use of a number of different baskets.

The aim of such regulation, from time to time combined with rate-of-return targets, was to prevent incumbent operators from increasing prices in the light of a lack of competitive constraints. However, rate regulation became mired in a quasi-political morass of cross-subsidies and universal service concerns which resulted in many rates, especially those for access, being below costs.

However, these blunt regulatory instruments, while potentially useful, were only poor proxies for competition and did not provide the right incentives for rapid innovation. Despite the torrid pace of innovation in electronics and PCs, the paradigm of the integrated, slow-to-innovate phone company remained intact for many years. Today, the fixed-line phone is still remarkably unintelligent. Advances such as integrated contact lists, e-mail capability and pre-paid service are innovations of mobile operators, yet could have been done far earlier by fixed-line carriers.

The introduction of competition in telephony required a suite of new wholesale services and mandated interconnection. A central debate was over the type of competition - facility-based or retail service competition over the incumbent telco's own infrastructure. Carrier pre-selection services opened the door to a form of competition for long distance and international telephone calls. Oftel in the UK originally raised concerns about the investment incentives of such types of regulation but the EU decided that it would provide more choice of providers in the short-term, and won the argument. By the end of the 90s, carrier pre-selection was the main form of competition in Europe. However, the introduction of these 'indirect-access' services required a new regulatory focus on wholesale prices charged for facilities used by these new long distance operators.

It proved very difficult for new entrants offering voice calls or simple data connections to replicate local loop infrastructures in many markets, partly because this segment of the network exhibited many of the characteristics of natural monopolies and partly because the retail prices for access were regulated to be below-cost.

Even countries where very significant investments in alternative infrastructure (cable) were made - such as the UK - ended up with a largely fragmented industry focused on broadcasting that also needed to consolidate to be able to benefit from the economies of

7. *From Open Networks to Open Markets: How Public Policy Affects Infrastructure Investment Decisions*, Martin Taschdjian, November 2000, Centre for Information Policy Research, Harvard University

8. RPI is the economy-wide retail price index and x, the 'productivity' offset.

scale required. The alternative infrastructure, which regulators hoped would provide competition with the telcos, concentrated on broadcasting partly because of the regulatory-induced subsidy to telephony access. This can be seen best in Canada where a strong and nearly ubiquitous cable network did not offer any telephony (i.e. voice services) until the IP revolution which allowed them to offer VoIP over cable broadband.[9]

The regulatory model in the 90s was still very much concerned with managing the transition from a legacy network-centric, largely incumbent-controlled environment to a more competitive, multi-operator market. The pace of regulation was slow and steady; change was incremental, and the regulatory paradigm remained intact. Regulatory concerns, as noted, shifted, albeit slowly, from retail regulation to wholesale regulation. Regulators appeared to have the largesse of time – after all, the telecoms sector was about voice calls and leased line services which themselves also changed slowly.

However, the semiconductor and other revolutions began to permeate American industry. The ICT paradigm shift was beginning. In 1995, the then Chairman of the US Federal Reserve Bank coined the phrase 'the new economy', referring to the leap in economic growth and productivity enabled by ICT. Academics began to see sharp shifts in US productivity performance and attributed this rise to the use of ICT.[10] Firms such as Dell in computer assembly and delivery, Zara in clothing production and retail, Walmart in retail and others were using ICT to revolutionise their industries in a way called by Eric Brynjolfson 'the Digital Organization'.[11]

But telecom regulators did not move in tandem; they did little to ensure that telecom networks would utilise the latest advances in computers and to ensure the necessary investment in telecom networks – the crucial infrastructure on which computer-enabled production moves. In the US in 1996, a year after the beginning of this new economy, politicians designed an incredibly complex compromise between the might of AT&T and the might of the RBOCs.

The Telecommunications Act of 1996 ushered in a decade of litigation and increasingly constraining regulation of the telecom access providers, in order to ensure me-too service competition. But politicians and regulators lost sight of the crucial importance of telecom investment and pricing. Thus, the central importance of ICT to US economic growth and productivity may, suggest Jorgenson and Stiroh, even have diminished by 2001. With the 2005 ruling ending the UNE Platform in the US,[12] Crandall (and others) see resurgence in US telecom investment.

In Europe, with the exception of the UK and a few other countries, competition in telecom services did not begin until eight years ago, in 1998. The EU has moved quickly to ensure

9. Very small cable systems also exist for example in Kingston-upon-Hull and the Isle of White. The dozens of UK licensed cable operators that were operating in the early 90s have all merged and ntl, the new refinanced cable entity, is now the only significant cable player in the UK.

10. Dale W. Jorgenson and K.J. Stiroh (1999), *Information Technology and Growth*, American Economic Review, Papers and Proceedings, Volume 89, Number 2

11. Eric Brynjolfson (2003), *The IT Productivity Gap*, Optimize, July, Issue 21

12. D.C. Circuit Court, 2 March 2002 in United States Telecom Association v FCC. US Supreme Court decided not to review this decision on 15 June 2004

similar policies across Europe, beginning with interconnection and basic Framework directives to open up former monopolies. The 2001 Framework (actually introduced in July 2003, and which began to be drafted in the 1990s) mimics much of the US experience, albeit without the US's political and adversarial overtones. Hence, today the state of EU (and UK) regulatory policy needs reconsideration. There are four key principles of current regulation that I address:

- interconnection;
- retail price constraints;
- wholesale mandated offers based on a notion of the 'Ladder of Investment'; and
- equivalence.

1. Interconnection

Interconnection in general is of more value to smaller networks than larger ones. Indeed, large networks can limit entry and growth of rivals by foreclosing interconnection[13], which is why interconnection was a vital regulatory tool in the 1990s.

However, mandating interconnection in competitive markets is not necessarily a socially-beneficial policy. If the larger firm has innovated and invested, its size reflects these investments. Mandating interconnection then can allow free-riding, deterring those innovation and investment incentives in the first place. Take Instant Messaging (IM) as an example: AOL and Microsoft were the first ISPs to offer IM, AOL purchasing Mirabalis ICQ and Microsoft purchasing Hotmail in 1998. AOL was the predominant IM provider, using IM to grow its ISP business through network effects and scale economies. Microsoft always trailed as an ISP – and attempted both to interconnect with AOL's IM network as well as to use antitrust policy to force interconnection.

Today IM is a service offered by many firms including a highly innovative US wireless provider Nextel, the latest entrant into the US cellular market and which used IM as its strategic differentiator.

What if Microsoft had been able to use competition policy or regulation to access AOL's IM service? AOL's IM would have become the incumbent offering and with mandated interconnection, its competitive advantage would have been eroded. Nextel may, as a result of regulation, not have considered IM to be a core strategic advantage, assuming that, given what had occurred to AOL, its own IM service could be viewed as exclusionary.

This is not to say that the interconnection policies of the 1980s and 1990s in telephony were wrong – they weren't. But today, we must be careful in taking legacy regulatory instruments into a competitive world without legacy networks. We must evaluate the costs and benefits of mandating interconnection, as well as the costs and benefits of the alternative *ex post* competition policy.[14]

2. Retail price regulation

Most regulatory agencies are abandoning retail price controls in telephony. This is a good thing, but should have occurred earlier. Some regulators are still, however, fighting old battles. For example, in 2003 the CRTC attempted to

13. Mark Armstrong, Chris Doyle and John Vickers 1196, *The Access Pricing Problem: A Synthesis*, Journal of Industrial Economics, Volume 44

14. Here I part company with Shelanski, 2006, most which I agree with. He considers interconnection necessary. I think we need to evaluate its pros and cons.

raise regulated retail price floors, arguing that entrants could not compete with the incumbents' low prices.[15] The obvious retort is that no consumers are hurt by low prices, and it follows that the welfare gains of keeping prices high are non-existent. If an incumbent somewhere were actually to engage in predatory retail price behaviour, then competition policy could intervene.

3. Wholesale price regulation and the 'ladder of investment' theory[16]

This theory (best articulated by Cave in 2005)[17] suggests the following:

- access regulation can encourage entry by setting low access charges for entrants who would otherwise find insurmountable obstacles associated with replicating incumbent networks;
- entrants taking advantage of low access charges initially occupy the lowest rung of the ladder, often doing no more than reselling the incumbent's service;
- they then grow their customer and revenue bases, while regulators gradually raise the price-of-access products so that those entrants that have gained sufficient financial strength can start moving up the ladder – initially into other access products such as IP Bitstream and then into using fully unbundled local loops; and ultimately;
- entrants start investing in their own significant infrastructure, and facilities-based competition takes hold.

European regulators are firm believers in this model. The European Regulators Group (ERG) states:

"The…concept of a ladder of investment links a pro-competitive regulation with investment which in turn is pushing broadband penetration, in other words [sic] a virtuous cycle has been created." [18]

A variety of other countries may beg to differ, where evidence fails to provide conclusive support for the ladder of investment. While available European data does show some migration up the ladder (reflected by the growing importance of 'higher-rung' offerings based on fully-unbundled loops and shared access), there is also some evidence that the share of cable (the leading infrastructure alternative to the incumbent fixed-line network) has suffered. For example, the broadband scorecard of the European Competitive Telecommunications Association (ECTA) found that between the last quarter of 2003 and the first quarter of 2005, the share of cable as a proportion of all broadband connections had fallen to 17 per cent from 21 per cent.

There is therefore a risk that instead of developing healthy infrastructure-based competition, poorly-designed access regulation results in locking entrants to a business model where they rely on (arguably) subsidised wholesale access to the incumbents' networks.[19 20]

15. CRTC Telecom Public Notice 2003-8

16. This section here and in the paper written for France Telecom are quite similar.

17. Martin Cave 2005, *Investment Incentives and Local Loop Prices*, mimeo

18. See ERG *Revised Common Position on Remedies, Explanatory Memorandum*, ERG (06) 19, January 2006

19. Pindyck (2004), *Mandatory Unbundling and Irreversible Investment in Telecom Networks*, NBER Working Paper 10287, Cambridge MA, February, shows that accounting for opportunity costs or 'option values' but also adjusting for the fact that sunk costs are a barrier for entry, wholesale prices for network access are too low and thus subsidise entrants.

20. The European Regulators' Group (ERG) has explicitly endorsed the ladder of investment theory and it is enshrined in the European Commission's New Regulatory Framework.

The best illustration of the substantial negative impact of subsidised access to 'rungs on the ladder' comes from the United States. There, an unsustainable model of entry developed, based on leveraging the margin between the regulated rates for unbundled network elements (UNEs) and the rates that the Incumbent Local Exchange Carriers (ILECs) were able to charge in order to satisfy requirements on predatory pricing. In particular, AT&T and MCI-WorldCom made heavy use of the UNE-P platform, which offered them the unbundled network elements plus switching. Entrants utilising the UNE-P were essentially the same as simple resellers. There was no innovation involved, and entrants using UNE-P did not branch out into investing in their own facilities.

Between December 1999 and December 2002, UNE- prices decreased by a median of 28%, while the share of UNEs as a percentage of CLEC switched-access lines increased from just over 20% to over 50%. (See Ingraham 2005).[21] The FCC's announcement in 2003 which abolished the mandatory provision of UNEs had a major effect on the strategies of many CLECs. In August 2004, AT&T announced its intention to stop promoting its local and long-distance services. The major motivating factor behind this was that it could no longer use the UNE-P/resale platform and could not sustain its local entry strategy. MCI was subsequently incorporated into one of the largest incumbent carriers, Verizon. Both AT&T and MCI largely abandoned the residential local voice services market, once the UNE-P was no longer available.

The example of UNE-Ps in the United States illustrates the fact that access regulation can result in distorted competition, where entrants fail to innovate or move up the ladder of investment. When the regulatory decision to scrap the UNE-P was made, the biggest entrant into the market (AT&T) chose not to migrate to the next rung of the ladder – it simply withdrew from the local services market altogether.

The 'core capabilities' (and the 'core rigidities') of a firm that 'buys' network facilities from the incumbent and then relies on repackaging incumbent services are potentially different from those of a firm that invests in its own network facilities and offers the full range of services that its network technology enables. Once they have developed a business model based on regulatory arbitrage (the generous US regulatory regime afforded UNE-P entrants discounts of between 50 to 60 per cent off retail), it may prove very difficult for firms to then make the leap to competition based on developing their own facilities.[22]

4. Equivalence

Ofcom has invented the new regulatory paradigm in Europe: equivalence. Equivalence is essentially non-discrimination in wholesale services. That is, the incumbent telco – in this case, BT – cannot discriminate at the wholesale level to the benefit of its retail arm. As a principle, non-discrimination is hard to argue against. However, one must be careful to distinguish regulatory-mandated discrimination from competition policy findings of anti-competitive discrimination.

21. Allan Ingraham (2005), *Impact of Mandatory Unbundling at TELRIC Prices on Telecommunications Investment*, Powerpoint, Criterion Auctions

22. For a definition of core capabilities and core rigidities, see Leonard-Barton, D. (1992), *Core Capabilities and Core Rigidities: A Paradox in Managing New Product Development*, Strategic Management Journal, 13, pp. 111-125.

The difference, I think, is clear: should we as a society impose *ex ante* rules on service provision? Or should we rely on competition policy to sort out those discriminatory policies which are anti-competitive from service discrimination which is innovative and welfare-enhancing? The calls for net neutrality (see below) push regulatory non-discrimination principles too far. We are better off as a society in addressing anti-competitive discrimination when and if it appears, rather than relying on *ex ante* across-the-board regulatory mandates.

The evidence appears clear that in ADSL service provisioning, BT did give its own retail arm an advantage in terms of time, quality of service and so forth.[23] However, dealing with the sins of the past should not prevent us from questioning whether a current regulatory policy is sensible for a new era.

The importance of telecom investment and its primary drivers

Investment is a significant contributor to economic growth. Most empirical studies of the determinants of economic growth find a strong, positive and statistically significant correlation between overall economy-wide investment and growth.[24] A growing body of literature finds that there are significant returns to investment in public infrastructure, including telecommunications. Roeller and Waverman (2001) found that as much as one-third of German growth between 1970 and 1990 could be attributed to the growth in the penetration rate of fixed-line telephones. Further studies such as those by Greenstein and Spiller (1995) and Correa (2003)[25] find that the diffusion of telecommunications products and services into the wider economy has a significant spillover and impact. This impact seems especially pronounced for the services sectors in general, particularly sub-sectors such as FIRE (Finance, Insurance, Real Estate) which drive the growth of the modern economy.[26]

Telecom networks are the key enabling infrastructure for the information economy. While existing studies suggest the impact of telecom is quite high, such studies might well *under-estimate* the indirect impacts of telecom networks. These arise from the role of these networks as the critical infrastructure over which data, voice and video content is being transmitted. It is the vastly increased ability to exchange information that drives the growth of the high-tech knowledge-intensive service sectors around the world; thus the role of telecom networks has grown ever-more critical as the diversity of uses for the network has grown greatly in the past 15 years. There is no indication that these uses will diminish in the future, either. Indeed, the advent of Next Generation Networks will greatly increase network speeds and also the multiplicity of uses for the network.

23. See Ofcom

24. The work of Summers and DeLong (1991), *Equipment Investment and Economic Growth*, QJE, May established that certain kinds of investment – for example, investment in machinery and equipment – provide an even greater spur for productivity and economic growth than other types of investment.

25. *The Economic Impact of Telecommunications Diffusions on UK Productivity Growth*, Queen Mary College, Working Paper 92, June

26. Greenstein and Spiller's, *Modern Telecommunications Infrastructure and Economic Activity: An Empirical Investigation*, Industrial and Corporate Change, Volume 4, Number 4, econometric results suggest that a doubling of fibre optic cable leads to a more than 10 percent increase in the level of economic activity in the high-tech knowledge-intensive sectors. Correa's input-output analysis of the impact of telecommunications investment eliminates some of the biases that previous macroeconomic studies of the impact of telecommunications sometimes suffered from; yet her work also suggests a very high impact of telecommunications investment on service sector productivity.

ICT contribution to labour productivity differential 2003					
	USA	EU-14	France	UK	Sweden
Output per hour worked (in 2000 USD at PPP)	$37	$35	$43	$32	$33
Non-ICT capital per hour worked (in 2000 USD at PPP)	$54	$63	$87	$44	$58
ICT capital stock per hour worked (in 2000 USD at PPP)	$6.3	$3.6	$3.4	$3.7	$4.9
Telephone penetration (main lines per 100 population plus mobiles per 100 population)	118	140	126.03	150.69	174.62
Personal computer penetration (PCs per 100 population)	69	38	41.74	44.95	68.78
Total hours worked (in 000s)	249,634	249,616	33,961	45,345	6,723

Source: Fuss and Waverman 2006

Table 1: Europe, the USA and UK compared

The European Commission recognises the importance of telecommunications networks and, therefore, of safeguarding incentives for telecommunications investment. The Commission states:

"The telecoms sector…[accounts] for 44.4 per cent of the total value [of the overall European ICT sector] in 2005….ICTs drive about 40 per cent of productivity growth and one-quarter of overall growth in Europe. The contribution of the telecom sector to Europe's growth, jobs and prosperity is therefore crucial."[27]

Europe has an ICT problem. In general, Europe invests less in ICT, and uses this lower-level ICT less intensively, than the United States. While some European countries are good performers on measures of ICT usage and penetration, the core countries of Continental Europe lag behind the USA and countries such as Japan and South Korea. Between 1995 and 2001, productivity growth in the United States (as measured by growth in GDP per hour worked) was 2.2 per cent per annum, compared to 1.3 per cent per annum in the 15 pre-accession European nations.

Table 1 illustrates that Europe is lagging behind the United States in investing in ICT (as shown by the substantially lower levels of ICT capital stock per hour worked), but also in one means of ICT adoption: the penetration of PCs. PC penetration can be seen as an indicator of the pervasiveness of ICT in the wider economy, and the fact that Europe trails the US on this measure suggests that European businesses and households are less technology-friendly than their American counterparts.

The UK's labour productivity is 15 per cent below US levels and below the EU average, as well as that of France and Sweden. An average

27. European Commission, 11th Implementation Report on the Telecom Regulatory Framework (complete citation)

worker in the UK has less non-ICT capital and far less ICT capital at her/his disposal than the average US employee. A number of studies show that the salient difference between productivity in the UK and the US lies in the much higher contribution to productivity growth of the ICT-using sector in the US.[28] No one doubts that factors such as managerial resistance, some remaining labour market rigidities and other characteristics of the UK's macro-environment contribute significantly to the ICT 'problem'. However, it is clear that there is a productivity and ICT gap, and regulatory policies which impede investment incentives in MIP networks will only exacerbate the UK's ICT and productivity problems.

The growth of bandwidth, investment in MIP networks and new service offerings increase the challenges on regulators. We are moving into a new digital world, and thus the costs of getting regulation wrong are high. As Commissioner Abernathy has noted in the United States:

"Most economic regulation comes from the rule-making process and the notion that we, as regulators, can always make things just a little better, no matter how competitive the market may already be."[29]

Regulation of PSTN in the past did cause some spillovers to other sectors but any regulatory error was largely subsumed within telecom itself. But today telephony can be provided over multiple paths by multiple players, hence regulating the previous telecom sector as a silo produces wide-ranging deleterious spillovers to a variety of sectors. *The costs of regulatory mistakes in PSTN telecoms or in cable are now magnified well beyond the areas where regulators act.*

Regulation that was designed to kick-start competition in the telecom sector in the era of fixed-line voice service monopolies runs the risk of ignoring the rapidly multiplying possibilities for substitutability across services and platforms. When ignoring the possibilities of demand and supply substantiation, regulators send the wrong signals to incumbents and entrants alike, and these policies will likely result in a sub-optimal allocation of resources.

It is also crucial to note that not all investment is equal. The investment that should concern us most is investment in core network assets. Investment in switching, cell centres or even intelligent network components such as DSLAMs certainly counts as useful, but ultimately it does not transform the network's core characteristics in terms of the services that can be provided. Investment in new core network capabilities has transformational capabilities and thus generates greater economic benefits. Regulation should particularly encourage investment in new core network assets and encourage inter-modal competition between different core networks.

The ladder of investment theory (LOI) recognises that some investments are more worthwhile than others. However, a more careful analysis is required not only to determine if entrants are climbing the rungs on the ladder but also to determine if the ladder stretches high enough! The set of incentives induced by the LOI approach may produce the wrong investment, i.e. not the investment which maximises social returns, growth and productivity. There is another serious concern with regulation that aims at one form of

28. See Melvyn Fuss and Leonard Waverman 2005, *Canada's Productivity Dilemma: The Role of Computers and Telecom*, Appendix E-1, Bell Canada. Submission to the Telecommunications Policy Review Panel

29. As quoted in ETNO Reflection Document RD-235, January 2006

infrastructure. Regulating one MIP network – that is, telephony NGN alone – runs the risk of promoting service competition on these facilities but at the risk of lowering competition between infrastructures. The original focus of Oftel was infrastructure competition and that's something we need to return to. The best way to be technology-neutral is to treat all technologies equally. This does not mean to regulate them all; rather, where possible, to use competition policy tools equally.

Regulation and competition policy

The justification for regulation in the telecommunications industry has been the legacy investment of the incumbents that began as monopoly providers. Regulation began by allowing entry into national and international calling, activities which had been the prerogative of a nationalised public monopoly. Entry into the 'last-mile' local telecommunications was much more difficult – the market presented several challenges for new entrants.

Most notably, replicating the incumbent's network was both costly and inefficient for new entrants, especially as public policy held retail prices below cost. We tend to forget the importance of pricing. Even today, we continue to have national geographically-averaged rates for both access to the PSTN network (the last-mile) and for calling. It was unrealistic several decades ago to believe that new entrants could build their own networks to compete with the incumbents when they had no market presence or customer base to begin with.

Today, however, we do have ubiquitous competing networks (mobile) and competing broadband in at least half the UK (cable). In addition, competitors now have billing systems, customers, brand names and deep pockets.

Regulatory policy evolved in the 1990s with the aim of encouraging progressive competition in local services. Mandated access to incumbent networks encouraged the growth of carriers who were either simple resellers or made use of local loop unbundling (LLU) to offer their services over incumbent networks.

At this point, regulation of access was arguably justified as a means of kick-starting competition. However, regulation of access – most specifically, granting entrants access to incumbent networks at regulated rates – rests on the presumption that the networks are inherently not replicable. Once this presumption is removed, the rationale for regulation diminishes.

The greater danger is that regulation, particularly regulation pertaining to access to essential infrastructure, could induce too low a level of investment – a tragedy at a time when such investment is ever-more socially useful.

Laffont and Tirole note that: "There is in general a trade-off between promoting competition to increase social welfare *once the infrastructure is in place* and encouraging the incumbent to invest and maintain the infrastructure".[30] [My italics]. A potential analogy is the issue of patenting in the pharmaceutical (or other R&D-intensive industries). In the R&D case, it is generally agreed that patents protect the firms' *(ex ante)* incentives to innovate by allowing them to earn monopoly rents on their innovation for a limited period. However, *ex ante* regulation of incumbents' new networks does the reverse. It undermines the incentives for innovation by threatening to expropriate part of those returns from innovation.

30. p7 Jean-Jacques Laffont and Jean Tirole (2000) *Competition in Telecommunications*, the MIT Press, Cambridge Mass., London England

To return to Laffont and Tirole's point, the relevant trade-off that access regulation must make is between promoting competition once the relevant infrastructure is already in place and protecting the incumbents' investment incentives. For example, Telstra, the Australian incumbent, recently announced plans to abandon its proposed high-speed broadband network since it could not agree with regulators over whether competitors should have access at wholesale rates to this new network.[31]

The disagreement between the European Commission and the German Government about whether to grant Deutsche Telekom's new VDSL network a 'regulatory holiday' also captures the tension between foregone investment and the desire to prevent the new network from becoming a bottleneck that thwarts future competition.

Simply assuming that new incumbent networks automatically represent potential bottlenecks ignores several important aspects of investment in new networks.

First, new MIP networks offer the potential for vastly improved service offerings at lower prices to greater numbers of consumers than existing networks. Thus the pure consumer welfare gain (i.e. households and businesses alike) from such networks is potentially very large. The social costs of delay are also large. Hausman 1997 showed that FCC delays in licensing US cellular systems cost some $20 billion in foregone annual consumer surplus. This is not to say that there is a direct comparison between not licensing cellular networks and imposing *ex ante* regulation on some MIP networks (i.e. investment in NGN telecoms). However, regulatory action or uncertainty can clearly delay investment and such delay has high social costs. The costs of delay in the case of NGN in the UK and Europe are, in some respects, higher than in the case of early cellular systems in the US. NGN networks are a vital component of firm-to-firm computer aided exchange, and hence are important to raise productivity and growth.

Secondly, the new networks require significant risky capital outlays. Investors in new networks, whether incumbents or entrants, assume considerable risks, as we saw from the wave of telecom-related bankruptcies in the United States, the UK and Europe during the late 1990s.

In the 1960s and 1970s, the period of rapid expansion of the existing PSTN in Europe, investments were far less risky simply because the incumbent operators were public monopolies entitled to recover their regulated cost-base through their rates. Today, incumbents do not enjoy that luxury to take care of any unanticipated cost shocks, and face the additional risk that their rivals can either choose to build out their own new high-speed networks or can wait for the incumbent to move first and then 'buy' facilities from the incumbent.

The *ex ante* imposition of regulated access on new incumbent networks creates significant asymmetries between entrants and incumbents. Incumbents and entrants alike could build new networks, with the only difference in cost being the (presumably) lower costs of financing the investment for the better-heeled firm. But the threat of mandated access to the incumbent's network raises (asymmetrically) the cost of capital to incumbents by reducing the expected return on invested capital.

31. Wall Street Journal, Europe, August 8, 2006

Thus, in a purely *ex ante* context, the threat of regulation reduces the likelihood of incumbent investment. Furthermore, option values are not generally incorporated into regulatory rate-setting or in LRIC calculations as they should be. Alleman, Darby and Rappoport (2005)[32] point out that when incumbent firms make large-scale irreversible (sunk) investments in their network infrastructure they forgo their option to delay rolling out this network infrastructure.

The current regulatory paradigm allows entrants two choices: one, to wait for the incumbent to build its network and then either invest in their own facilities or lease facilities from the incumbent; and two, simply to roll-out their networks regardless of the incumbent's decision. Thus the entrants enjoy the 'option value' resulting from the greater variety of choices they have relative to the incumbent. This option value is not factored into the Long-Run Incremental Cost (LRIC) models that are used to price network access. Unfortunately for incumbent and society, well-functioning financial markets will take these opportunity costs into account, altering the cost of capital for financing entrant projects and jeopardising incumbent investments.

None of this is to say that regulation is always a bad thing for investment. Where there are genuine bottlenecks, without regulation, the long-term consequences of an unchecked monopoly could be just as bad for investment and long-term welfare as the deleterious consequences of inappropriate regulation that we outlined above. Inasmuch as regulation boosts competition, it helps to bring competitive discipline to bear on the behaviour of the dominant firm. In the absence of competitive pressure, the dominant firm might well operate inefficiently, and offer consumers a limited variety of services at high prices.

However, access regulation that favours a particular model of entry runs the risk of distorting competition in such a way that the dynamic outcomes resulting from this distortion (where competitors base their business models around mandated access to incumbents' infrastructure) are decidedly sub-optimal.

'One-size-fits-all' national policies no longer suffice. One way forward for regulation is to adopt a more case-by-case approach to regulating access. For example, the underlying cost conditions faced by a new entrant operating in the centre of London are very different from those posed in the Lake District or the Highlands of Scotland. In the urban areas, the incumbent network (even the existing PSTN) could be replicated by an entrant investing in its own facilities (fibre-rings, for example). In such cases, it is appropriate to make geographical distinctions in the nature of the regulatory regime since, otherwise, entrants could simply base their strategies on the best opportunities for regulatory arbitrage.

Perhaps the more formal use of an 'essential facilities' test is the way forward, wherein no wholesale regulation is imposed where there is a compelling alternative infrastructure (such as cable). In these regions/areas, the pre-conditions for healthy inter-modal competition between services offered over multiple infrastructures already exist.

32. Alleman, Darby and Rappoport (2005), *Optimal Pricing with Sunk Cost and Uncertainty*, in Cooper and Madden, eds., The Economics of Information, Communications and Technology Networks

New challenges

There are a number of new challenges for regulators in a MIP world, three of which are addressed briefly here:

- net neutrality and discrimination;
- margin squeeze; and
- content as a two-sided market
  - 'normal' regulatory tests – predation, cross-subsidy are not applicable.

Net neutrality

Telecom infrastructure providers now offer connectivity for more than a person-to-person voice call; with falling voice revenues, such infrastructure providers are considering new ways of 'distinguishing' prices. The old model of 'value-of-service' pricing, where long-distance callers and business customers generally paid higher mark-ups, is no longer possible. Hence new 'value-of-service' models are being examined. These models include charging intermediaries (such as ISPs or search engines) for network access or uses. These intermediaries who have not, to this point, expressly paid for transmission or differentiated quality argue that they should *not* be discriminated against. Hence net neutrality is proposed – that infrastructure providers treat all content equally.

This principle of net neutrality should not, however, be enshrined. As Larry Darby puts it:

"A key element of net neutrality put forth by proponents is the proposition that of all stakeholders in the value chain, only end-users may be charged for network access and uses. Other beneficiaries of broadband networks, including providers of software, applications and content would by statute be insulated from paying providers of broadband networks."[33]

The proposition is exactly as Darby puts it. *Ex ante* regulation binds us by limiting the ways in which access and usage costs are allocated through prices. Such limitation is neither efficient nor equitable. In the past we relied on end-users paying for all costs of connection and transmission – but with circuit-switched calls, the calling and receiving parties did 'bear' infrastructure and usage costs. Thus 'end-user pay' made economic sense. However in an IP packet switched world, there is no simple correspondence between a user and its associated costs. Hence there is no simple cost causality.

Moreover, neither customers nor suppliers invest in broadband for one service only such as voice calls; the demand for broadband comes from the desire for access generally, and the desire by users to multi-cast content. Consider as an analogy the broadcasting world. The old model of free-over-the-air broadcasting did not involve end-users (viewers) paying. However, advertisers did not ask for regulatory intervention to end this 'discriminatory' treatment and to force end-users to pay.

Telecoms and Broadcasting are two-sided markets but we have done little to recognise this basic fact in our analyses or polices (see below).[34] Hence having advertising-based 'free' voice calls, or having content providers paying for network costs, is neither unfair nor inefficient. The issue is not one of determining,

33. *Consumer Welfare, Capital Formation and Net Neutrality: Paying for Next Generation Broadband Networks*, Darby Associates, June 6, 2006

34. In two-sided markets, simple correspondence between prices and costs (i.e. price at LRIC) needs to be adjusted to reflect the externality on the other side.

ex ante by regulation, who should pay what. The issue surely is this: is the infrastructure provider acting anti-competitively by their charging mechanisms? We cannot rely on *ex ante* regulatory prescription to answer this question.

Two-sided markets

There is a growing literature in the economics of two-sided markets. A two-sided market is one where 'platforms' enable two or many 'sides' to interact.[35] Above, I discussed broadcasting briefly. The provider of the platform (be it cable or satellite, or indeed ADSL) enables content providers to reach an audience. Viewers want programming, and advertisers want viewers or eyeballs. So we have three, at least, sides to the market – viewer, advertiser and content provider. Rochet and Tirole show that a market is two-sided if:

(i) the structure of prices and not just their level affects demand; and

(ii) the two sides cannot transact an efficient division of the price.

Broadcasting is a two-sided market: charging viewers, instead of advertisers does alter the level of demand. Telecom is also a two-sided market. For a voice call there is a calling party (C) and a receiving party (R). Clearly, both benefit from the call yet in fixed line networks, while both calling and receiving parties pay for access, the calling party traditionally pays all the usage costs of the call. Mobile networks differ across the world in their charging mechanisms. In some jurisdictions – the USA, Hong Kong, Israel and others – the receiving party pays part of the cost of the call, as does the call's recipient. Here in the UK and in Europe, the calling party pays all the costs of the mobile call. These differences in charging policies appear to affect the volume of calling.

We do not as yet know all the economic theory of two-sided markets, but we do know:

(i) efficient prices can be below, equal or above marginal cost. That is, setting prices at LRIC is not necessarily optimal!;[36] and

(ii) this is because of the 'opportunity cost': the impact that adding one user to one side has on those on the other side. That is, there can be an externality on the other side of the market. This externality, say, to the R side, lowers the required price to the C side and can lower it below marginal cost.

A typical example is a night club. The platform (i.e. the club) wishes to maximise revenue and can charge any price it wants to the two sides of the market – men and women. Setting price at marginal cost which is gender-neutral requires equal prices to men and women. However, many clubs offer free entry to one side of the market while charging the other side a price that covers all costs. This appears to be discriminatory but need not be once the externality is taken into account.

A recent paper by Hausman and Wright 2006[37] is one of the first to address the setting of telecom prices using a two-sided theory. They use a model where callers can be both fixed and mobile subscribers in order to address the issue of whether fixed-to-mobile termination rates are set too high. They calibrate this two-sided model to Australian data for 2003. They find that equilibrium fixed to mobile termination rates are well above estimated costs

35. See J Rochet and J Tirole 2004, *Two-sided Markets, An Overview*, mimeo, IDEI, Toulouse

36. See J Tirole, *Industrial Organization of Two-sided Markets*, UCL Two-sided Markets Conference, UCL London, 11 May 2006

37. J Huasman and J Wright, *Two Sided Markets with Substitution: Mobile Termination Revisited*, mimeo, April 2006

and that regulating termination rates to costs lowers welfare. The reason is that raising the fixed to mobile termination charge makes mobile subscription and usage more attractive. Since most customers use both fixed and mobile phones, welfare effects must be analysed by capturing these two-sided effects.

Conclusions and recommendations

This paper highlights the importance of investment in advanced telecommunications services as a key determinant of economic growth. There is a growing concern around the productivity gap between the UK and Europe, and the USA. Many academic studies point to the greater level and use of ICT in the US compared to the UK as one of the causes of the productivity gap.

Other studies point to a recent acceleration of US investment in NGN following the dismemberment of wholesale access regulation of incumbent local telco access providers. There is a risk that the continuation of existing regulatory policies in the UK and in Europe could end up undermining investment in the very core telecommunications networks that will be key enablers of the Information Economy.

Just as regulators insist on 'fit-for-purpose' services from incumbent telcos, all current regulatory instruments should be fit-for-purpose and assessed to see if they increase social welfare. Current regulation was based on the problems of dealing with monopoly provision of services over legacy infrastructures with little or no demand or supply substitutability between services. The new MIP world is very different. There are no legacy investments and we are just beginning to see the many services that will emerge.

We cannot, as a society, risk maintaining legacy regulation for a new converged world without a complete review of the bases for such regulation. This does not mean that all current regulatory instruments should be instantly disbanded. What it does mean is a re-evaluation of the costs and benefits of a policy framework of *ex ante* mandatory regulatory intervention compared against the costs and benefits of relying on an alternative framework – *ex post* competition policy instead.

We have come from a world of decades of retail price controls on voice carrier incumbents, and a decade of mandatory wholesale access to these incumbents' legacy infrastructure. The principles we developed, which basically revolved around non-discrimination by the incumbent telco, needs rethinking. Discrimination, innovation and investment are the ways in which markets function and firms add value.

The European Commission's regulatory stance on mandated access (as embodied in a recent speech by Commissioner Reding) needs to provide greater clarity. As part of her analysis of Deutsche Telekom's argument for a regulatory holiday for its new VDSL network (an issue which we do not discuss in this paper), the Commissioner wondered whether the products and services offered over the proposed network were not substitutable with products and services offered over the existing network.

However, the justification for extending access regulation to such new networks must rest on evidence that the new networks are truly essential facilities which only the incumbents could develop – which appears not to be the case. It is far too narrow a test to rely only on the substitutability of products and services offered over these new networks with products and services offered over existing infrastructure by these same incumbents.

Regulation that encourages facilities investment by incumbents and entrants alike permits a greater variety of services (existing ones and new ones) to be offered by a greater variety of competitors over a greater variety of competing technological platforms. Regulation that dampens such competition may have the short-term merit of leading to greater services competition based on existing platforms, but the innovation, investment and dynamic efficiencies it forgoes more than outweigh the benefits of more effective services-based competition.

The investment community remains concerned about the European Union's regulatory stance. For example, Merrill Lynch comments that:

"Although we expect some fibre build particularly...for VDSL in European countries with cable competition, we still see the regulatory backdrop as unsupportive of a sustained increase in capital intensity."[38]

Credit Suisse First Boston remarks:

"We...see little incentive from a regulatory perspective for incumbents in Europe to pursue FTTP [Fibre to the Premises]."

Regulators must always balance investment incentives with the promotion of effective competition. However, when considering the application of access regulation to new networks, regulators should pay attention to the possibilities for entrants to replicate these networks, and to the possibility that incumbents will severely delay or even abandon their investment in new networks under the threat of having to share them. A keen recognition of the economic importance of high-speed telecom networks as an enabler of the Information Economy should inform regulatory decisions.

- Regulators should consider adapting regulation to suit different geographical and competitive circumstances. For instance, densely-populated urban markets offer greater entry possibilities for entrants of all varieties (including facilities-based entrants); in this instance, there does not seem to be a justification for access regulation on the grounds of access to essential facilities. Similarly, where ubiquitous cable networks exist, they provide alternatives to incumbent facilities (new or old) and in this instance, regulators should also rethink mandated access to new networks on the grounds that they are not an essential facility. In these instances, asymmetrical wholesale regulation of incumbent facilities could prevent necessary investment and facilities-based competition. Such case-specific regulation could be complex to implement, however, but as a theoretical precept it does provide guidance about the merits of wholesale regulation and its likely impact on investment.
- We need to rethink the central tenet of non-discrimination, on which current regulatory policy including equivalence is based. Discriminatory treatment, discriminatory offerings and differential prices all exist in competitive markets. Indeed, the need to discriminate offerings is the key to competitive success. We then need to take care that the old tenet of non-discrimination is applied wisely, and not simply as a core legacy instrument.

38. Source: Indepen, *The ICT Sector and Productivity in Europe: the Challenge Ahead*, Presentation to Global Communications Consortium, London Business School, June 29, 2006

Market-based alternatives or complements to regulation

Jonathan Cave

Department of Economics, University of Warwick

If a market is seen as a form of democracy, the co-evolution of dominant firms and dominant regulators might be said to miss the maxim: "the cure for the ills of democracy is more democracy".
Jonathan Cave

Introduction

Many economic regulation regimes are based on some concept of market failure – a misalignment of power, information and incentives that damages technical, allocational and/or dynamic efficiency. More broadly, key 'stakeholders' affected by market functioning do not have an appropriate and efficacious voice in market governance and hence may not participate appropriately.

This governance gap has led to three broad classes of regulatory intervention: general economic (competition), sector-specific (industry, including technical) and social, with their own objectives, stakeholder relations and favoured tools, but a shared focus on the market. A changing environment and the lessons of experience suggest that regulated markets may fail as well and that structural rebalancing of roles and responsibilities between markets and regulators can be enormously helpful.

The basic message is simple: regulators balance the interests of firms (dominant, fringe and potential), customers and societal stakeholders. Dominant firms tend to have the most intimate contact with regulators while key societal interests have a political or policy voice. Information about other parties may be harder to come by, which can lead to mutually-reinforced structures of regulation and industry, which may not adapt well or swiftly to changing conditions. If relevant information is not elicited or efficient, arrangements do not materialise, the combined system fails, and it is reasonable to seek to (re-)design markets to fix failures in regulatory systems devised to fix markets.

Technological changes, new regulatory challenges to traditional command-and-control regulation and advances in the theory and practice of mechanism-design have raised hopes of market-like mechanisms to perform or assist regulation; reduce or eliminate regulatory costs and distortions; and address new regulatory challenges while relinquishing obsolete ones and rebalancing regulatory roles and responsibilities.

This paper considers the prospects for what David Salant calls the "re-engineering of regulation". The concept is not new – concrete examples of new tradable property rights to 'carry' externalities (marketable discharge rights), mechanisms for allocating access to congestible public goods (spectrum auctions) and markets to ensure efficient capacity creation and utilisation in the face of long- and short-term risks (gas and electricity markets) have helped to fix ideas and drive the integration of theory and practice. Many have been narrowly conceived in ways that produce unintended side-effects or deepen our ability to improve the regulatory function of markets themselves, realign administrative regulation and combine or unbundle economic, sectoral and societal regulation.

Why regulate markets?

Economic regulation arises out of the 'failure' of markets to ensure technological, allocational and dynamic efficiency by means of long-term contractual relationships as well as investment and innovation incentives. These hoped-for consequences of competition (rather than competition itself) are the usual objectives of economic regulatory regimes, but only rarely can regulators identify specific efficient behaviour.

It is often easier to reinforce competition or replicate market outcomes. This paper considers the use of markets to perform regulatory functions. Although primarily concerned with economic regulation, it is important to recognise that other forms of regulation also use market mechanisms and that this use may have implications for economic regulation.

Sectoral or industry regulation addresses problems that reflect specifics of industrial structure, technology and demand and are therefore poorly-served by general antitrust (pricing and merger) regulation. The objectives include facilitating entry and exit via open-access and/or standards requirements, ensuring satisfaction of consumer needs by quality-of-service (QOS) and related rules and advancing general sectoral economic goals of competitiveness (as distinct from competition), growth, structure (especially SME participation), risk allocation and management, and so forth.

Societal regulation involves objectives outside the immediate matter of commercial operations, including equity, macroeconomic impacts such as employment, and by-products or externalities ranging from privacy to environmental impact.

Regulators pursue these objectives via a range of functions embedded in different institutional arrangements. These include:

- selecting and regulating monopoly/oligopoly providers.
- infrastructure provision of:

  a) investment;

  b) access, pricing;

  c) operation (esp. QOS); and

  d) horizontal and vertical interconnection.
- universal access/service provision.
- controlling and facilitating entry and exit.
- controlling access to congestible local public goods to provide incentives, etc; and
- exploiting regulatory purchase to advance indirect or general objectives e.g. content regulation, security, information assurance (integrity, authenticity…), privacy, IPR, environmental impact, etc.

These functions cross the borders of economic, sectoral and societal regulation.

So too does the replacement of these functions by market alternatives. In addition to learning from experience in other regulatory domains, this also suggests that economic regulation may, in effect, substitute for other forms under some circumstances. In general, economic regulation by means of markets entails trade in complementary property rights, or unbundling those already traded, to ensure internalisation of externalities in market transactions; counterbalance concentrations of market power; or enable efficient bypass of failed markets. But markets created to advance sectoral or societal regulatory objectives may themselves fail or exacerbate failure of existing markets[1] and thus involve more, rather than less regulation. This is justifiable if the required (economic) regulation is more accurate or effective – and less burdensome or distorting – than the sectoral or societal regulation it replaces.

In determining appropriate outcomes and ensuring that they are realised, regulators face two fundamental strategic problems. The necessary information is not always available to them, and must be gathered from interested parties – who themselves may be partially informed or, indeed, disagree. Even if the activities can be determined, they may lie within the competence of unregulated parties, or outside the scrutiny of the regulator. These information and incentive problems are the subject of the game-theoretic approach to regulation sketched in the following section.

Game theory in relation to regulatory design

This section sketches the main concepts of game theory as applied to the analysis and design of regulatory institutions. Further details can be found in any modern game theory text, and especially Tirole.

A game is a general model of interacting rational choice, defined by a set of *players*, who choose *strategies* in order to maximise their expected payoff or return. The *rules* of the game determine the outcomes (market and other) resulting from strategic choice. Players' incentives are captured in the *payoff* function, and the expectations that guide their behaviour reflect their *information* about likely outcomes, other players' identities, strategies, incentives and information, and so on. In tackling problems with substantial information asymmetry or those where linkages among players are complex, 'optimal' strategies may be extremely difficult to compute or implement and thus lack credibility.

To introduce a further note of realism, games used for policy purposes sometimes include 'non-strategic' players who follow behavioural rules rather than optimising each choice (e.g. consumer behaviour summarised in a demand curve) or alternate models of rationality (such as what, if anything, they maximise, and how much information they take into account). Regulation in a market game can affect each or all of these elements.

1. For example, the inclusion of airlines in the European Emissions Trading Scheme – or the creation of a separate scheme dedicated to airlines – may further enhance the market power of incumbent operators (e.g. national champions), especially when initiated by grandfathered endowment rights that institutionalise existing dominance and/or when used in combination with geographically localised slot markets. While access to a dominant airline's slots, for instance, could be 'purchased' by an entrant for emissions rights, a favourable slot allocation could equally encourage predatory or collusive bidding.

The game theoretic apparatus analysis is completed by a *solution concept*. This summarises the information in the game description and a qualified prediction of the outcome in terms of the way players 'trust' others to behave or think. It also reflects whether players behave myopically or play a 'long game'.

We can then further divide the analysis of games according to the ability of players to make binding commitments and the degree to which the rules of the game can be changed by design. We start with the positive analysis of non-cooperative games, in which players cannot make binding commitments and the focus is on describing what will happen if they behave rationally. We then discuss the positive analysis commitment and co-operation before concluding with an illustrative example of normative analysis or mechanism design, in which the rules of the game are adjusted to implement a desired outcome.

The simplest class of non-cooperative games involve *perfect information* – players' identities, strategic possibilities and payoff functions are all *common knowledge* (each player knows them, and knows that the others know them, and so on *ad infinitum*). The key solution concepts are dominant strategies, Nash equilibrium and perfect equilibrium.

A strategy is said to be dominant if it is the best strategy regardless of what the other players are expected to do. A simple example is adhering to, or deviating from, a collusive output agreement in a Cournot (quantity competition) market. If the rival firm is expected to restrict output, the best reply is opportunistically to expand beyond the cartel level. If the other firm has already deviated, the best reply is (defensively) to expand.

The 'game' interaction does not affect the strategy choice by the players, but does matter to society, regulators and other 'public-interest players', and affects entry/exit decisions. A further example is bidding in a second-price, sealed-bid auction with independent private values, where truthful bidding is dominant.

Many games lack dominant strategy equilibria. The most common alternative[2] is Nash equilibrium, in which each player adopts a best reply to the others' strategy choices. Almost all games have at least one Nash equilibrium. Implementing it does require players to know each other's strategies, so it should be seen as an *ex post* stability concept of mutually-consistent expectations.

It is often inefficient, and its stricter concept of rationality makes it a less compelling benchmark than dominant strategies, but it is easier to achieve. It is also significant to note that it may not be unique.

An example is provided by equilibrium bidding in an English (ascending oral) auction. According to the Revenue Equivalence Theorem[3], Nash equilibrium gives the same result in this type of auction as in the second-price sealed-bid auction in the independent private values, risk-neutral case, but in general may have additional equilibria which may be complex to compute. Roughly, each player should bid the expected second highest *bid* conditional on his *value* being the highest among the bidders.

In some games, expected payoff depends on contingent response behaviour of other

2. Not the only one – rationaliseable strategies are a viable alternative.
3. Ref: Klemperer

players in the future. Such games typically have many Nash equilibria, and the concept of perfect equilibrium reduces this multiplicity. It is similar to Nash equilibrium, but requires all contingent threats and promises to be 'credible' (i.e. it is in the players' interests to carry them out).

An example is price-matching behaviour in a cartel. In general, price competition is unfriendly to collusion. Each firm has an incentive to cut price unilaterally in order to stabilise a cartel, so any promises to hold up the price umbrella lack credibility. But if the firms publish 'price matching' policies, a firm knows that price cutting will not increase market share, but will only reduce prices for all firms.

In games with imperfect information, the players do not necessarily know each other's payoffs – or even their own. In such games, we must take account of the behaviour of players with different information and treat them as separate players. The analogue of Nash equilibrium in such games is Bayesian equilibrium, where each 'type' of each player (in other words, each possible combination of a player and what he might know) is expected to play optimally given his knowledge and others' (type-dependent) strategies. This builds in inferences from observing others' behaviour.

One example is provided by spectrum auctions. If a bidder thinks a given licence is valuable, his knowledge of the experience of other players may tell him something about the correlation of his valuation with the valuations and bidding behaviour of others. In particular, this allows him to revise his estimate in the light of others' observed bidding behaviour in an ascending oral auction.

The analogue of perfect equilibrium for games of incomplete information is sequential equilibrium. All contingent actions ('threats' or 'promises') must be rational, given what a player should infer from the prior history.

In many applications, one player knows strictly more than another and the analysis divides into two special cases. In signalling games the informed player moves first, but if the uninformed player moves first, we have a principal agent game – this type of model is used to analyse optimal contracting and regulatory design. A well-known example is price cap regulation, in which the regulator allows the monopolist flexibility in setting individual basket prices.

In some games, the players have the further power to 'escape from' non-cooperative solutions by negotiating binding agreements. In bargaining games, players choose among agreements based on individual threats or 'reserve prices'.

For example, in imposing content regulation on the audiovisual media services industry, the burden of regulatory compliance can be reallocated among the parties in various ways.[4] The agreement that will be reached in practice will differ depending on whether the *de jure* regulatory burden falls on content providers, service providers or end-users.

In co-operative game analysis the emphasis shifts to groups – threats may involve changes in alignment and co-ordinated predatory or defensive behaviour, and some solution concepts even predict which groups will form and what (if any) side-payments are used.

Examples include the Shapley formula for allocating joint costs in a multi-product firm,

4. RAND Europe (2006)

and its application to, for example, access pricing. This can be used to indicate outcomes in markets for network capacity if there were 'serious' externalities and possibilities for shifting alliances and mergers. A second example is the reformulation by Sharkey of the 'contestability' criterion for arguing that an apparently non-competitive outcome (e.g. with one or a few firms) is actually efficient (using the core of the associated cost game).

The normative or mechanism design analysis looks at how to design a game that will achieve the regulator's desired outcome without obliging the regulator to know and use information the parties may not wish to provide, or to monitor and enforce detailed conduct prescriptions. The main analytic results are that only a few regulatory objectives can be attained as dominant strategy outcomes, but more interesting alternatives are available if Nash/Bayesian equilibrium is enough – in other words, if firms can be relied on to think strategically about each other. Further results relate to the existence and structure of optimal (regulatory) contracts and auctions, and the consequences of requiring that regulatory markets be robust to collusion.

One simple example is the design of a mechanism for choosing a common standard to be used by many firms. To keep the example simple, assume that there are two standards (A and B), that standard x will cost C(x) to implement and that the consequences to the firms are given by incremental profits i(x). Consider two decision mechanisms:

1. The regulator could ask the firms to report their profits and implement the standard with the greatest excess of total profit over cost, dividing the costs in proportion to reported benefit. In this case, firms will attempt to 'free-ride' on each other by under-reporting their benefits – there is no guarantee that the appropriate standard will be chosen or the burden reasonably divided.

2. The regulator could fix cost shares $ci_i(x)$[5] for each standard and ask firms to report their *net* profits $\square_i(x)$ - $ci_i(x)$. Again, the standard with the greatest total net benefit will be implemented. In this case, firms whose profit exceeds cost will over-report their gains, and those whose net benefits are negative will exaggerate their losses. Once again, there is no guarantee of an efficient outcome.

3. A third alternative is to proceed according to option 2, but add a 'pivot tax': if a firm's announcement changes the selected standard, it must pay an additional amount equal to the aggregate change in net value for the other firms. (In other words, it must pay the full marginal external cost of its announcement.)

It is easy to show that under the third mechanism, firms have a dominant strategy to report truthfully – that is, it is in their best interests to do so regardless of their expectations about others' behaviour.

A further example is provided by the well-known Baron-Myerson mechanism for regulating a monopolist with unknown costs, in which the regulator designs a contract that induces the monopolist to price in the way the regulator would prefer, regardless of his marginal cost.

We conclude this section with an example related to the 'net neutrality' debate. Suppose

5. Where $c_1(x) + c_2(x) + \ldots = C(x)$

that there are two types of firm (content providers P and service providers S); three regulatory regimes (a – to prohibit charging based on content; b – to allow unrestricted charging on the basis of content; and c – to use a 'heavy' rule of reason/administrative procedure to evaluate cases and/or allow parties to sue each other). The firms' preferences depend on the effectiveness of competition between the Ps, the Ss and possibly integrated players and on transactions costs.

Consider these two possibilities that are known to the firms but not to the regulator.

In state 1, the Ps prefer no charging to unrestricted charging and unrestricted charging to 'rule of reason', while the Ss prefer unrestricted charging to 'rule of reason' and 'rule of reason' to no charging.

In state 2, the Ps prefer no charging to 'rule of reason' and 'rule of reason' to unrestricted charging, while the Ss prefer unrestricted charging to no charging and no charging to 'rule of reason'.

The socially efficient decision is option a (no charging) in state 1 and option b (unrestricted charging) in state 2. This cannot be implemented by any 'purely' market-based game (i.e. without fines or penalties).[6] But if the regulator can threaten fines, there is a game that implements a socially optimal solution without – in equilibrium – collecting fines, no matter which state prevails.

The regulator begins by asking the service providers whether they want unrestricted charging (option b). If they say no, the regulator bans charging (option a).

If the service providers say yes, the regulator asks the content providers whether they want charging prohibited (option a). If they say no, the regulator allows full charging (option b).

Otherwise, the regulator bans charging, imposes a small fine on service providers and imposes on content providers a fine F that is large enough so that they would rather get their preferred outcome (no charging) and pay the fine (compared to full charging and no fine) if – and only if – the market is in state 1.

It is easy to see that the unique outcome of this game is the socially efficient outcome, no matter what the state is.

The example illustrates a different sort of 'market alternative' in that the two parties to the vertical relationship compete for the regulator's attention.

Evolution of the regulatory relationship

In many arenas, initial regulatory developments were driven by market power. This was most visible – and most easily addressed – in the case of monopolies or single dominant enterprises. Potential market failure in sectors dependent on bottleneck facilities or shared network infrastructures such as transportation, energy and tele-communications was initially managed in many countries by creating public utilities to operate in society's interest. In the event, they became increasingly profit-driven and much of the early development of regulatory economics was devoted to reconciling socially optimal behaviour (e.g. Ramsey pricing) with reasonable reward (rate-of-return regulation, etc.).[7]

6. The optimal rule is *non monotonic* – in moving from state 1 (when a should be chosen) to state 2 (when the choice should be b), no player has increased his preference for b over a.

7. Sherman (2001)

Both empirical and theoretical economists pointed out that the optimal behaviour might not even be feasible. Ramsey prices depend on cost information that the public utility should possess, and on demand information which it might not. Regulators typically have worse information about both, and thus are vulnerable to informational capture. Simple mechanisms to ensure commercial viability and continued access to capital were shown to distort investment behaviour even as they 'policed' pricing behaviour.[8] As the positive game theoretic analysis of regulatory tools was extended, 'smart' regulation attempted to balance prevention of rent-seeking and deadweight loss with innovation incentives (e.g. price-cap regulation).

Use of theoretically-derived mechanisms for specific regulatory tasks was extended to the normative domain by models of optimal mechanisms (or games for firms to play). These models addressed problems of hidden information as well as problems of hidden action and provided a strong early indication of the analytic similarity between the problems facing regulators, those they regulate and those with whom regulated firms compete. Simultaneously, corporate finance literature on optimal contracting was solving similar problems by means of similar models.

Of course, they also had similar weaknesses: they did not handle (imperfect) competition very well; made strong independence and even asymmetry assumptions about information; and – perhaps above all – were based in a principal-agent world where one (or more) well-informed agents behaved optimally in a game or contract proposed or imposed by a strictly less-informed principal.

Often these contracts or games could be regarded as markets. Examples include menu contracts, in which the principal specifies different options from among which the agent can choose, or mandatory use of market mechanisms by the regulated entity.

Perhaps the best-known example is the 'essential facilities' doctrine, which applies to natural monopoly suppliers of bottleneck assets such as key parts of transmission or transportation networks. The 1912 *Terminal Railroad* case[9] prohibited owners of such facilities (in this case, a bridge) from discriminating among firms who wished to use it. This established the principle that such facilities cannot be used to foreclose downstream markets but must be offered on the same terms to all-comers, and thus gave a legal basis for access and interconnection markets. The essential facility idea can be generalised to multi-layer markets.

It is now commonplace that economic regulators should strive to introduce (or mimic) competition where feasible (for technological, cost, structural or demand reasons), and protect those layers from competitive failure arising elsewhere by means of common carrier, open access, interconnection, easement (use or share, use or lose) access rights, and so on. Such rules constrain market activity and can themselves distort competition through the second-best principle.[10]

We can illustrate this point by the example of an infrastructure owner who offers

8. Averch-Johnson (1962)

9. US v. Terminal railroad Association of St. Louis, 224 US 383 (1912)

10. The observation that 'fixing' one of a linked set of markets may reduce overall efficiency.

interconnection on non-discriminatory terms to downstream entrants with different (U-shaped) marginal and average cost curves. An infrastructure provider who charges a possibly non-linear, but non-discriminatory price for interconnection may create a 'winner-takes-all' condition in the downstream market and deter all but the incumbent or a subsidiary – even if it is not technically the most efficient.[11] Alternatively, the access price may be set at such a level that only the most efficient downstream firms can survive, transferring all surplus to the infrastructure provider and removing those downstream firms' incentive to innovate or invest in efficiency gains.

Infrastructure providers may even seek to foreclose downstream markets by keeping a set of downstream firms in close competition with one another. In doing so, they accomplish two valuable tasks. The mere fact that many providers buy access under changing terms may convince antitrust regulators that the downstream layer is effectively competitive. Perhaps more importantly, such a strategy may give the infrastructure provider not only the downstream firms' producers' surplus but their information rents as well[12].

As public utility sectors matured, dominant firms were gradually privatised and vertical market power was eroded – initially via external regulation, but eventually by market forces related to inputs (e.g. finance, fuel supply or content), outputs (e.g. reselling) and substitutes (bypass markets). However, in most cases the principal-agent relationship persisted, albeit adjusted to multiple layers of agency (a regulator dealing with more than one infrastructure provider), multiple agents and even (especially for international interconnection and capacity-sharing[13]) multiple principals. But all were seen as games in which players moved sequentially and dynamics could be handled by the tools of repeated or (rare) dynamic game analysis. This perspective led to some unexpected and undesirable results, and at least one major redesign of a market substitute for administrative regulation. In what follows, we consider the evolutionary symbiosis between regulators and regulated firms and the regulatory and market implications of changes between client-server and P2P architectures.

While utility industries progressed from public firms assumed to operate in the public interest to monopolists regulated at arm's length, and eventually to dominant hubs in more-or-less star-shaped network industries, the single *regulators* who oversaw this development typically underwent no such liberalisation and were to some extent preserved by it.[14] Long-term relations between regulator and regulated created durable links and barriers to access and information. Some regulators found it difficult

11. A potentially more efficient firm may not be able to 'jump' to a scale of operations where its cost advantage was realised.

12. This can be accomplished easily if the downstream firms' cost information is affiliated, by using a Cremer-type two-stage mechanism. In the first stage, the bidders effectively reveal each other's information, and in the second stage they are charged the infrastructure provider's optimal access price.

13. Newbery (2003)

14. The point could be elaborated to argue that the resulting regulatory creep led to gaps, duplication and working at cross purposes, especially between economic and sectoral regulators.

15. Two supporting footnotes are possible – the first would refer to Ofcom's experience with BT in areas such as introduction of ADSL, LLU and international settlement, while the second would refer to sectoral asymmetries in the 1996 US Telecommunications Act – such as cross-ownership – or even the AVMS, which attempts to extend broadcast-style regulation to the internet and, by implication, to telecommunications providers.

to act vigorously to restrain their charges and encourage 'serious' entry, particularly across international and sectoral boundaries.[15] By the same token, some well-regulated firms (particularly those making the transition to the private sector in a more globally competitive world) found it hard to respond to market challenges and sought instead to exploit regulatory relationships, in some cases ceding more power than was necessary.[16] Regulators, in short, came to regard themselves as uniquely competent to allocate the right to operate, determining and enforcing qualifications and regulating pricing and other market conduct in order to approach more closely an analytically-determined efficient outcome. This sometimes took predatory form (regulatory capture or foreclosure) but in general seemed a natural co-dependence. Suitable markets in which even regulatory functions may be traded could at least open these relationships to external change.

An effective regulator is an attractive target for a market participant and can sometimes be captured by informational asymmetry or the sheer complexity of the information or computations involved. This seems to play to the strengths of markets, including decentralisation and informational efficiency.

In reality, regulators are agents as well as principals; they act in their own interest in ways that can be predicted or even manipulated. Their legal and operational bases are set by many principals including firms, customers and society's political representatives, among others. But regulators tend to listen to some parties more than others. So they are not simple first-movers, and their powers of commitment are limited. While some of their structural interventions can be thought of as design, the very mixed experience of wholesale adoption of spectrum auctions shows that the goalposts are constantly shifting.

Because no player is wholly 'outside' the market system or has perfect powers of commitment, both structure and conduct are driven by *variation* (innovation and shocks), *selection* (conscious design and market outcomes) and *heredity* (persistence, imitation, standardisation…). These three factors taken together comprise evolution.

This is not a steady drift or cycle towards or away from free markets. Markets are regulatory mechanisms in their own right and innovative forms have arisen to address particular problems, including:

- sectoral regulation of market access, especially those reached via commons. Licensing creates a tradable right and auctions are used to allocate the right efficiently, as well as to elicit information relevant to other commercial, regulatory or policy decisions and capture some returns from use of the public good;
- societal regulation of externalities has been dominated in recent years by trade in rights linked to the problem – discharge rights, liability permits, landing slots, etc.;
- the use of yardstick competition for economic regulation of competitive behaviour implicitly bases regulatory decisions on market signals;
- economic regulation of infrastructure provision and other 'natural monopolies' by franchising; and

16. Examples here might include the California combined regulatory regime for power, in which separate organisations (with industry agreement) were set up to regulate long-term load balancing and short-term power exchange.

- efficient discharge of public service obligations, attained by broadening the market choices of potential providers through 'pay or play' provisions, auctioning the right to provide universal service or through redesigned public procurement arrangements.[17]

Not all of these have passed the tests of selection and heredity but variation continues.

We should not make the mistake of thinking that evolution is the same as improvement; it does not always operate at the expected level. Some regulatory objectives cannot be implemented in the strict (dominant-strategy) sense of a competitive market and must either be sought through less certain Bayesian equilibrium in market-like games, or compromised. Hybrid regulatory schemes may attempt to 'steer' the natural evolution of market structure, conduct and performance.

An example is provided by marketing boards used to allocate a number of commodities between more- and less-elastic markets.[18] Typically, a voting regime is used[19] to choose a 'pro-rate' – a common maximum proportion of capacity that can be sold on the less-elastic (more profitable) market. In a concentrated industry, introducing this kind of board improves allocational efficiency. Smaller participants both drive the selection of a relatively generous pro-rate and are more likely to sell up to the maximum allowed, so they are more profitable and gradually accumulate capacity.

This reduces concentration and the risk of structure-induced market failure. At this point the regulatory arrangement facilitates collusion and should 'sunset.' In much the same way, behaviour in networked markets is strongly influenced by the geometry of connections among participants.

Regulation *can* change that behaviour, but the required intervention may be substantially distorting compared to a more modest approach that exploits evolution by encouraging beneficial cascades and possibly changing the network structure.[20] This is reflected in, for example, competitive effects merger analysis, which relies implicitly on markets for corporate control.

Regulatory challenges

The introduction of new markets is – or should be – a likely development in the evolution of utilities which support the dynamism of the economy as a whole and are inevitably dragged across regulatory boundaries. But specific challenges give the development of regulating markets special urgency and specific shape.

Infrastructure provision requires a number of factors: ongoing investment by owners only partially able to appropriate returns; non-discriminatory interconnection; availability of essential facilities; and prevention of foreclosure. Short-run pricing, co-location or technical requirements 'solutions' create future regulatory risks that discourage investment or favour investment in exclusion.

17. Notable among these are compulsory competitive tendering of public service contracts, PFI arrangements and the redesign of the Procurement Directives to increase the market-like character of tendering (by means of, e.g., competitive dialogue, electronic markets and Most Economically Advantageous Tender rules).

18. Cave and Salant (1995)

19. The same result can be achieved through a secondary market in capacity tied to voting rights.

20. E.g. by means of divestiture or merger policy.

In analysing repeated franchise auctions, Engel, et. al. (1997) argue that their limited success stems in large part from fixed franchise terms. They demonstrate that necessary flexibility can be achieved if the regulator fixes the infrastructure user fee and potential operators bid the required present value of revenue in order to operate the facility. This reduces dependence on detailed demand forecasts, and thus the risk premium attached to bids. Suitable auction forms can also reveal the income flow required for competitive rates of return and thus the scope for *ex post* renegotiation or hold-up. Finally, contract terms allow either side to terminate on payment of the residual value, thus preserving the regulator's flexibility if demand changes and the firm's incentive to invest in cost-reducing innovation (and thus accelerate the end of the 'first' contract).

In utility markets with variable demand capacity utilisation (and capacity), risk management can be complicated by specific market features. Perhaps the best-known example is electricity; it cannot be stored, so demand and supply must be kept in fairly close balance – otherwise the network as a whole may 'go down'. Not all of this risk can, or should, be placed on individual generators or the (typically public) network operator without distorting forward contracting, which is one reason why regulatory regimes typically opt for load-balancing *or* power-trading, but not both. There are also externalities on the generator side, particularly for coal- and gas-fired power plants: continued operation (e.g. as gas turbine 'spinning reserve') is costly, but stopping and starting imposes capital as well as operating costs. Where *aggregate* demand is relatively predictable, this can be addressed through contingent contract trading, but this has perverse effects when demand shocks are correlated or when security of supply concerns dominate.

The problem occurs with transportation and communication networks and can be exacerbated by network externalities. Newbery (2003) discussion of auctions for electricity network capacity – in particular grid interconnection – identifies conditions under which they might be preferred to price-cap, yardstick and other regulation. He notes the possibility that an auction resulting in insufficient capacity may trigger a reversion to more onerous regulatory intervention and greater competition. This regulatory risk can discourage investment and trigger a self-fulfilling collapse.

Secondly, Newbery considers the implications of the regulatory stance (in the Electricity Directive) against long-term contracting, which would provide revenue assurance over a term commensurate with the capacity investment payback period. If long-term rights allocated at auction could be resold as shorter entitlements, it would be possible to auction them without endangering competition. The alternative of synthesising long-term contracts from short-term contracts bought at auction may be less likely to accomplish the twin goals of efficient investment and capacity utilisation.

A third challenge is the problem of finding optimal access 'size'. Marketable rights have to be delimited: spectrum rights are defined in terms of frequency (range), power and location. If the rights are 'wrong' the market may be too thin on one side, create too much market power or provide insufficient commercial purchase to cover investment.

To achieve an efficient outcome optimal collections of 'idiosyncratic properties' must be identified either inside the market or before it operates, but bidders have only limited incentives to reveal relevant information. The dynamic challenge may be even greater: how can we adapt the allocation of rights to fit demand or technology changes? Incumbents can use endowments to foreclose entry, supporting, for example, technological collusion and distort innovation.

To date, economists have concentrated on identifying correct initial structures via a modified Vickery combinatorial auction. This is enormously complex and costly to operate, vulnerable to strategic behaviour and does not always produce a core outcome; for this reason, various proxy auctions have been devised.[21] These solutions do assume that bidding firms are able to evaluate different combinations. Like most market access auctions, design is complicated by risks or changes arising in response to, but also outside, the auction: failure of regulatory commitment (for example, a price so high that the 'winner' was allowed to default); the bankruptcy or merger of some or all winners; or a shortage of capital needed to realise anticipated licence values.[22]

In these and other cases (such as electricity auctions) the returns expected by bidders also depend on the fate of other bidders. A firm that fails to win (or even enter) an auction may be strongly affected by the identity of the eventual winner(s) and the price(s) paid. This suggests some form of menu auction where participants bid on different allocations of licences across the entire bidding population.

In ICT-rich sectors, traditionally regulated activities are increasingly important complements to unregulated activities and both are subject to rapid and often disruptive technological change. This sharpens challenges to conventional regulatory paradigms. Dynamic efficiency (incentives to invest in innovation) may be as important as short-run technical or allocational efficiency.

Necessary investments are risky and often protected by intellectual property rights. Recent US merger policy puts control of key IP on the same footing as control of 'real' goods and services markets. The same considerations apply to regulation of access to protected facilities or the granting and licensing of key patents. This does not invalidate regulatory procedures, but enormously complicates asset and market valuation. Markets in access rights – including IP access – can at least simplify the regulatory problem.

Two further implications follow. First, Ramsey pricing theory and the analysis of non-linear prices and bundling[23] suggest that 'light-touch' regulation of part of a mesh of complementary markets can have a better overall effect than regulating all connected markets or full unbundling. Second, the emergence of substitutes for (some) services offered by a regulated firm may weaken its market power

21. Ausubel and Milgrom (2004)

22. Or debt overhang problems associated with distorted financial structure.

23. See e.g. Armstrong and Vickers (2006) and Mason and Valetti (2001)

24. This has been argued in the case of fixed-line telephony, gas pipelines (USDOJ (1991), etc.

25. In this way, for instance, ONP provisions enforced under a dominant-firm relation with BT effectively created a market for interconnection that fostered cable and mobile alternatives and thus allowed a weakening of price regulation. Note that the expansion of a competitive market at one part of the value chain may, in turn, create the conditions for extending competition vertically, as with LLU.

to the point where it no longer needs to be regulated[24] – indeed, regulation of part of a value chain may encourage or forestall 'bypass' substitution, producing dynamic effects quite different from the immediate response.[25]

Access rights – and regulatory liability – allocations that are needed for short-run efficiency may have to be adjusted as markets evolve. Administrative proceedings tend to understate the 'real options' aspect and there is no *a priori* reason why a market in regulatory risk might not simplify the construction of adaptive markets capable both of rebalancing regulatory pressures and enabling 'make-or-buy' decisions about self-regulation.

For network industries, changes in legally-required regulation affect the interacting ecologies of legal postures and regulatory strategies. For instance, as discussed in RAND (2006), content regulation of internet audiovisual media services may fall on a layer of the market that is naturally regulatable, but which may not be in a position efficiently to monitor behaviour or enforce compliance. The net effect will depend on the degree to which regulatory costs can be monetised and transferred to content users and/or providers, and on resulting changes in behaviour. These in turn reflect the magnitude and nature of those regulatory burdens and the willingness to pay for outsourced protection by those potentially liable.

Regulatory strategies range from heavy and literal enforcement on those currently regulated to 'light-touch' regulation that makes maximal use of market forces (but may not keep pace with subsequent changes) to a strategy that relies on lawsuits to impose and continually re-examine the application of (if not the need for) regulation, albeit at heavy administrative and risk cost. Simplifying market options, service providers (ISPs) provide users with access to third-party content. If the roles of content user and content provider are clear and fixed, the ISP can offer the protections of a closed environment at both financial and connectivity cost or a riskier 'open access' environment. In all regulatory environments, the 'walled garden' protection offsets regulation cost. It is not, in general, efficient to regulate content providers or peer-to-peer content hosts, so internalising costs through service markets may be the only way to change their incentives.

To achieve reasonable outcomes in the face of incentive, information and communications problems may require structural intervention (such as limiting contract type and duration, or by ISPs imposing content-based charges on content providers). This is less a matter of mechanism design than guiding the evolution of the distribution of ISP models and thus the internalisation of regulatory externalities among content providers, content users and redistributors. Finally, the globalised nature of ICT-rich markets – at least compared to the distribution aspects of other utilities – means that regulatory stance will be chosen via a game among regulators mediated through the market.

Many of these markets will involve explicit or implicit auctions. The most notable recent successes in the economic access regulation have come from auction theory, and many other problems could, in principle, be addressed the same way. This might be as simple as replacing detailed access price regulation with a requirement to provide access using a certain type of auction. Alternatively, the regulator or other public body could design and implement a specific auction to replace or accompany regulation in the usual sense. Auction design involves, at a minimum:

- what is auctioned (e.g. size, scope, duration, subsequent tradability, etc.);
- initial conditions (choosing an auctioneer or initial distribution of rights);
- analysing the identities, exploitation possibilities, information, incentives and strategic sophistication of potential bidders;
- technical aspects – the form of bids (amounts of money, shares, contract terms, etc.), entry qualifications and requirements (e.g. deposits), bid qualification rules (e.g. reservation prices, activity rules) and 'auction form' (open- or closed-bid, inter- and intra-round information, provisional allocation rule, ending rule and subsequent negotiation);
- provision for risks of inadequate or inappropriate entry, predation, collusion, hold-up etc.; and
- specifying the conditions prevailing before and after the auction.

There is a vast literature on auction design. Some assumptions of conventional auction analysis are questionable in industries subject to strong technological change, network externalities and/or extended value chains. Bidders' values are likely to combine common and private elements (especially when access value depends on access to content, finance or customer loyalty) and will be to some degree affiliated.

Auction theory as generally employed also views values as unaffected by the auction outcome – while there may be externalities (such as distribution of property rights among rival firms) and bidders may know more about values after the auction closes, the underlying values themselves do not change. Experience has generally shown that specific auction forms (such as simultaneous ascending, Japanese, package and modified Anglo-Dutch auctions) tend to work well under specific circumstances – but the same experience suggests that learning about auction forms strongly affects their performance.[26]

As mentioned, debt arising from auction prices may affect winners' investments; this line has been taken strongly by some commentators on the German and UK 3G spectrum auctions.

In addition to foreseeable financial effects that could be taken into account via real option analysis of bidding strategies[27], public-sector or regulatory risks may arise from the auction mechanism itself. These include retroactive charging of non-participating incumbents; 'beauty contest' allocation of unsold licences; additional spectrum made available as a direct result of auction revenues; or changed competition policy in response to concentration via the auction or subsequent merger, takeover and exit.

These resemble the regulatory risk identified by Newbery (2003, see above) and could be addressed by inclusion of option arrangements in the licence contracts or by increasing the freedom to resell, subject to rule-of-reason competition policy.

26. Klemperer (2005)

27. As pointed out implicitly by Binmore and Klemperer (2005).

These endogenous changes exemplify larger knock-on effects. Consider a network-based service provided by several suppliers via client-server architecture and delivered over an infrastructure requiring long-term investment. Markets have to reallocate services in the short-run to meet demand, supply and price shocks while providing investment and capacity balancing for the network as a whole. Balancing is critical in networks without much storage slack (such as congested transport networks and electricity networks).

A large literature has analysed the contractual forms and novel assets arising, and various regulatory institutions have been used to assist their function. But contingent contracts, novel derivatives and the like are adapted to eliciting and aggregating information to reach unique outcomes and work best with relatively exogenous shocks.

But there may be multiple equilibria as in Newbery (2003) which gives rise to a 'co-ordination game' problem: if bidders believe the auction will lead to weak infrastructure investment, they may anticipate regulatory efforts to secure additional critical infrastructure investment, further weakening bidding and leading to a low-investment equilibrium with an ongoing need for regulatory support.

Alternatively, strong bidding can lead to higher direct investment and returns that reflect limited supply and lower regulatory distortion and cost. The inefficient equilibrium could be headed off by regulatory commitment but this seems difficult, especially when the 'client-server' roles of supplier and user become blurred (by peer-to-peer networks, combined heat and power units, and so forth), shifting the revenue-generating activity to the periphery and reducing overall private investment incentives.

Newbery identifies conditions under which capacity auctions will avoid the inefficient equilibrium and recommends regulation for other cases. But the choice of equilibrium – like the choice of ISP strategy in the previous example – may be an endogenous consequence of evolutionary forces and network geometry.

Reduced to its simplest form, a co-ordination game has two strategies – aggressive or timid bidding – either (but not both) of which could prevail in equilibrium. Because the auctions run repeatedly, firms may adjust their strategies as circumstances change. They will not do so every time; necessary computations are complex and depend on the expected behaviour of other bidders who may in turn be learning.

In addition, participants do not always face the same bidders, as demand and supply fluctuate. We can model the situation using a variant[28] of Young (1993), in which players periodically re-optimise their strategies in the light of opponents' current behaviour. Due to imperfect information, they do not always succeed and result in a Markov process whose states are distributions of bidding strategies.

In a fully-connected network (where the set of bidders is the same and comprises all users of the rights on offer) the long-run stable outcome behaviour is the 'risk dominant' bidding

28. See Young (1993) for the basic model, and RAND (2006), Annex A for an application to the ISP game.

29. This analysis deals only with long-run outcomes. The short-run behaviour may display sudden jumps or cycles, but the same policy interventions or market conditions that encourage efficient behaviour mitigate these problems as well.

behaviour that would be best against a randomly-bidding opponent. Quite which equilibrium meets this criterion (or its multi-strategy generalisation) depends on out-of-equilibrium outcomes. Conditions for evolution in the direction of the efficient equilibrium are milder than those under which capacity auctions always work – and the corresponding light-touch regulation may be relatively modest.[29]

There are also additional challenges for regulation of network industries arising from the network layering. Regulation of parts of *extended value chains* (such as essential facility and access price) exploits competitiveness in one layer to mimic market outcomes in others, with suitable vertical coordination and cross-ownership controls. Similar principles should apply in *layered network industries* but the dynamics and bypass possibilities may be greater. Bypass is not harmful in itself. Proposals for auctioning universal service provision rely on incentives to develop more efficient ways of meeting universal service needs. Problems arise from persistent isolated pockets of inefficiency and the threat that bypass might pose to 'converged' regulation.

One particularly difficult aspect is cross-sector convergence, when entrants with superior network reach can achieve predatory market power across other sectors. Another concern is the ability of differently-configured network or network-enabled service providers to 'unbundle' economic, sectoral and societal regulatory objectives. An example is the current evolution of VoIP services and emergent business models for exchange of copyrighted content.

Conclusion

The discussion above considers the forces driving regulatory change. Technology, for instance, creates new issues for regulation but also creates new possibilities for attaining regulatory objectives. These range from the relatively trivial (for example, using ICTs for electronic mechanism implementation) to the more profound – such as the emergence of layered structures and sector convergence or splitting, that allow regulation 'at a distance'. We noted that regulatory and market structures tend to co-evolve, and that inappropriate outcomes on one side can be stabilised on the other. New forms of regulation, including self- and co-regulation, make greater use of commercial interactions to address governance issues. Finally, advances in mechanism design experience and analysis have led to better (and better-understood) options for meeting new challenges and recognising and relinquishing obsolete ones, as well as a new suite of candidate market designs and business models. We conclude by describing two specific possibilities.

One is a market for regulatory liability, access and/or compliance. This could concentrate compliance around a regulated firm or take the form of compliance outsourcing. The latter market may show a 'tipping' reversion from oligopoly regulation to a co-regulatory relationship with an industry representative. The only difference would be market-mediated participation, which would certainly lower compliance burdens and distortions and could also mitigate hold-up and foreclosure.

The second is a market for regulatory services. Given partially overlapping jurisdictions and regulatory agendas, there seems little *a priori* reason why sectors or individual firms should be regulated based solely on the jurisdiction of establishment or on changeable 'principal lines of business'. There is some evidence that local oligopoly provision of regulation has led to regulatory flight, forum-shopping and even the distortion of business models and investment flows. These may have as much to do with specific regulators' policy objectives and tools as with their regulatory remit. A system whereby firms – especially those operating in multi-jurisdiction markets – could choose among competing regulators or provide acceptable self-regulation – would seem eminently feasible and could produce a more proportionate allocation of regulatory burdens, mitigate 'unforeseen consequences' of broad directives and provide efficiency incentives for regulators themselves.

Mechanism design tools, traditionally and sparingly applied to firms, can be applied to a population that includes all stakeholders (including regulators). The result will be transfer of some command-and-control functions to suitably-regulated markets and a greater dynamism of regulation itself.

References

Armstrong, M. and J. Vickers (2006) *Competitive Nonlinear Pricing and Bundling mimeo* at: http://else.econ.ucl.ac.uk/papers/uploaded/220.pdf

Arozamena1, L and E. Cantillon (2004) *Investment Incentives in Procurement Auctions* Review of Economic Studies, 71(1): 1-18

Arrow, K., and R.Lind. (1970) *Uncertainty and Public Investment Decisions*, American Economic Review, 60: 364-78

Ausubel, L. and P. Milgrom (2004) *Ascending Proxy Auctions*, Stanford Institute for Economic Policy research Paper, 03-035

Ausubel, L. Cramton, P. McAfee, R. and J. McMillan (1997) *Synergies in wireless telephony: evidence from the broadband PCS auctions*, Journal of Economics and Management Strategy, 6(3): 497-527

Averch, H. and L. Johnson (1962) *Behaviour of the firm under regulatory constraints*, American Economic Review, 52(5): 1053-1069

Bacon, R. (1996) Competitive Contracting for Privately Generated Power, in Public Policy for the Private Sector: Infrastructure. Washington, D.C. World Bank Group

Baron, D. and R. Myerson (1981) *Regulating a monopolist with unknown costs*, Econometrica, 50(4): 911-930

Cave, J. and S. Salant (1995) *Cartel Quotas Under Majority Rule*, American Economic Review, 85(1): 82-102

Cave, J. Georghiou, L. and R. Wilkinson (2006) *Public Procurement for Research and Innovation: Developing procurement practices favourable to R&D and innovation*, European Commission

Cramton, P. and S. Stoft (2005) *A Capacity Market that Makes Sense*, Electricity Journal, 18: 43-54

Crew, M. and P. Kleindorfer (2002) *Regulatory economics: 20 years of progress?*, Journal of Regulatory Economics, 21(1):5-22

Demsetz, H.(1968) *Why Regulate Utilities?*, Journal of Law and Economics, 11, 55-66

Engel, E., Fisher, R. and A. Galetovic (1997) *Revenue-Based Auctions and Unbundling Infrastructure Franchises*, Inter-American Development Bank, Infrastructure and Financial Markets working paper, IFM-112

Joskow, P. and J. Tirole (2004) *Reliability and competitive electricity markets*, NBER Working Paper No. 10472

Laffont, J-J (1994) *The new economics of regulation ten years after*, Econometrica 62(3): 507-537

Mason, R. and T. Valetti (2001) *Competition in communication networks: Pricing and Regulation*, Oxford Review of Economic Policy, 17(3):389-415

Milgrom, P. (1996) *Procuring Universal Service: Putting Auction Theory to Work*, Lecture at the Royal Swedish Academy of Sciences in honor of William Vickrey's Nobel Prize

Mougeot, M. and F. Naegelen (2003) *Using Auctions for Contracting with Hospitals When Quality Matters*, International Journal of Health Care Finance and Economics, 3(1): 5–23

Newbery, D. (2003) *Network Capacity Auctions: Promise and Problems, Utilities Policy*, 11(1): 27-32

Salant, D. (2000) *Auctions and Regulation: Reengineering of Regulatory Mechanisms*, Journal of Regulatory Economics, 17(3): 195-204

Shleifer A (1985) *A theory of yardstick competition*, Rand Journal of Economics, 16(3): 319–327

Sorana, V. (2000) *Auctions For Universal Service Subsidies*, Journal of Regulatory Economics, 18(1): 33-58

Stavins, R. (1998) *What Can We Learn From The Grand Policy Experiment? Lessons From SO2 Allowance Trading*, The Journal of Economic Perspectives, 12(3): 69-88

United States Department of Justice (1991) *Comments of the United States Department of Justice in Response to Notice of Proposed Rule-making*, Docket No. RM91-11-000, for In Re Pipeline Service Obligations and Revisions to Regulations Governing Self-Implementing Transportation Under Part 284 of the Federal Energy Regulatory Commission's Regulations at: http://www.usdoj.gov/atr/public/comments/200509.pdf

Williamson, O. (1985) *The Economic Institutions of Capitalism*, New York: The Free Press

Young, H.P. (1993) *The Evolution of Conventions*, Econometrica, 61(1): 57-84

Communications policy, economic policy. The inextricable link

Reed Hundt

Senior Advisor, Telecommunications and Technology, McKinsey & Company

For any country, its communications policy is one of the most important aspects of its overall economic policy.

In fact, the microeconomics of communications markets drives much of the macroeconomic performance of national and regional economies. It follows, then, that if you want to determine the best communications policy, start with selecting the goals for economic policy as a whole.

The desired result for most countries will be the same: to create ubiquitous broadband access for which public users such as schools and libraries pay nothing, private users pay only a little more than nothing, and virtually everyone is on the net all the time.

Economic growth is the best social policy, preached Robert Rubin, and indeed while he was President Clinton's Treasury Secretary Americans of all income classes saw their wages rise. Coupled with a judicious tax policy, that economic growth produced a strong American dollar, a national government surplus, a stock market boom, and increased investment in capital. Even when the stock market inflated into a bubble and popped into a bust, American growth tempered only slightly. From 1998 to 2003, the United States accounted for 98 per cent of all global economic growth, as China's growth was offset by declines in other continents and the American machine kept generating new wealth.

But where did all that growth come from? The surplus was an effect, more than a cause. If investment was the fuel of this growth engine, what provided the ignition plug that caused it to roar into life? The answer, according to Paul Romer and the other new growth theorists of the late 1980s and early 1990s, was technology. And nowhere did technology show its suzerainty more dramatically than in the communications and information sector.

At the core of not only American but global economic growth since at least the early 90s was a trilogy of technology advances, identified by cable tsar John Malone as the microprocessor, digitisation and fibre optics. In more general terms, hardware, software and the networks to carry data were responsible for the economic boom that started in the United States and now reverberates throughout China, India and the rest of what was formerly, and rather derogatorily, known as the Third World.

Yet technology is just technology. Alone, it could have done nothing to improve standards of living, had entrepreneurs been unable to translate its breakthroughs into goods and services generated, distributed and sold in ever-increasing volumes and at ever-lower prices per unit.

The entrepreneurial revolution in information and communications policy deserves as much credit as scientific discovery for the fact that the United States became the epicentre of the growth earthquake. That development had its roots in the American culture. Its three architectures were the openness of technological knowledge, the strength of the leadership gene or character in American business, and the commitment to open markets of American law and regulation.

Genetic entrepreneurship

However, in the 21st century, the United States has tended to dismantle all three architectures. Secrecy has cast a veil over openness, as intellectual property rights have become excessively complex and extensive. Too many leaders have contributed to a culture of deception, and law has been used to close markets and limit freedoms.

Nevertheless, even when judged by its shortcomings, entrepreneurship lies in the DNA of American culture. Indeed, some scholarship suggests that this trait is literally genetic – those people who migrated from their origins in eastern Africa to Europe and, ultimately, to America, brought their exploratory genes. Even if this is true, surely nurture as well as nature breeds entrepreneurship, and the particular nurturing force for information and communications rivalry in the 90s was the set of policies embodied in the Tele communications Act of 1996.

It was a law aimed at breaking up monopoly and opening the communications and information markets to new entrants, in two dimensions. First, the law used unbundling and other network sharing provisions to encourage competition in the physical medium of communications. Second, the law, at least as applied by the Federal Communications Commission (FCC) in the 90s, encouraged competition in services or applications developed in that physical medium.

That medium can be wire-based or wireless; it can be Malone's famously identified fibre, Alexander Graham Bell's copper or Marconi's electromagnetic waves vibrating in the non-existent ether at distinct frequencies which were capable of coding messages through modulation of many kinds. Until the 90s such media were owned by governments and assigned either to state-owned or state-regulated firms organised into monopolies or oligopolies.

The tremendous change in the 90s was the imposition of the concept of communication across networks of networks. Instead of one firm operating a single network from a client device (such as a computer or telephone) to all other client devices, many firms might build and operate pieces of networks that could interconnect. Any particular packet of communication might trace a route that, like a pathway traversing the property of different landowners, comprised portions of many different networks. To make this new model practical, the new legal paradigm granted every network owner the right to lease and connect to other networks.

Just as important, although far less clearly identified in the early years, this new paradigm also granted users of the networks the right to be free from commercial restrictions imposed by the networks. The most important example, by far, was that American telephone companies could not charge extra, much less deny the use of their networks, to those who leased the network for the purpose of carrying bits to and from computers participating through common shared software, in the vast voluntary consortium called the World Wide Web.

To extend the metaphor, if some wanted to use the pathway over the network of networks, for singing and dancing, the network owners could not charge for, or otherwise limit, that activity – even if they were also in the business of selling songs and dreaded losing revenue to all the new collaborating singers. So it came to pass from the early days of the new paradigm that the far-sighted could see that the web would destroy the voice business. If client devices could use software to send messages over the physical medium of the networks, then the network owners would not be able to charge money for the service of carrying such messages.

The purpose of the new paradigm was to open up the market to entrepreneurs to build networks and generate new services. The desired effect was to lower the cost of transmission of such information and to expand the potential for value creation in the form of web-based businesses and social activities. New goods and lower transmission cost led to expanded economic growth and a higher standard of living.

In this new world, communications networks were divided between conduit and content (or bits carried), and the so-called triple play of voice, video and data (a.k.a. web material) was not then necessarily provided by the owner of the conduit.

The new paradigm was anathema to the American network owners. Meanwhile, they faced the additional threat of the fourth addition to the triple play: wireless services. Under the auction methodology of the 90s, the FCC in the United States created a multi-firm competitive wireless market that drove prices down and penetration up. Wireless prices by the turn of the millennium fell fast enough to enable some consumers to substitute wireless for wire.

Under the Bush administration, the policies shifted back toward permitting network owners to use their physical medium in order to obtain competitive advantage in the services or content markets that used the 'pipe' to gain access to consumers. Similarly, the Bush 'reforms', championed by Congress and the regulators at the FCC and in the Department of Justice, with crucial help from the White House and friendly judges, encouraged wire-based businesses to consolidate with each other, and to acquire wireless firms.

Yet the web kept escaping the grasp of both regulators and consolidators. The new phenomenon of the web was social networking, meaning that end-users connecting to each other fashioned their own content, and demanded low-price, high-speed connections to make it happen. Services such as Skype, Vonage, Facebook, MySpace, YouTube and millions of individual blogs represented intensely creative user-managed and user-crafted content of both voice and video dimensions. They placed new demands on physical network owners while, at the same time, almost casually supplanted at least some of the value-capturing strategies of traditional content generators, such as movie studios in the visual arts and voice connectivity by the traditional telephone companies.

Content vs. conduit

In current times, then, the battle between content and conduit has taken on a new shape. It now appears that the pro-content side carries the banners of individual freedom and creativity. The content side now seems to embody much more of the entrepreneurial spirit that in the 90s, under most aggressive regulatory intervention on behalf of market entry, could also be found in the conduit markets.

We saw recent proof of content's power in the rise of the net roots – frequent visitors to popular blogs, especially on the left of the political spectrum – as a source of some significant opposition to the new telecommunications bills moving through Congress. In general, these bills focused on enhancing the power of the telephone companies, somewhat at the expense of cable, and definitely to the disadvantage of social networkers, content firms and consumers. The net roots denizens rallied support for open networks, under the code name 'network neutrality' and, at the time of writing, at least temporarily halted the congressional progress.

What, then, is the new economic policy of the current one-party government? No one knows. There is a distinct lack of co-ordination in the United States Government on policy issues of all kinds. No one seems to have decided what to do about North Korea; the calamitous events in Iraq, Israel and Lebanon; or issues at home, such as the federal deficit and the widening income gaps among Americans of all classes. Not surprisingly, the crucial topic of communications policy also draws less focused attention than at any time in the 90s or even the 80s.

Yet policy-makers deserve some charity because, in fact, good communications policy at this point in the development of markets and technology is particularly difficult to determine. It does seem plain, for instance, that if a big firm or millions of individual users want to send large video files over physical networks, they must expect the builders of those networks to be reimbursed for the cost of transmission.

Suppose telephone companies in the United States installed a fibre access network to my house. In general, they would have to spend a couple of hundred dollars on the electronics, about $400 on labour, and about $200 on overhead (and these numbers could be on the low side). But who would pick up the tab? Either the Government has to fund it from taxpayer's money, or shareholders have to be willing to make the investment in return for the prospect of profit. There is no other way to pay for the build of that fibre. Yet without it, I cannot watch the video, much less send my own breakthrough-contribution to YouTube.

But what is the national economic policy in the United States – or indeed anywhere? According to Michael Porter in the *Competitive Advantage of Nations*, it is to generate a high and rising standard of living for the country's citizens. Although Professor Porter does not dive into the details of communications policy, that standard of living goes up, generally, as the cost of transmitting bits falls and as the number of people reached by the means of transmission grows. In other words, a major, perhaps essential, contributor to a high and rising standard of living is ubiquitous and truly high-speed broadband at very, very low cost.

To this end, every nation needs a network and, even better, a network of networks, in which case a country can depend on competition to generate new technologies that increase the speed and extent of broadband access. Particularly vital to this end is the creation of unlicensed and licensed spectrum in vast amounts, so that a licence is as easy to obtain as a franchise to open a Starbucks coffee shop.

This need is always underestimated when national policies are shaped: in country after country, too few wireless licences are granted with the inevitable effect of diminishing welfare gains. There is no reason why the market for spectrum should not be *flooded* with licences, just as there is no reason why a mall should not have as many stores as there is demand by retailers for space. In addition, unlicensed spectrum can be an ideal testbed for innovation: WiFi has grown to be a major new technology precisely as a result of using unlicensed spectrum to experiment.

Similarly important are pole access rights; easy and rapid licensing of municipal WiFi networks; accommodating zoning authorities eager to permit wireless towers to be erected; reliable regulation of very cheap interconnection; and some unbundling of essential facilities, if any exist.

However, self-interest dictates that any owner of a physical medium will want to provide content and, at the same time, give their own content an advantage against rivals. If the owner of a road wanted to go into the taxicab business, it would of course deny the road to its rival cabs, or charge them an amount equal to their firms' profits to use the road. This impulse could motivate competitors to build their own roads, but meanwhile it's the passenger who suffers, either by paying for the service offered by the road owner or simply by not being able to afford to travel.

Similarly, as social networking is seemingly inventing new forms of collaboration in all spheres every day, particularly in entertainment and politics (a distinction that is often blurred), the loss to a nation's standard of living from the lack of access to very cheap and ubiquitous communications networks appears greater all the time.

Defining the policy

What, then, should a nation adopt as its policy for communication? On the one hand it wants to make sure that network builders have enough profit to encourage them to build high-speed networks. On the other hand, it wants the input of communications services – the transmission of bits – to make a very small addition to the total cost of generating the fascinating new services that produce a high and rising standard of living in today's information societies.

The first of these goals – incentivising the network builders – is tricky. A nation wants its networks to provide rapidly increasing capability at a rapidly decreasing cost per unit of information carried. It wants the performance and function of the microprocessor industry to be replicated in physical networks: namely, every year or two, the performance for constant cost should double, and the cost of existing performance should halve.

To this end, a nation's networks, as Malone said in the early 90s, should borrow from and build on the microprocessor and its related technologies. That implies the use of common and open protocols, software that is internet-compatible from end-to-end, and easy interoperability with applications software of all kinds, ranging from voice replication to search tools.

The design of the physical networks, in short, should reduce the advantages of proprietary, first-mover installation and translate semiconductor advances into ever-cheaper transmission costs. Ultimately semiconductor technology will bring light waves, instead of lower frequency radio waves, through networks right into client devices, enabling ubiquitous visual computing. The user will literally and metaphorically be able to see everything in the world that ever existed or does exist. That will be the fulfilment of what James Joyce referred to in *Ulysses* as the "ineluctable modality of the visible".

In order to achieve this rapid expansion of networking capability, any nation needs to have its networks become combinations of wire and wireless networks. It is quite acceptable for any nation to stimulate network build-out by providing taxpayers' money, whether by letting contracts for networking rural areas or granting tax breaks to network builders.

The economists tend to recoil at such programmes, but roads and public (state) schools have not served the nation all that badly for Americans. People should not ignore the lesson that some public money can provide some useful public services. The cost of a few months in Iraq, for an actual example, would suffice to close the gap between what consumers are willing to pay, and what network builders have to invest, to fund broadband networks to most of the United States.

The essential step in all cases, however, is to assure that such networks be open to the fundamental design efficiencies of semiconductor technology. That is a kind of openness that is vaguely, although poorly, reflected in the debates over network neutrality.

To advance technological openness, a nation should designate at least one of its major networks – whether wire, wireless or a combination of both – as its chief open network. Its owner, or owners, ought to be required to increase continually the available bandwidth and should obtain government subsidies for this purpose, awarded to the low-cost provider by bid, with the contract renegotiated every three years. This network would be open, or neutral, as to content and all other software. It would be the public highway, in a landscape that might also include many private highways.

In addition, this open network would have the assigned task of reaching everyone in any particular political jurisdiction, in return for which it would obtain the aforementioned partial subsidy of public funds. Its owners could charge anything for providing connectivity – but could not offer any service, such as voice. The owners would have to obtain at least 90 per cent penetration of the high-speed connectivity, a driver that would limit what would be charged better than any regulation. The owners could price-discriminate; that is, charge each connected party a different price for connection. But, to repeat, the owner of the one partially publicly supported network would not be able to compete in service provision.

As for all the private networks – those who are not providing the public 'network-neutral' connection – they could connect to anyone and provide any service, whether theirs alone or the services of others. By analogy, Sony is free to offer and sell PlayStation, with proprietary, closed, non-internet content; Apple can introduce the proprietary and unconnected iPod.

One, very low-cost, very high-speed, public network ought to be enough to assure that the others would either offer distinctive services of a proprietary nature that had their own value, or would provide 'open access' to all websites simply in order to be competitive. In this paradigm there would perhaps be no investment in private networks. Yet private schools exist side by side with public schools; cellular services exist next to wire-based services. There is always some demand for the special or distinctive services of private networks with private content as against the public network. And if that demand ever dissipated, then the country would still have the public network.

As I explain in my book, *In China's Shadow: The Crisis of American Entrepreneurship* [1], the world is not flat. The dramatic drop in communication

1. Yale University Press: October 2006

costs and the equally dramatic increase in capitalism have not created equality in trading relationships among nations, nor competition among people and firms. Instead, advantages obtained by some are even greater than they were in less competitive eras.

Within nations, winning classes obtain even more income than those in the comparatively poorly performing classes. Between any two nations, the principle of free trade does not assure increasing equality of wealth, but rather tends to produce big winners in firm-to-firm competition coupled with the real possibility that most of the winning firms will be in one country rather than in another. The fans of teams in the American League of Major League Baseball, which typically wins All-Star games and interleague contests, simply enjoy a higher standard of baseball. Similarly, the citizens of countries that host winning firms will enjoy a higher standard of living than the citizens of countries hosting losers.

Workers, of course, also compete in local and global labour markets. They win if they work in winning firms; they lose, on a relative basis, if they work in losing ones. Losers can look for some support from national governments but the best social policy, it will be recalled, is a fine economic policy. The key to such economic wisdom is the creation of a microenvironment for any particular industry that promotes productivity gains, vigorous competition, one or more big winners of that competition, and easy entry by rival firms so as to keep the next cycle of competition rolling quickly through time. In turn, that easy entry, as taught by Tim Bresnahan of Stanford, is dependent on divided technical leadership in the particular market, a cycle of significant technological change (many "10x inflection points", in Andy Grove's terminology), and adjacent market entry.

These three attributes of a dynamic, productive, wealth-creating economy would all be enhanced by ubiquitous physical wire and/or wireless networks that rely on open semiconductor technology and open software protocols. A good economic policy depends on that underlying fabric of communication just as, in the great boom of 1845 to 1870, the underpinning was railroads, and in the great boom of 1945 to 1970 it was highways. But railroads and highways had to be opened by regulation and government spending; so both these tools, wisely deployed in sparing but surgically smart ways, are necessary for national economic success in the current potentially long boom of the 21st century.

section 4

Utilising the airwaves

The future of regulation – not

Peter Cochrane

ConceptLabs, Co-Founder

Pre-amble

Take a look back over the development of international and countrywide communication systems, and they all have one thing in common: they have needed some form of control in order to orchestrate resources and achieve the highest efficiency. And, once they attain a certain scale, they have tended to be overseen, regulated and managed by one government body or another.

This has certainly been the case for radio technology and a spectrum dominated by analogue technologies for over 100 years. Which begs a question: with the emergence of radically new digital technologies marking a wholesale revolution in broadcasting, do we actually need regulators, or government interference, at all?

Even a quick look at how we use radio spectrum today points to a future where everything has fewer wired connections. Just compare, for example, how many wireless transmitters you'll find in a typical home compared even to your parents' time. There'll be the inevitable congregation of remotes – perhaps at least four, for hi-fi, TV, set-top boxes and DVDs.

It's odds-on there'll be a bunch or two of car keys with wireless transmitters, and perhaps even a remote control for the garage door. And then of course, there could be burglar alarms and security devices operating wirelessly; maybe a local area network for computers; and the ubiquitous cordless phones, mobile phones and perhaps walkie-talkies, and so on. If you're in any doubt as to the rate and pace of change, just try adding up the number of transmitting devices you own. (I gave up at 70.)

The fact is, it would have been extremely rare to find any kind of transmitting device in the analogue home of even 50 years ago. Those who did own one would have been an amateur radio enthusiast, or perhaps an official who worked for the Police, the Fire Service, military and the like. (Or, indeed, a spy..) The vast majority of transmitter sites were also owned by the BBC, the General Post Office, emergency services, military and utilities.

Interestingly, the real proliferation of transmitting devices has taken place very quickly, in the last 20 years, as a result of the digital revolution. There'll be more of the same over the next 20 years as well. I think it's safe to say, then, that wireless devices are here to stay – and that, to some extent, their spread is already out of control, and will become increasingly so.

As someone who worked as a wireless engineer in the analogue days, I find it nothing short of miraculous that these millions of wireless systems can coexist, and apparently work interference-free. The operational problems that used to plague us have magically disappeared with the arrival of digitisation – yet this is no happy accident. It is the fruit of the labours of small armies of engineers and scientists, who spend their lives designing systems that can work together in harmony and, most importantly, free of human intervention.

All of this work is orchestrated by government agencies, regulators and standards bodies across the planet, in concert with industry and users. But new technologies are arriving that call into question whether these traditional roles, shaped and formed in an analogue world, are relevant any longer. In the same way that typing pools and reprographic departments are relics of another era, so governments and regulators also face extinction in their role as the guardians of the airwaves. In a digitally-dominated future, what do we actually need now?

The fact that we now have the ability to dispense with the old dedicated bands and channels, devoted to specific use and users, changes everything. New forms of digital modulation and signalling allow us to layer wide-band signals, one on top of the other, without significant interference problems. At the same time, we can dynamically limit the transmit energy to what's required to meet our Quality-of-Service (QoS) needs. And it gets better: we have the power of microprocessors to control our systems to the extent of antenna steering and signal adaptation, in order to achieve continual optimisation – and without human intervention.

In addition, we have rafts of new technologies that are opening up the radio spectrum above 30 GHz to release more frequency space than we have used before. Add to this the need for smaller and smaller cells of concentrated capacity; a web of optical fibre that is being spread across countries; and the move towards DIY radio systems as the norm – and we are witnessing a wholesale revolution in wireless.

It's very likely that we'll need to retain some form of band and channel structure for many applications, but when it comes to interference, modulation and coding scheme control, it looks as though we are heading towards a hands-off regime. This all pre-supposes that, at the design stage, the companies producing the equipment are fully up to speed with the true operating mode, the environments in which their equipment will be placed and used, and what it will really be used for. Providing this is so, it seems entirely possible that the role of government and regulator will at least be diminished, and most certainly changed.

Therefore, I address these likely changes in this chapter, and plot the most probable course for the future.

Some near-invisible forces

How did we get here? It's tempting to think it is all down to technology, but in reality the changing behaviour of people, society, companies and institutions have been powerful drivers also.

- During the 1970s, the population of the UK developed a thirst for commercial radio, and for the freedom to communicate using the citizens' band. However, the government said no. So it happened anyway! Pirate commercial radio stations were established, people imported cheap CB radio equipment from the USA, and the government lost control. They had to bow to the will of the masses and make both legal.

- During the 1970s and 80s, the retail climate was driven by suppliers and companies deciding what we would buy.

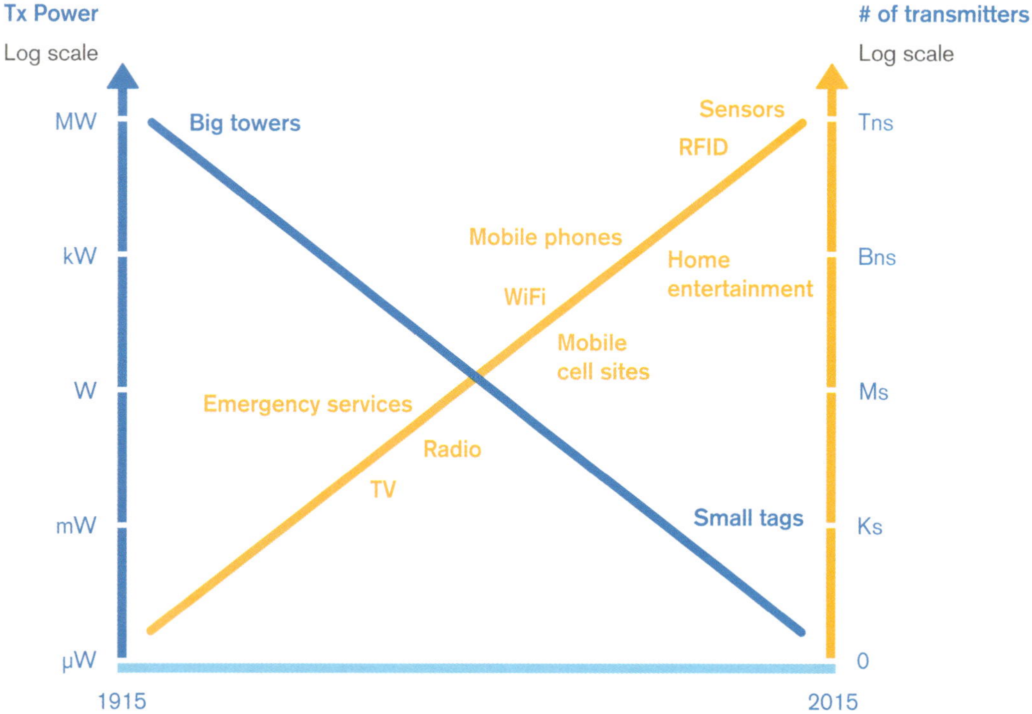

Figure 1: Growth in the number of transmitters

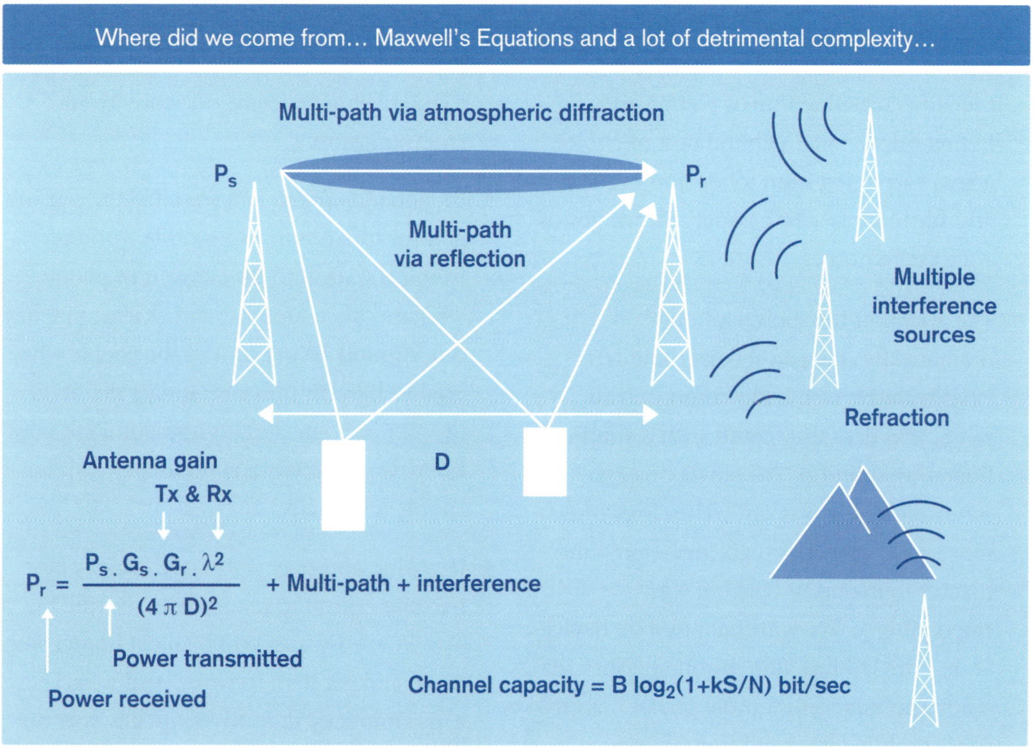

Figure 2: Where have we come from – dumb wireless systems

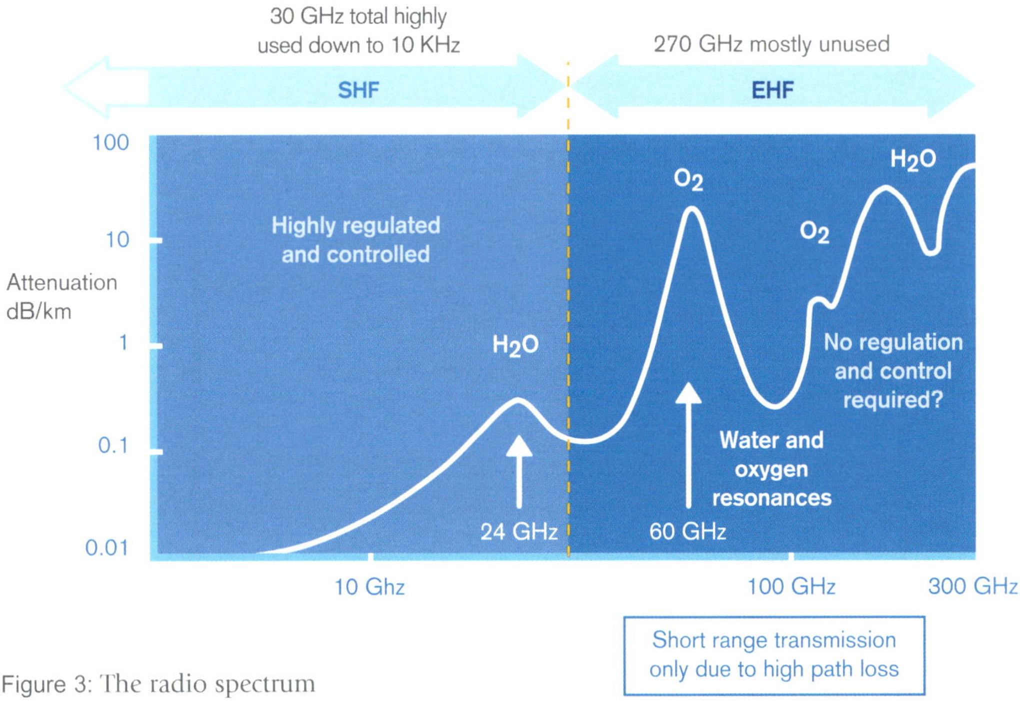

Figure 3: The radio spectrum

However, by the 1990s and into the new millennium, the power had shifted to customers, as we drove what they should supply. Globalisation and commoditisation together created a time of plenty with falling prices and an abundance of choice across everything from food, clothing, cars and furniture to white goods, brown goods – and even homes.

- The arrival of the mobile phone dramatically changed the user mindset. People always wanted to communicate freely, and now they could with a simple, liberating solution. The single biggest innovation here was not technology but marketing: 'pay as you go' accelerated the market to the point where it was impossible to live without a mobile device. Today, the UK has 90m mobile phones for a 60m population. And the planet has far more mobile than fixed-line phones, at more than 1.4bn.
- Computers and skills migrated from office and school into the home with the advent of the PC. ICT was now demystified: everyone could use a PC, and we all began to establish new forms of creativity and communications.
- Broadband arrived and sounded the starting gun for true home and mobile working. In the UK, around 80 per cent of people who are office-based spend at least one day out. Around 30 per cent of the people who used to have an office no longer do, and over 25 per cent of the population are now home workers. Work is no longer a place – simply an activity.
- In a plug-and-play world, technology has become central to all our lives. People buy TVs, PCs, Hi-fi and Wi-Fi, take it home, plug it up, and get it all to work. And it isn't only simplicity that drives uptake: year-on-year it costs less and less, whilst providing more and more, with even simpler

interfaces and lower cost of ownership. It's a seductive proposition that sells itself.

- Of course, it is Moore's Law – the exponential growth in transistor packing densities at an exponentially falling price – which is powering all of this. Just about everything we now buy is packed with digital electronics, and gradually this will include wireless technology. Cars, TVs, Hi-fi, PCs, laptops, PDAs, MP3 players and indeed homes are all, increasingly, being delivered with a wireless capability built-in and mostly for free. It isn't an extra or an afterthought – it's a necessity.
- Mobility and mobile working have had a profound impact on the way we view networks and the way they are designed and operated. Fixed line telephone networks tend to be governed by randomness, with little correlation between calls. Mobile networks, in contrast, are governed by far more chaotic behaviours. There are strong correlations between people-clustering, events, device and call-events. And in the case of IP networks, eMail, IM, VoIP and net downloads, the correlations are even stronger. We see the peak to mean traffic ratios of each class of, typically:
  - Fixed line telephone = 3 or 4:1
  - Mobile telephone > 50:1
  - Internet > 1000:1

Of course, central to all this are our mobile devices and wireless systems. And it is all a very long way from the old analogue world of BBC-only, and no mobility. Some of the key changes and forces are shown in figures 4-6.

Technology to come

Why did we go digital? The reason seems to have been lost in the mists of time, but it's simple: economics. As telecoms started to expand in the 1960s, it became clear that switching and transmission standards would be increasingly expensive, if not impossible, to maintain transmission standards in a truly global network.

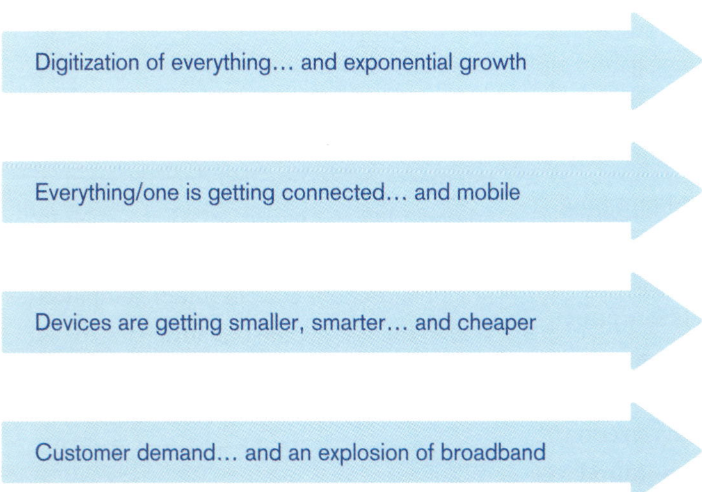

Figure 4: Disruptive forces

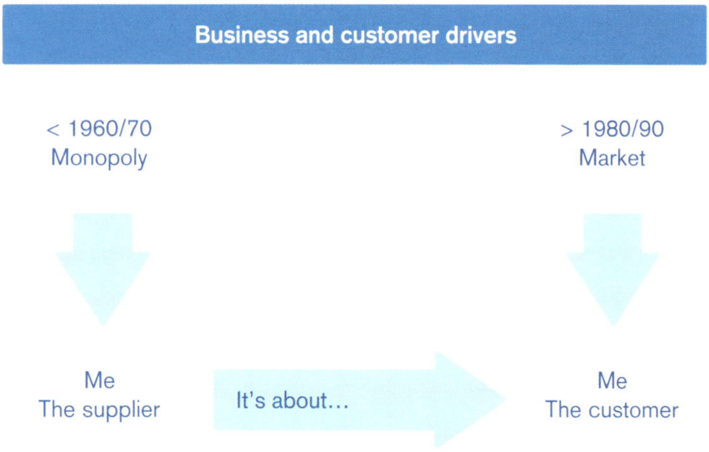

Figure 5: It's about me, the customer

It also became rapidly apparent that the variability of analogue solutions and systems could be removed from the equation by going digital. And so the decision was made country-by-country: go digital for all transmission and switching systems, and so reduce costs and increase inter-working compatibility.

What was not contemplated at the time was the impact this decision would have when overlaid with the decision to go all-digital in computing. Well, as we know now, the space race and the integrated circuit changed everything. A revolution duly followed with the PC, internet, IP networks, mobile phone, electronic games and digital content. So what happened to radio? It got smart.

Instead of simply being static and pumping out a fixed amount of power in all directions, the communication and control between mobile transmitters and receivers saw power levels continuously adjusted to the minimum required to maintain a given Quality of Service. Next, a degree of dynamic antenna steering (or, more precisely, 'sectoring') was introduced to beam the energy between two or more devices in order to further reduce unnecessary radiated energy.

The early mobile digital systems (2G) used relatively narrow channels, allocated in bands (with speech seen as the primary service) that were strictly controlled and regulated. To some extent they look very similar to the early (1G) analogue mobile systems that used even narrower band channels, the only difference being the modulation mode. But when the latest technology (3G) arrived, it started with the objective of delivering digital content on the move, and to that end it looks truly digital.

Instead of using conventional (analogue-like) channels, 3G systems use a Spread Spectrum mode that sees coding as the differentiating mechanism. This is why signals are very broad and can be layered on top of each other without interference. This was first demonstrated around 1945 as a means of hiding signals from the enemy in WWII, but at that time involved not a handset but a room full of computers! Within the last decade our advancing digital knowledge has allowed us to accelerate this technology to become a commercial proposition; we can now do the necessary digital signal processing on a fraction of a single chip.

So what lies in the technology pipeline that will cause another giant leap in utility and

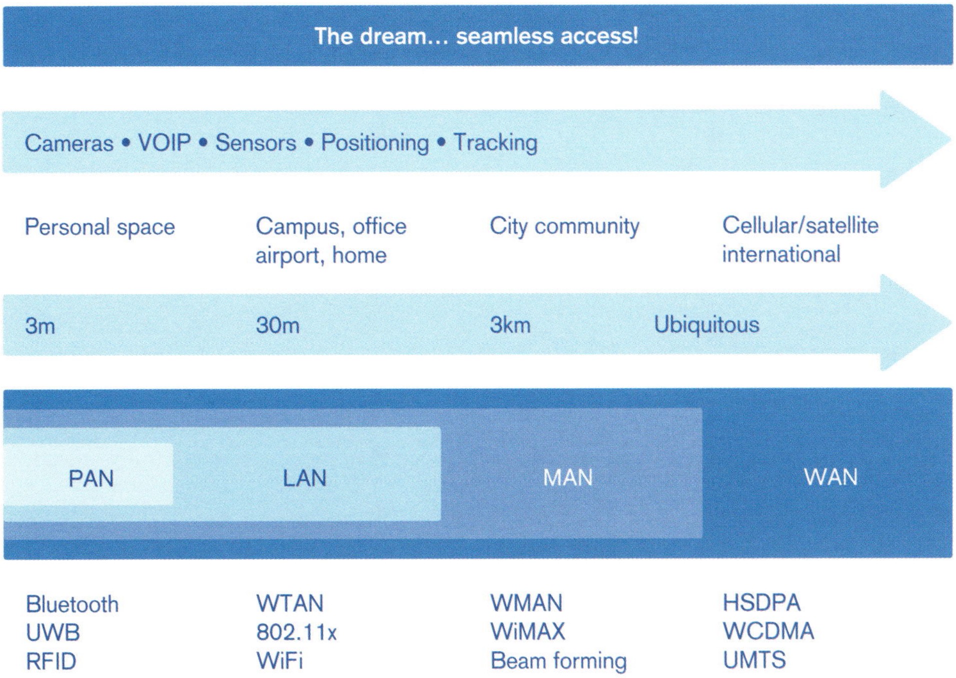

Figure 6: The dream of seamless access

usability? In a phrase: dynamic adaptability – in every dimension. Computing power allows us to search for unused spectrum, adapt our signal to the available channel and comms needs, and just get on with it in a dynamic manner we have never seen before. The spectrum then looks more like a freeway than a railroad; space is a variable rather than a fixed and rigid quantity.

The false perception of today's regulated airwaves is that every Hz of space has been allocated and is in use, and that there is a severe lack of spectrum. Nothing could be further from the truth! Recent studies across the USA have shown that even very 'wireless-dense' environments such as the city of Chicago, very seldom exceed 17 per cent of the available spectrum in use today (up to 30 GHz). So, of the total spectrum theoretically available (up to 300 GHz) this equates to around only 1.7 per cent. Clearly, there is considerable opportunity for further exploitation, should our technology allow.

Beyond the raw wireless technology, it is the management software and systems that offer another big gain. If frequency, space, power, time and code can be managed dynamically, then so can the communication session. Real-time and delayed messages/signals can be mixed with real-time channels held open and delayed, and becoming opportunistic as shown in figures. 9, 10 and 11.

As terminals are increasingly mobile then the use of beam steering antennas also becomes more attractive. Being able to harness scattered energy, as opposed to trying to cancel or limit its effects, is a really positive move in terms of maintaining QoS and minimising transmitter power. Again, this is only possible with low-cost computing power.

Note that digital coding in the spread spectrum sense implies dynamic changes in time, frequency and power distribution

	Frequency	Space	Power	Time	Code
Analogue 1	✔				
Analogue 2	✔	✔			
Analogue 3	✔	✔	✔		
Digital 1	✔	✔	✔		
Digital 2	✔	✔	✔	✔	
Digital 3	✔	✔	✔	✔	✔
Digital 4	Dynamic/ adaptable	Dynamic/ adaptable	Dynamic/ adaptable	Dynamic/ adaptable	Dynamic/ adaptable

Figure 7: The migration from analogue to digital systems

So, there we have it. Humans are taken out of the loop, control is given to the machines and self-organisation wins over hierarchy. Which begs a question: why will we need control and regulation of the spectrum?

A clean sheet of paper

Given the incredibly complex situation we now find across the board in the wireless space, it is worth starting with a fundamental question. If we had a clean sheet of paper, with no analogue past, no history, only digital systems, and no spectral occupancy anywhere – what would we do?

The first observation to make is that analogue and digital systems still have to obey the basic laws of physics. Information and field theory – Shannon's Law and Maxwell's Equations (figure 2) – still apply, as do the Radar Range Equation and the propagation conditions for radio waves. Add man-made and natural noise, reflections and interference, and we have all the fundamental elements that dictate the ultimate performance of all wireless systems.

There is no magic, nor short-cuts to success; it is down to physics and engineering. And in this regard the analogue era has taught us a lot about propagation, noise and interference below 30 GHz. Broadly speaking, the segmentation by frequency (and, of course, wavelength) of the spectrum into global, national, regional and local reach systems (by way of the ELF, VLF, LF, MF, HF, VHF, UHF and SHF) has worked well and seems to form natural divisions for technology, applications and performance.

The reality is that we don't have a clean sheet of paper below 30 GHz where we have a growing mix of analogue and digital systems with a gradual migration to all-digital. It is fair to say that governments and regulators are going to be engrossed in the progressive mode change

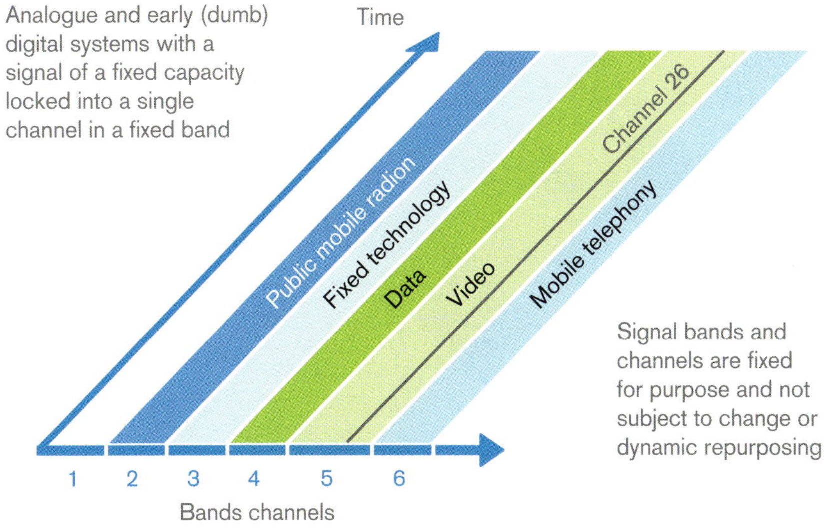

Figure 8: Analogue bands and channels

over, reallocating and repurposing of these bands for a decade or two yet. So let us look at what might be done above 30 GHz and see if we can establish some basic principles; we can then return to the sub-30 GHz region and look afresh.

Above 30 GHz

Frequencies above 30 GHz have largely gone unused because of the high path attenuation due to atmospheric resonances (figure 3), and the difficulty of building the necessary electronic circuits. But now, we have both the technology and the need. Electronic devices clocked at speeds in excess of 350 GHz at room temperature are in the lab at the leading-edge, whilst devices operating in the 100-180 GHz range are available. In addition, we have to hand smart antennas, coding and modulation technologies, and the processing power to control and optimise our future systems.

As people cluster with their electronic devices, the bit density (bits/unit volume) to be communicated accelerates to the point where LANs and mobile cell-sites fall over due to overload (figure 12). Moreover, with the advent of VoIP, even more traffic is adding to the concentration at wireless LAN nodes, but with the added requirement of no packet loss and minimum delays. The only way this can be generally accommodated is through the provision of more capacity, which translates to ever-smaller communication cells.

Then, entering stage left, come two extra competing areas: home entertainment and games. Several manufacturers are looking to employ 60 GHz wireless systems to link all of their HiFi, TV, DVD, Satcom, Games and domestic appliances with links into all computers, laptops and servers. This will see the wireless home and office with integral broadband connection, WiFi, Bluetooth, 2.5 and 3G and VoIP.

The design philosophy is simple: due to O_2 absorption in the atmosphere, operating under the 60 GHz attenuation peak will ensure very little room-to-room and house-to-house interference through the natural filtering and attenuation provided by Mother Nature. So does this mean we might see the same philos-

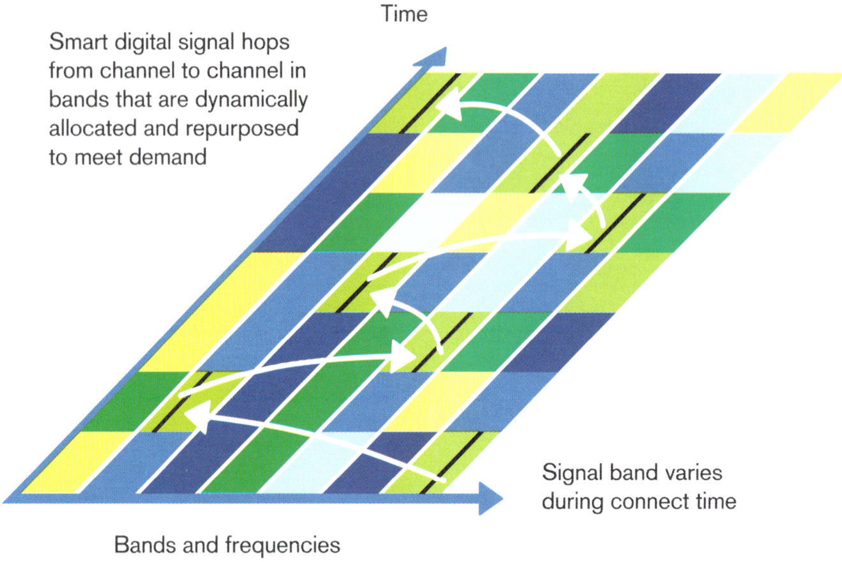

Figure 9: Dynamic channel hopping

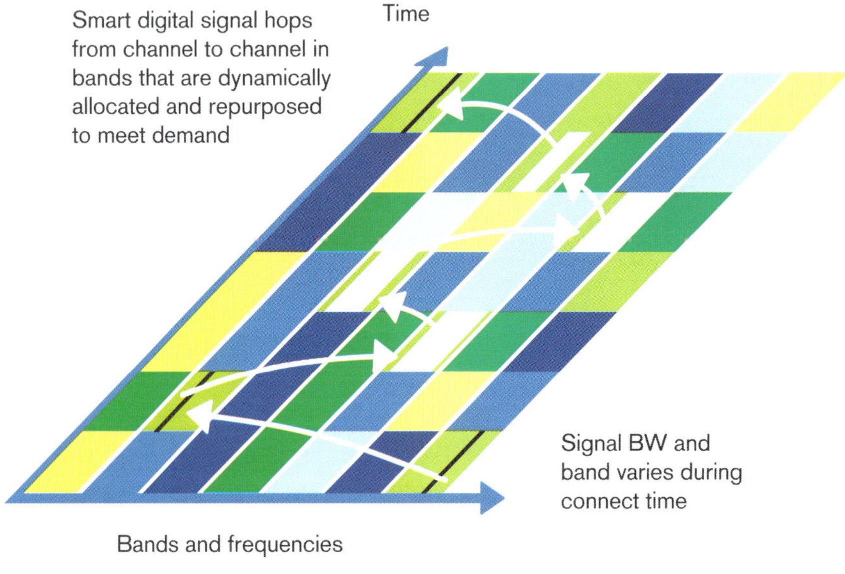

Figure 10: Dynamic band hopping and capacity allocation

ophy applied to WLAN design across this region of currently unused spectrum? I think it's inevitable.

Regulating the impossible?

When wireless systems were few and far between, numbering a few hundred, and then a few thousand; when transmitters and receivers, towers, masts and antennas were big and highly visible; and when just about every element of a wireless system was expensive and beyond the pocket of Joe Public – then, yes, it was feasible to licence, control and regulate everything. But that was then.

Now, when every device we buy, own and use has an integral wireless device; when they

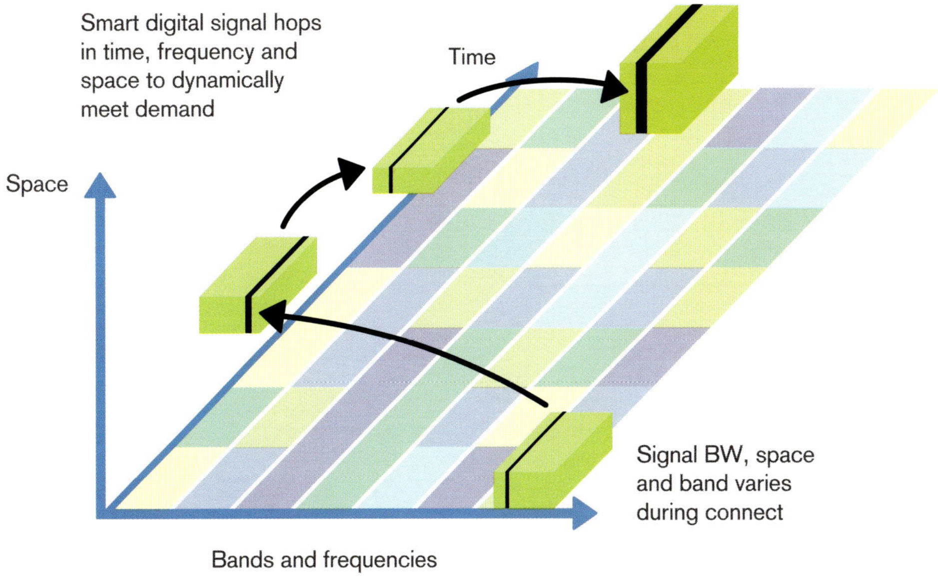

Figure 11: Dynamic hopping in time, frequency, space and capacity allocation

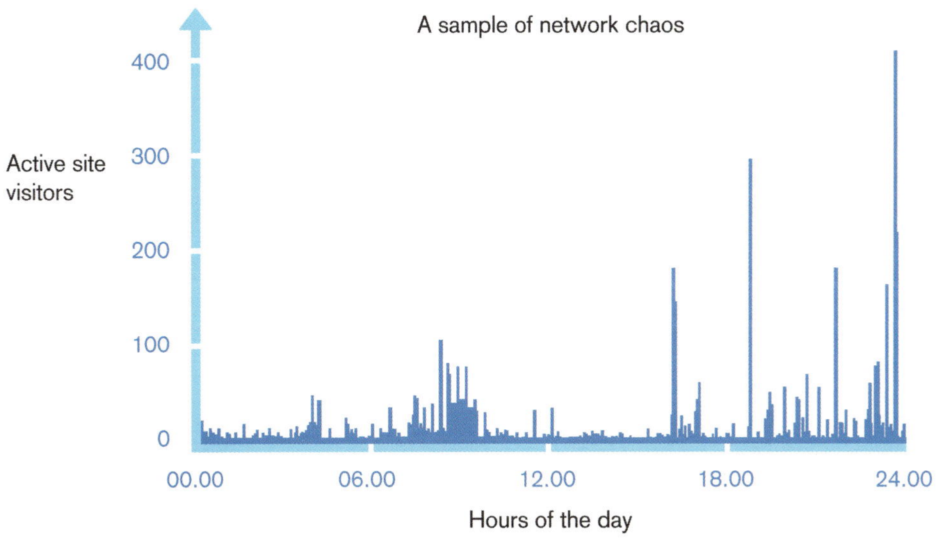

Figure 12: Network chaos – uncontrolled demand

number in their millions; when they flood in from abroad and cost virtually nothing; when they're so ubiquitous that Joe Public buys them even without knowing – then there is little or no chance that any form of regulation can be applied. It would be like trying to squeeze toothpaste back into the tube.

So what to do? Purchasing and operational freedom seems to be the *de facto* choice. Just let industry and the public get on with it. Open up the wireless spectrum and see what happens! Could it possibly work? I believe so. Just look at the licence-free bands supporting WiFi and so on. We might have predicted total chaos –

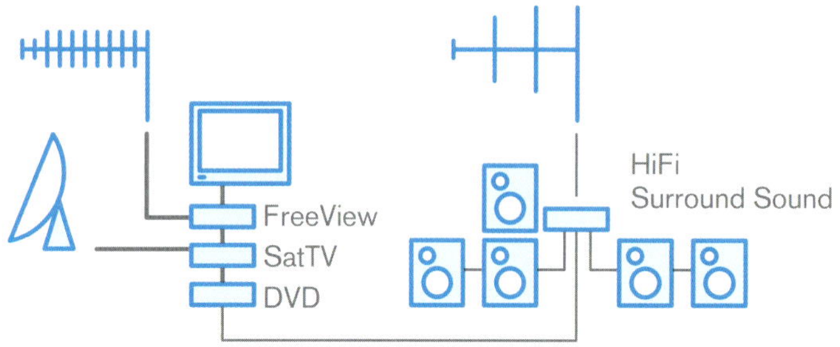

Entertainment and IT disconnected, but configured using hard wired solutions. Nothing inter-works or talks together together with ease, and there are wires everywhere!

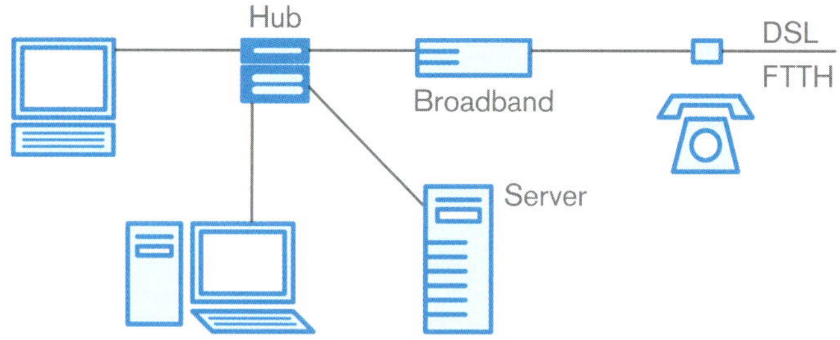

Figure 13: Disconnected IT in the home and office

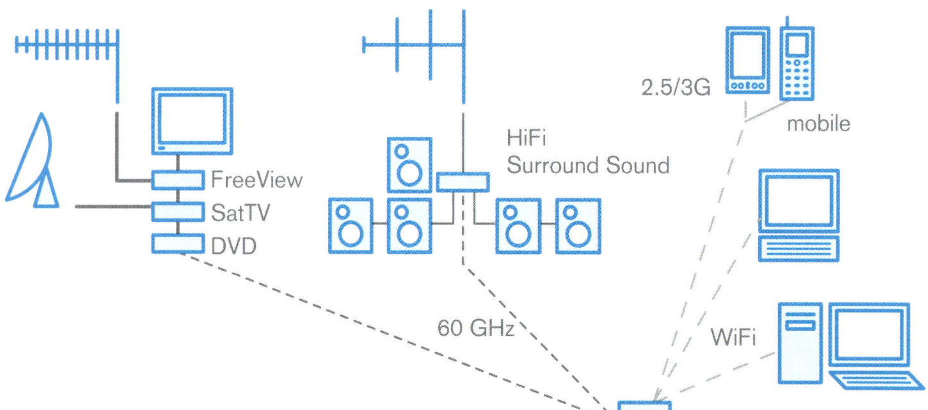

Entertainment and IT connected and configured using 60 GHz wireless solution. Everything inter-works and talks together with ease, and wires have mostly gone!

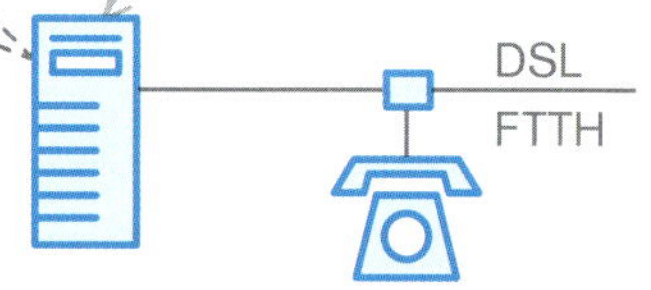

Broadband server + back up store
+ CD and DVD + movie store
+ Integral 60 GHz + WiFi
+ Bluetooth + 2.5 + 3.0G

Figure 14: IT connected at 60 GHz in the home and office

actually, many did – but it works simply because of the operating modes and embedded smarts. So would total freedom in spectrum and use be a benefit to society? Without a doubt. More freedom of communication always enlivens and regenerates an economy, and there are very strong links between GDP and communication infrastructure, connection availability and cost.

What needs to be done?

Total freedom in anything is rare, and the wireless spectrum will be no exception.

The first essential is to establish some broad-brush rules and guidelines to ensure that all participants in this new wireless space can co-exist and stand a reasonable success of operating without problems. If the future is truly to plug and play, and be far less problematic than today, it is essential that the ground rules are thoroughly hammered out early with all the participants. Agreeing modulation, signalling, power and antenna modes for cells might well follow the trend of licence-free bands currently used for WiFi, cordless phones and so on, in a broad range (see table 1).

While the above is simply a straw-man and may raise concerns regarding which precise mode of ultra-wideband (UWB) and spread spectrum (SS), I suspect that we can operate in 'don't-care' mode. All the technical studies and trials to date seem to indicate a high level of tolerance between co-habiting systems, and the emphasis really should be on design, manufacture and operational simplicity. Also, because of the geographical separation provided by the isolating feature at these frequencies, a 'don't care' mode ought to be (naturally) assured. BUT: *an industry and government feasibility study into the full depth of this concept would be well advised.*

What would happen?

All the practical evidence suggests that such a scheme will work – just as it does, for example, in the unlicensed WiFi bands worldwide. And this is at 2.4 GHz and similar UHF frequencies. If there are problems, companies and people simply get together and sort them out. After all,

Range/Cell Diameter	Frequencies	Modulation
1cm = Chip-to-chip and device-to-device	200 - 300 GHz	UWB
1m = PAN, box-to-box, and device-to-device	200 - 300 GHz	UWB
10m = PAN, box-to-box, LAN pico-cell	60 and 120 GHz	SS
100m = LAN Node/micro-cell	60 and 120 GHz	SS
1km = LAN/mobile node + point-to-point	40 and 90 GHz	SS
10km = Super LAN/ mobile node + point-to-point	40 and 90 GHz	SS

UWB = (ultra wideband) A direct sequence modulated bit stream generating huge spread factor

SS = (spread spectrum) A sequence spread signal about a centre carrier and not a simple frequency or channel hopping system

Table 1: Modulation, signalling, power and antenna modes for cells

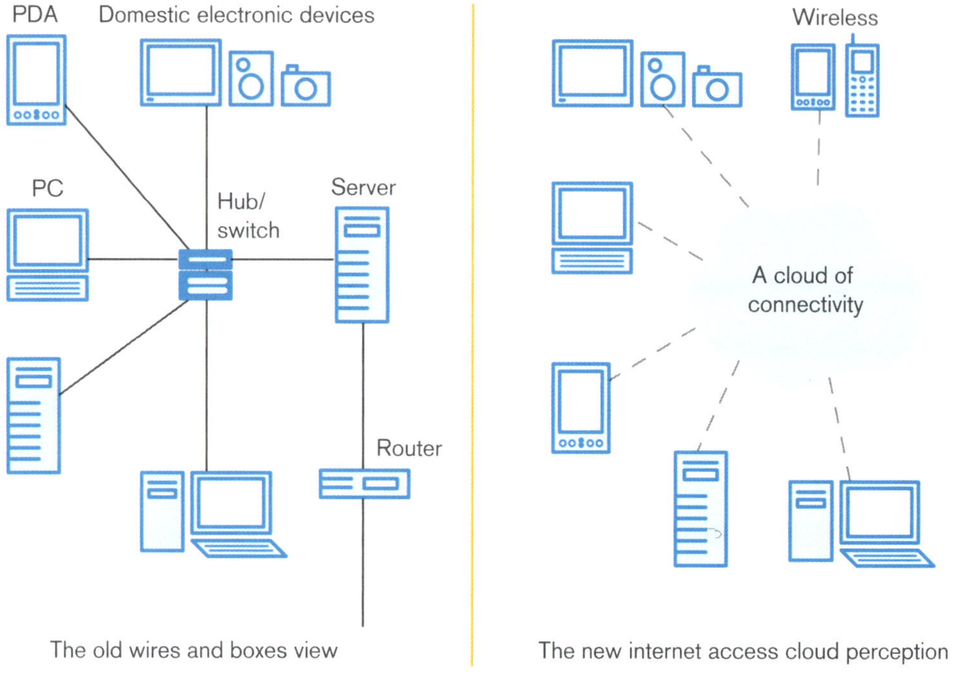

Figure 15: Our future network perception – wires or clouds?

Radio, TV, Hi-Fi, DVD, Satcom boxes and car keys all co-exist and work, so why not a free-for-all above 30 GHz? It seems that all the physics and laws of nature which govern wireless communication are even more in our favour, compared with the depths of 2.4 GHz. So my engineering guess is that it would indeed work.

Below 30 GHz

In today's world, 30 GHz seems like a small amount of bandwidth, but it is all we have for long and very long distance wireless communication. It is occupied, extensively used and yet rarely exceeds 17 per cent or so density. So what should we do? The frequencies below 30 MHz are the most problematic, since they support a great deal of legacy analogue global radio communication and broadcasting. It is likely that, for the next 20 years, any change here will be very slow, with international agreements very hard to reach. And we might expect digital to creep in here!

More interesting are the VHF, UHF and SHF bands occupied by users such as TV and radio broadcasters, emergency services, utilities, taxis, mobile phones, telecom carriers, companies, aeronautical and the military. These will be going digital far faster, but because of their strategic and commercial importance they will also require some sensitive handling. But there is no reason why digital prioritisation cannot be built-in to differentiate users and give urgency states. In short, spread spectrum, and all the other techniques detailed in this chapter, are applicable in this space.

It should be possible to give the 30 MHz to 30 GHz spectrum a higher degree of operational freedom than ever before. The real challenges are not technological at all, but established mindsets, commercial gain and political power including taxation. After all, it proved impossible to get 3G operators to share bands and base station sites and cut their individual install and operating costs by more than 80 per cent – such is the analogue mentality that prevails. Perhaps 4G will see this change? The reality is that change takes time.

The next step

The next obvious thing to do is for regulators and governments to try a hands-off approach for real. Just let go, and give individuals and companies the freedom of innovation in a space where little or no harm can be done. Open up some of the virgin 270 GHz spectrum to real product trials, get it out to customers and see what happens. Most importantly, think big. Don't allocate 0.01, 0.1, or 1 GHz – get things moving with a 10 GHz slot. Give spread spectrum and UWB some real space, and people some real freedom. I suspect that, like Pirate Radio and CB, it is going to happen anyway. It's just a question of time.

Mobile phones	>	Fixed line	=	12 years
Digital fax	>	Analogue fax	=	8 years
Colour TV	>	Black and white TV	=	11 years
CD	>	Audio tape	=	9 years
DVD	>	VHS tape	=	7 years
Digital nets	>	Analogue nets	=	15 years
Optical fibre	>	Copper and microwave	=	20 years

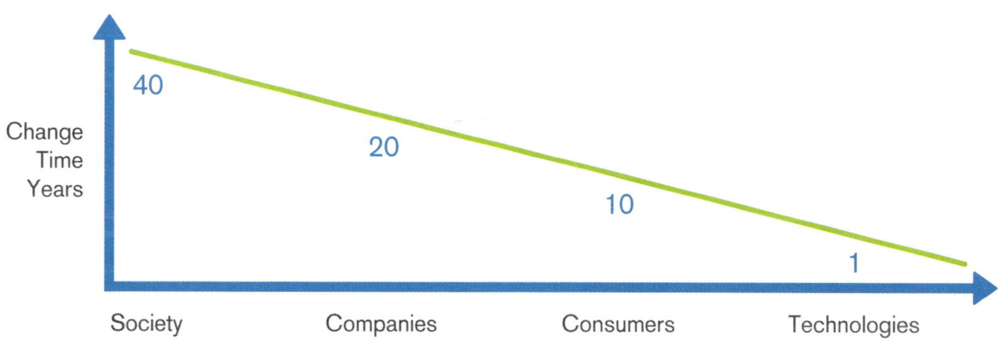

Figure 16: Change takes time...

New spectrum-using technologies and the future of spectrum management: a European policy perspective.[1]

Martin Cave

Professor and Director of the Centre for Management Under Regulation, Warwick Business School

This chapter is concerned with the two-way relationship between new spectrum-using technologies, such as ultra-wideband or software-defined radio, and the future of spectrum management, especially in Europe.

After a brief discussion in Section 1 of how spectrum fits into the value chain and of the benefits of promoting spectrum access, Section 2 sets out the case for a market-based spectrum management regime building on the argument that spectrum should be treated like any other input. It also sets out the need for smooth implementation of flexibility in the currently prevalent management regime for spectrum, which is based on discrete (unshared) assignments of spectrum to particular licensees. I call this 'first generation flexibility'.

Section 3 discusses how licence-exempt spectrum or commons can be fitted into this market-based framework. Section 4 discusses new spectrum-using technologies, many of which can operate on disaggregated components of frequencies, and Section 5 considers how spectrum management can be adapted to achieve flexibility in spectrum use in this richer context. I call this 'second generation flexibility'. Section 6 suggests spectrum reforms for the EU for the period following 2010, implied by the previous analysis, for which new legislation is now being considered.

1. Spectrum in the value chain

The goal of spectrum regulation and economic regulation should be the same: to pursue the long-term interests of end-users. Most observers agree that competition is the most effective instrument, although the nature of competition in the value chain can be very varied. Figure 1 shows a typical vertical structure of production of wireless broadcasting and communications (to the extent that a distinction between them can be maintained).

Different degrees of vertical integration or separation can allow a number of sharing opportunities, for example:

- the same physical assets and transmission capabilities can allow a range of programming to be sold or shared; equally, content can reach end-users via many platforms;
- access to spectrum can be achieved in numerous ways via commons, direct licensing, underlays or overlays (see below),

1. This is a shorter version of a paper prepared for Thinktel, an international think tank for telecommunications.

Value Chain		Examples
Content		Programmes, file-sharing
Physical assets		Towers and masts
Spectrum	Wholesale	Licensees, intermediaries, commons
	Retail	
Transmission		MVNOs, broadcasting transmission
Reselling		Airtime
Retail (end users)		Mobile telephony, broadcasting services

Figure 1: Spectrum in the wireless service value chain

via intermediaries such as band managers or operators of real time access regimes;

- physical assets, spectrum and other resources can be used to provide entry or access points nearer the end-user – for example via MVNOs or resellers; and
- end-users themselves – 2.4bn mobile subscribers – are skilfully brigaded by their operators into a pattern of shared spectrum use.

But should regulators have a preference among the points of entry? This depends to a considerable degree on the availability of competing platforms (for example, competing end-to-end mobile operators, cable, satellite, ADSL and so forth).

If this end-to-end competition is lacking (as it is for broadband in un-cabled areas), regulators, including spectrum regulators, should be interested in inserting competition in infrastructure as deeply into the value chain as possible, in order to sharpen commercial rivalries and promote service differentiation. Given the relatively low level of fixed sunk investment required of wireless access networks, replication of assets, including spectrum access, is fairly easy to achieve and spectrum policy should encourage it as far as possible.

2. Spectrum markets based on current technologies

Rather than set out the pros and cons of alternative spectrum management regimes in this section, I prefer instead to set out the case for a major extension of market forces. (For extensive reviews of markets vs. administrative methods, see Cave 2002, FCC 2002, Cave and

Webb 2003a, Ofcom 2004 and European Commission 2005).

The context of the discussion in this section is a management regime in which licences are assigned for the exclusive use of one firm or organisation. Under a market system, this is subject to change of ownership and/or change of use. The goal is thus 'first generation flexibility', based on an unrivalled access by each licensee to its full frequency.

The key underlying insight is that spectrum, like other natural resources such as capital and labour, is an input. The starting proposition in a market economy is that these inputs are best allocated to produce particular outputs by a market process, which co-ordinates the interests and information of millions of agents, in a way best calculated to encourage efficiency and innovation.

This approach is quite consistent with decisions about final demand being taken by the private sector (such as households buying goods, or firms buying investment goods); or by governments (such as a local government organising emergency services, or a national government pursuing the defence of the realm).

Nobody finds this proposition particularly challenging in relation to other inputs, such as skilled engineers, land, financial capital, capital goods and so forth. But it's necessary to check whether spectrum has some relevant significant difference which would suggest it should be treated differently. The main candidates for this are:

a) **'Spectrum is a resource which should be available to all.'** This proposition is fine in respect of spectrum used for certain applications where low power and short range allow users to co-exist peacefully; these needs are appropriately met by limited 'commons'. However, high-powered uses do demand exclusivity of some kind – which entails a mechanism to achieve it.

b) **'We need command-and-control (or administrative) methods to minimise police interference.'** Liberalising countries have shown that this does not pose a greater problem than other inputs (such as land). We can draw on experience from New Zealand, Australia, the UK, the USA (in relation to certain bands) and Canada, as well as Guatemala and one or two other less-developed countries.

c) **'Spectrum licence markets will lead to monopolisation, speculation and hoarding.'** In fact, the chief source of this problem is restrictions on *spectrum use* which have the effect of narrowing markets and creating market power. In any case, competition law should be able to deal with such problems. As with other markets, intermediaries ('speculators') and anticipatory purchasers ('hoarders') can smooth demand and supply, and enhance long-run market efficiency.

d) **'Administrative methods are needed to achieve harmonisation.'** On the basis of the success of GSM, there's an argument that a common allocation of spectrum across the EU (harmonisation) will encourage the adoption of a standardised technology and thus reduce costs, benefit consumers and achieve the aims of European industrial policy. As with command-and-control spectrum policy in other contexts, this argument assumes that the regulators are capable of choosing the correct frequency to adopt in common. This is clearly a gamble, with the risk increasing because of the uncertainties over which technology, and associated frequency, will triumph. For example, in the case of mobile broadcasting an array of technology and spectrum choices is still under examination, and the risk of regulatory failure is very considerable. In such circumstances, it is better to delegate the decision to a competitive process involving equipment manufacturers and operators, than to rush to judgement. However, a better bet is an alternative form of harmonisation: to harmonise spectrum management procedures across member states in ways that will first be likely to encourage competition, and then to facilitate convergence around the identified winner. As noted below, the European Commission now appears to favour this kind of meta-harmonisation, with occasional signs of support for old-style harmonisation.

While none of the above provides a good basis for rejecting markets as the basic method (which may be subject to minor exceptions) of allocating and assigning spectrum, restrictions created by treaties made under the International Telecommunications Union are a different story.

These establish primary and secondary uses for specific bands, but the uses identified tend to be broad (as well as increasingly obsolete) so that national regulators have considerable freedom of action: for example, the spectrum used for analogue television transmission has 'broadcasting' as its primary use but a Declaration signed by the vast majority of European countries at the Regional Radio Conference in June 2006 (RRC06) has the effect of authorising its use, following analogue switch-off, for other purposes.

The only provision is that they do not impose more interference than a broadcasting application, and require no more protection than broadcasting. Such strategies are less than ideal, but they show the scope for combining a substantial degree of liberalisation and flexibility while adhering to treaty obligations. This does not, however, make it unnecessary to seek changes in ITU procedures and spectrum management instruments.

To summarise, first generation flexibility in European spectrum management should be directed to an end-state in which (with small exceptions) market mechanisms including *change of ownership* and *change of use* of licences should operate to the fullest possible extent in respect of:

- geographical area (initially the EU, and later globally);
- frequency bands (as many as possible to ensure that spectrum use is optimised by market exchange over as full a range of usable frequencies as possible);
- form of licences (including freehold, leasehold etc);
- identity of licensees (which could be a direct public, or private sector, firm or organisation, a band manager or an intermediary of another type); and
- the nature of the service produced using spectrum (which could be a marketed or non-marketed good, a private or collective good, and so on).

3. Marketed licences vs. commons

Traditionally, a small number of frequencies sat alongside spectrum licences assigned by administrative methods to provide unlicensed access to users of particular apparatus. These include frequencies used for television remote controls and Bluetooth short-range communications, as well as spectrum for short-range broadband access using standards such as

802.11 or WiFi. In the UK, unlicensed spectrum of this kind amounts to 4-6 per cent of the total.

While several commentators have proposed a major expansion of the commons (Noam 1998, Benkler 2002, Werbach 2004), others (Hazlett 1998, Faulhaber and Farber 2003, Cave and Webb 2003b, Baumol and Robyn 2006) regard it as best suited to short-range applications where rivalries between operators for spectrum are more limited. (This does not exclude the development of technologies such as mesh networks, especially to exploit short-range transmission using sublicensed spectrum.)

However, drawing the line over time between the universal licensed and unlicensed spectrum is highly problematic, and historically has been done using administrative fiat in two dimensions: in the basic decision to assign a frequency for unlicensed use, and in the choice of restrictions imposed on its use (Hazlett 2006).

Figure 2 illustrates the allocation decision between licensed spectrum, some of which may be used for 'private commons'. This is an arrangement whereby users authorised by a licensee (for example, purchasers of the licensee's apparatus) have direct access to spectrum – and unlicensed spectrum.

In the past, spectrum regulators have made decisions on unlicensed spectrum on administrative grounds. But this is arbitrary and unsatisfactory. In a market environment, it would be better to introduce some form of market competition between the two modes of frequency management.

The difficulty is estimating, and making effective, the derived demand for spectrum from unlicensed spectrum users, in the same way that, say, mobile operators can express their derived demand. The root cause of the problem is establishing the willingness to pay of a large number of non-rival spectrum users – an illustration of the classic problem of establishing demand for a public good.

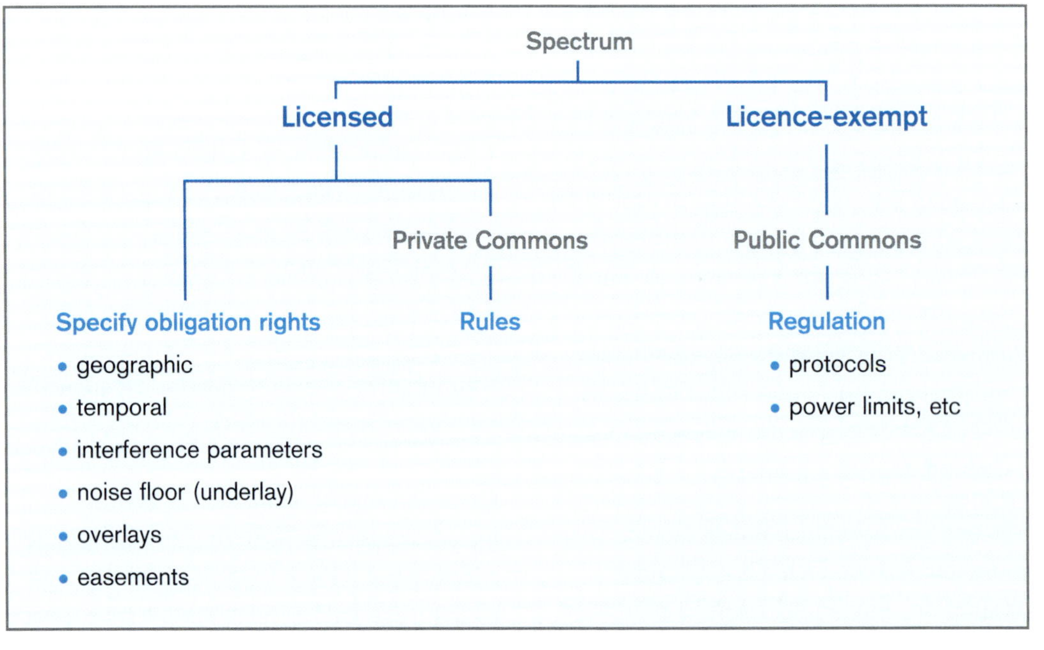

Figure 2: Licensed vs. unlicensed spectrum

In principle, individual levels of willingness to pay should be aggregated (vertically) to derive a social valuation. However, this brings with it the well-known problem that respondents have an incentive to falsify their estimates, in order to increase the supply of a resource for which they will not have to pay, or at least not the full cost.

A number of formal mechanisms have been developed to deal with the problems (for example, the Clarke-Groves mechanism reviewed in Campbell (1995) pp. 283-294). They have the feature that any respondent whose reported evaluation tips the decision to buy into the positive has themselves to pay a surcharge equal to the difference between all other participants' preference for the alternative option. This removes any incentive to report distorted valuations.

These mechanisms do, however, encounter problems associated with the need to make positive payments to some participants as well as to tax others. The budget may not balance, and other techniques such as conjoint analysis may be required to establish the aggregate valuation of unlicensed spectrum from willingness to pay for the services it can offer.

Since a spectrum commons is typically regulated existing services, a range of options may have to be established, in circumstances where consumer understanding of them may not be strong. This will obviously represent a significant challenge.

But let's imagine such an unbiased estimate were available. In this case, a proxy bidder seeking spectrum to be licence-exempt could compete against bidders for licensed spectrum, including those proposing private commons. This would require public financing, but for limited frequencies this might not be too problematic.

To summarise the argument so far: a case based on existing spectrum-using technologies has been made for the use of market methods, and an (imperfect) way, based on the market choosing between licensed and unlicensed use, has been proposed. It is pertinent to ask, however, if foreseeable technological developments support or undermine the case for the market. This is the task of the next two sections.

4. New technologies round the corner

New technologies show great promise in how to make more effective use of spectrum. Developments in spread spectrum technologies, software-defined radios and smart antennas have raised particular interest.

These technologies have the potential to increase spectrum efficiency in many ways, including a higher level of frequency re-use and sharing by means of both underlay and overlay techniques. Moreover, these technologies might be implemented together in mesh network architectures such as ad-hoc networks, thus offering a wide range of new opportunities for spectrum-based services.

Indeed, new technologies may have a relevant impact at different levels of the value chain of spectrum-based services, by enabling a more efficient use of spectrum either directly or indirectly.

In fact, on one hand, spread spectrum technologies appear to be able to bring about great benefits in increasing spectrum efficiency directly, by using frequencies more intensely. On the other hand, smart antennas promise better performance at both transmitter and receiver levels and generate opportunities for enhanced spectrum efficiency by building on the techniques to receive and send signals over frequencies without suffering harmful interference. Software-defined radio (SDR) technologies are expected to significantly change the way spectrum is used today, particularly as high-level cognitive radios will enable more frequency reuse and more flexible uses of the same hardware and infrastructure, delivering spectrum efficiency in a variety of ways.

Increasing spectrum efficiency directly: developments in spread spectrum technologies

A key feature of spread spectrum technologies is that they allow coexistence of multiple applications in the same spectrum, although the capacity of any one application will be reduced as others are introduced. In fact, these technologies spread power over a wide range of frequencies, so their signals only interfere to a limited extent with narrowband radios. Also, some technologies support spectrum overlay techniques, which offer secondary users the opportunity to access a primary channel when this is not in use on the basis of an 'intrude and avoid' principle.[2] Others, however, support spectrum underlay techniques that enable spectrum sharing between systems of either high or low power density levels.

Spread spectrum technologies include a multitude of legacy and new technologies (such as direct sequence spread spectrum, frequency-hopping spread spectrum and ultra-wideband). Their functionality is based on implementation of a code to send information over a much wider band than the actual bandwidth of the transmitted information.

The code, which might be known by other parties, is only used by any one party at one time. Therefore, the information can be coded and decoded by the intended parties at each end of the radio connection, giving, to a considerable extent, security, interference avoidance and high-speed connections. In fact, spread spectrum technologies have been used in military communications for several decades and they have more recently been moved into commercial applications.

Increasing spectrum efficiency by implementing new technologies along other parts of the value chain of spectrum-based services

The new smart (or advanced) antenna technologies have the potential to enable more efficient use of spectrum both directly and indirectly, according to the type of antenna and deployment in the radio-based system. In fact, smart antennas seek to increase coverage, capacity and reliability of a radio network by improving the ability to send and receive signals (i.e. by limiting interference). Actually, antennas are 'dumb', but can be made 'smart' if a digital signal processor enables an analysis of the spectrum environment, such that it is possible to either precisely determine and combine the sources of an incoming transmission, or to direct energy in a narrow beam towards the user.

Therefore, intelligence and greater (indirect) spectrum efficiency can be obtained in two fundamental ways along the value chain of spectrum-based services: on the receiver side, by improving the ability to listen (even with a high level of interference, sometimes known as the 'cocktail party effect'); and on the transmitter side, by transmitting a response only in the desired direction, whereas traditional omni-directional antennas transmit in all directions. Hence, smart antennas can save on spectrum requirements and also help reduce harmful interference, including multipath problems.

2. For instance, cognitive radios make use of this principle.

However, there are two main practical problems with advanced antennas: they are expensive (particularly when implemented on users' devices) and they don't work properly if there is not a line of sight (in which case, transmissions are easily reflected or blocked).

Making spectrum-based systems more flexible with software-defined radios

Software-defined radios (SDRs) are wireless communication devices that use software to implement almost completely their functionality on the physical transmission level. Microcomputers are installed in the transmitter and the receiver to handle a wide variety of waveforms[3] and their associated settings. Software is used to control high-speed signal processors, thereby performing functions that in traditional radios were carried out in the hardware.

SDR technologies can have a great impact on electronic communications, although their potential to improve efficiency along the value chain is mainly indirect with regard to spectrum. In fact, by implementing as many functionalities as possible into the software, SDR systems are able to adapt to a wide range of frequencies, bandwidths and transmission standards, without any hardware changes. It follows, then, that SDR can provide more flexibility. Also, SDR is an important enabler for advanced forms of dynamic spectrum access; i.e. any form of flexible spectrum use obtained by dynamically changing the set of transmission parameters.[4]

However, a key feature of SDR technologies is their novel way to carry out the radio functionalities. Indeed, SDRs represent a major technological advancement away from traditional radio devices for a number of reasons: with the exception of the antenna, physical layer functions are implemented in software rather than in hardware; there is no built-in waveform predetermined by the manufacturer and radios can be re-programmed on the fly, thus providing a high degree of flexibility by adjusting frequencies, bandwidth and directionality.

In this way, a radio becomes a generic device with its functionality defined by running software in a dynamic manner, allowing multiple uses of the same hardware and infrastructure.[5] These features of SDR technologies have generated some interest in the United States, as SDRs allow carriers to run multiple standards on the same network, and promise to reduce operational costs.

In the US, the FCC has recently approved Vanu's system, which commercially deploys SDR technologies in licensed spectrum. Although its commercial jump-start was with the rural market, the system is expected to gain popularity with large carriers, with WiMax potentially becoming part of the system.

Taking SDR technologies a step further: cognitive radios

Cognitive radios are a relevant and recent development in dynamic spectrum access technologies, based on SDR capabilities. These are smart devices that can perform a multitude of tasks autonomously, enabling more flexible and efficient use of radio frequencies: indeed, a high-level cognitive radio is able to "observe, orient, plan, learn,

3. Waveform is used to refer to the whole software application determining the behaviour of the system.

4. Early forms of DSA mechanisms can be found in modern car radio sets, DECT cordless communication systems and WLAN devices operating in the 5 GHz frequency band. They are all examples of automatic frequency selection mechanisms.

5. E.g. the same SDR could be used as a mobile or cordless phone, a pager or provide WLAN connectivity. Also, a network operator could offer his customers a set of waveforms including UMTS and IEEE 802.11.

decide, act". (Mitola 1999) Although their cognitive behaviour may take different forms, the focus here is on spectrum usage only.

Cognitive radios can acquire information on their spectrum environment and employ this information to decide on their transmission behaviour. Also, they are able to learn from their own behaviour and experience. Indeed, key features include their ability to learn user preferences, prevailing spectrum rules and operator charges. Particularly, cognitive radios can sense vacant frequencies and transmit over unused frequencies until another radio tries to use the same portion of the spectrum.

Therefore, one important advantage is their ability to transmit over temporarily unused frequencies, so considerably contributing to increase available bandwidth. Moreover, by exploiting their ability to acquire information on their physical environment in relevant dimensions – including time, location and frequency – cognitive radio technologies appear to have great potential since parts of the spectrum are often unused in at least one dimension.

Although a prototype military radio system, scheduled for August 2007, is being developed by the US Defense Advanced Research Projects Agency (DARPA), cognitive radios are still in development and it will probably take several years before they come onto the market. Meantime, a number of technological challenges still remain: they include wideband sensing, opportunity identification, interference prevention and dynamic co-ordination.

These issues come together in the so-called 'hidden terminal' problem: if two devices are out of range of each other (for example, if there's a building between the two), they might not spot any activity on a particular part of the spectrum by making independent measurements, thereby starting transmissions. However, the receiver (for example, a base station located within range of both transmitters) will suffer from interference as the spectrum was actually already in use.

Implementing ad-hoc mesh networks to introduce more flexibility in mobile spectrum-based applications

In mesh networks, radio nodes receiving information can also pass information along, so providing retransmission capabilities. Mesh networks can be divided in two groups: structured and ad-hoc. In the former, radio nodes are fixed and a relevant planning activity takes place in advance. In ad-hoc networks, mobile radio nodes are equipped with relay functionality, which provides a potential for great flexibility and ability to reroute transmissions along different paths by spotting another node within radio range.

While fixed wireless access systems are often included in the concept of structured mesh networks, ad-hoc mesh networks are in their early stages of development. Implementation of SDR technologies might facilitate deployment by allowing radio nodes in the network to become instantly flexible, according to the physical environment. High-level cognitive radios are candidates to be deployed as radio nodes in ad-hoc networks by creating their own transmission path in every specific situation, involving several hops of other radios in the network (perhaps using spread spectrum technologies such as Ultra-wideband (UWB) for high-speed, short-range communications).

What requirements do the new technologies impose on spectrum management?

The traditional regime for spectrum management is 'one frequency, one user' (who is bound to provide a particular service using individually licensed apparatus). Developments in the use of market methods, as recommended earlier in Section 1, have in some jurisdictions already achieved first generation flexibility, allowing change of ownership and use across discrete spectrum-using services and technologies in any given frequency. The next issue we consider is how to achieve the benefits of flexibility in the context of the more sophisticated technologies described above, which are capable of accommodating spectrum sharing, often dynamically – in other words, in ways which vary over time.

I am not suggesting that all these technologies will necessarily be applied on a large scale. Indeed, Ofcom's own analysis of the costs of SDR and cognitive radio casts doubt on their cost effectiveness, compared with the simple historical alternative of boosting efficiency by decreasing cell sizes and reusing spectrum more efficiently on a discrete-user basis. It is important, however, to give thought to their potential impact, if the spectrum management regime is to be made future-proof.

A simple dimension of spectrum use is its efficient scale. A technology such as cognitive radio relies on agility to pile more use into given frequencies. Much of the benefit flows from pooling intermittent demands to achieve a greater utilisation rate. Other things equal, this process occurs more efficiently on a larger scale, subject to the increasing cost and technical complexity of ranging over more spectrum. The least we can expect, therefore, is aggregation of demand under cognitive radio, which will capture the benefits of scale, possibly involving intermediation – for example, a band manager which sells access to a range of possible frequencies.

Mesh networks have the prospects of diminishing the power required of transmissions, by virtue of their use of multiple short hops at low power levels rather than one long hop at a higher power. To that extent, they enhance the scope for commons, provided that increases in equipment costs do not outweigh savings in spectrum use.

The two remaining, and more fundamental, issues concern underlays and overlays. Underlays are exemplified by UWB, which operates under the noise floor of other services.

In principle, UWB could be used in at least three ways. First, one or more separate geographical licences could be carved out beneath any existing noise floor and assigned on an exclusive basis. Or the same space could be carved out, and made licence-exempt. Or an obligation could be imposed on any prospective user of UWB to negotiate an arrangement with all licensees under whose noise floor it proposed to operate.

The last option would almost certainly fail because of the transactions cost incurred in negotiating with countless licensees. Choice between the licensed and licence-exempt modes of permitting UWB should hinge on a calculation of the risks of interference, either between UWB users or from UWB users to other licensees. It is necessary here to conduct a risk management exercise to establish the consequences in the future if the multiplication of UWB users ultimately led to either of the interference problems noted above. The conclusions reached so far by regulatory agencies (notably the FCC and Ofcom) favour creating a traditional commons with very strong power limitations.

Finally, there is the question of overlays, or access by users to spectrum licensed to others.

In principle, this could be made generally available. Indeed, the European Commission's recent proposals on spectrum reform[6] seem to contemplate just this kind of general right of access, when it says that:

"A new system for spectrum management is needed that permits different models of spectrum licensing (the traditional administrative, unlicensed and new marked-based approaches) to coexist so as to promote economic and technical efficiency in the use of this valuable resource. Based on common EU rules, greater flexibility in spectrum management could be introduced by strengthening the use of general authorisations whenever possible." (p 7)

This clearly raises fundamental issues of spectrum management and the design of property rights. Within one given frequency, transaction costs would not preclude bargaining between an original licensee and potential secondary users. There is also a concern that unlicensed entrants, lacking security of access, would not be in a position to make collateral investments, or to offer adequate assurances to their customers of continuity of supply, and quality of service. Alternatively, they might establish *de facto* some squatters' rights of a contestable nature, which would prevent the licensee from being able to exploit its asset to the full.

These questions have been widely debated in recent years. Notably, Faulhaber and Farber (2003) have been in favour of unlicensed non-interfering overlays or easements of this kind, arguing – *inter alia* – that absent imposition of such a regime, licensees will not be prepared to supply access to secondary users by entering into appropriate contracts with what might be their rivals in downstream markets.

Baumol and Robyn (2006) conclude the opposite, basing their argument in part on the analogy of a successful market for licensing intellectual property. They assert that:

"The argument favouring market accommodation of such emissions rests on the now familiar position that spectrum rights holders have an incentive to act in ways that result in (approximately) optimal use of spectrum space. There is reason to question whether a governmental rule without price and profit incentives will be able to match the performance of a market regime." (pp 61-2)

Absent any experience of the non-interfering easements regime, it is hard to discriminate on a *priori* grounds between the two approaches. The one economises on transactions costs, while the other eschews use of the price mechanism to ration access. As usual in economics, it is unlikely that there will be a corner solution, with the same regime optimal in all frequencies. Ideally, the choice of regime would be determined by a procedure akin to that described in Section 3 above, where some kind of quasi-market testing of a secondary commons would be undertaken. But the design of such a mechanism is very challenging.

The best way forward may be to undertake some limited testing of the non-interfering easement approach in likely-looking frequencies. This could provide a better empirical base for the European Commission's careful espousal of the method reported above.

Whatever application easements may have in the future, market methods will almost certainly provide the foundation for achieving flexibility in spectrum use of both the first and the second generation varieties. In the

6. Commission's Communication on the Review of EU Regulatory Framework for electronic communications networks and services (SEC(2006)816-7), on which more details can be found in Section 6.

next section, I turn to how this might be accomplished in the forthcoming revised regulatory regime for e-communications now being prepared in Europe to take effect from 2010.

What Europe needs to do[7]

There has been fairly widespread recognition that the current regime of spectrum management in most of the European Union is insufficiently flexible to achieve the Union's objectives of promoting industrial competitiveness and innovation. Despite this, the pace of reform so far continues to be slow. In the light of the previous arguments, urgent measures need to be taken to speed it up by extending spectrum reform in the leading countries to other Member States, thus generating the benefit of 'large area liberalisation'.

The European telecommunications market was fully liberalised in 2002 through the Full Competition Directive and, since then, the current regulatory position in respect of spectrum trading has been clear. By virtue of Article 9(3) of the Framework Directive, Member States are currently permitted, but not required, to introduce spectrum trading.

Article 9(4) of the Framework Directive sets out some procedural aspects of spectrum trading and makes it clear that the Spectrum Decision[8] is the principal means by which spectrum use is to be harmonised:

"Member States shall ensure that an undertaking's intention to transfer rights to use radio frequencies is notified to the national regulatory authority responsible for spectrum assignment and that any transfer takes place in accordance with procedures laid down by the national regulatory authority and is made public. National regulatory authorities shall ensure that competition is not distorted as a result of any such transaction. Where radio frequency use has been harmonised through the application of Decision No 676/2002/EC (Radio Spectrum Decision) or other Community measures, any such transfer shall not result in change of use of that radio frequency."

According to Article 1(1) of the Spectrum Decision, the aim of the Decision is to "establish a policy and legal framework in the Community in order to ensure the co-ordination of policy approaches and, where appropriate, harmonised conditions with regard to availability and efficient use of the radio spectrum..."

Any fresh decisions made under the Spectrum Decision must also be limited to 'technical implementing measures'. Although there is no binding interpretation of the Spectrum Decision that defines the permissible parameters of a technical implementing measure, it is clear from the past practice both of CEPT and the Commission that it can be used to adopt decisions on a band-by-band basis to harmonise the way in which a certain band is exploited – both practically and technologically – across Europe. Article 4 of the Spectrum Decision could be used to achieve freedom of use on stipulated frequency bands.

In September 2005, the Commission published a Communication on a market-based approach to spectrum management in the European Union (Com (2005) 400 final) which noted that a fragmented approach to spectrum reform would make it more difficult to achieve the Union's objectives. Accordingly, it proposed the co-ordinated removal of restrictions on spectrum

7. This section is taken from Cave and Crowther (2006).

8. Decision No. 676/2002/EC of 7 March 2002 on a regulatory framework for radio spectrum policy in the European Community.

use in all Member States in order to promote an open and competitive digital economy.

In practice, it was suggested that substantial amounts of spectrum, including roughly one-third of the spectrum below 3 GHz (the spectrum best suited for terrestrial communications) could possibly be made subject to tradable and flexible use by 2010. Clearly the Communication is a key document in which the Commission nailed its colours to the liberalisation mast. If the plan were realised it would represent a significant step towards the desired end-state set out above, even though a lot of non-communication related spectrum, which makes up much of the remaining two-thirds of spectrum below 3 GHz, would not be covered.

The Commission's Communication on the Review of the EU regulatory Framework for ECS (SEC(2006) 816-7), which proposes greater flexibility in spectrum management, could be introduced by strengthening the use of general authorisations whenever possible. Also, selected bands agreed at EU level via a committee procedure would become available for use under general authorisations, or subject to secondary trading across the EU. Common authorisation conditions for the use of the radio spectrum would also be enacted with this procedure in appropriate cases.

The proposals thus embody a strong attachment to the key concepts of technology neutrality and service neutrality (flexibility in use). They entail extensive use of the Spectrum Decision but little by way of thorough on-going reform of the Directives. Is this adequate for the task?

It probably isn't, so what's needed is a legislated change to the relevant Article of the so-called Framework Directive, the linchpin of the whole regulatory regime in Europe, which would mandate trading and flexibility of spectrum in the bulk of the key band, maintaining the option of exceptions for certain specified public sector uses (Cave 2005). A proposal along these lines is set out in Cave and Crowther (2006).

Actions by the Commission and the European Institutions are therefore required on two levels: use of the 2002 Spectrum Decision ad interim, and a more radical rewriting of the framework directive to allow significant harmonised spectrum trading and liberalisation in the EU. This would respond both to current and to future needs.

References

Baumol, W.J. & Robyn, D. (2006), *Toward an evolutionary regime for spectrum governance: licensing or unrestricted policy?* AEI-Brookings Joint Center for Regulatory Studies, Washington, D.C.

Benkler, Y. (2002), *Some economics of wireless communications,* Harvard journal of law & technology, vol. 16, no. 1, pp. 25-83

Campbell, D.E. (1995), *Incentives: motivation and the economics of information*, Cambridge University Press, Cambridge

Cave, M. (2002), *Review of radio spectrum management. An independent review for Department of Trade and Industry and HM Treasury*, London

Cave, M. (2005), *Independent audit of spectrum holdings*, London

Cave, M. & Webb, W. (2003a), *Designing property rights for the operation of spectrum markets,* Warwick Business School, papers in spectrum trading no. 1

Cave, M. & Webb, W. (2003b), *Spectrum licensing and spectrum commons – where to draw the line,* Warwick Business School, papers in spectrum trading no. 2

Cave, M. and Crowther, P. (2006), *Driving forward spectrum liberalization,* Vodafone policy paper series no. 5

European Commission (2002), *Directive 2002/21/EC of 7 March 2002 on a common regulatory framework for electronic communications networks and services,* European Commission, Bruxelles

European Commission (2002), *Decision No. 676/2002/EC of 7 March 2002 on a regulatory framework for radio spectrum policy in the European Community*, European Commission, Bruxelles

European Commission (2002), *Commission Directive 2002/77/EC of 16 September 2002 on competition in the markets for electronic communications networks and services,* European Commission, Bruxelles

European Commission (2005), *A market-based approach to spectrum management in the European Union*, European Commission, Bruxelles

European Commission (2006), *Communication on the Review of EU Regulatory Framework for electronic communications networks and services,* SEC(2006)816-7, European Commission, Bruxelles

Faulhaber, G. & Farber, D. (2003), *Spectrum management: property rights, markets, and the commons,* in Rethinking rights and regulations: institutional responses to new communications technologies, eds. F.L. Cranor & S.S. Wildman, MIT Press, Cambridge, MA

FCC – Federal Communications Commission (2002), *Report by the Spectrum Policy Task Force, ET docket no. 02-135*

Hazlett T.W. (1998), *Spectrum flash dance: Eli Noam's proposal for 'Open Access' to radio waves, Journal of law & economics,* vol. XLI, pp. 805-820

Hazlett (2006), *'Barbed wireless' & the vertical structure of property rights,* forthcoming.

Mitola J. (1999), *Cognitive radio for flexible mobile multimedia communications,* IEEE International Workshop on Mobile Multimedia Communications, pp. 3-10

Noam E. (1998), *Spectrum auctions: yesterday's heresy, today's orthodoxy, tomorrow's anachronism. Taking the next step to open spectrum access,* Journal of law & economics, vol. XLI, pp. 765-790

Ofcom (2004), *Spectrum framework review. A consultation on Ofcom's views as to how radio spectrum should be managed*

Werbach, K. (2004), *Supercommons: toward a unified theory of wireless communication,* Texas law review, vol. 82, pp. 863-973

Spectrum management and the achievement of policy goals – an independent regulator's perspective

Isolde Goggin

Chairperson, Commission for Communications Regulation, Ireland.[1]

Introduction – the importance and uniqueness of spectrum

The radio frequency spectrum – an invisible, intangible and ubiquitous resource – is used for numerous business and consumer applications. In telecommunications, radio frequencies are used for aviation, shipping, defence services, public safety, fixed links, and public and private mobile phone networks. Uses other than telecommunications include broadcasting, radar, radio astronomy, radio-location, space research and amateur activities. Wireless communications are increasingly used to provide broadband internet access, enabling online public services such as e-government, e-health and e-learning, as well as business and social uses.

Radio spectrum is unique and, in many cases, un-substitutable. It is unique in that it is an infinitely reusable resource with the ability to recover instantly from any effects of pollution or abuse. And although, for some applications, wired and wireless services are alternatives, there is no reasonable substitute for spectrum in mobile phones or radar tracking systems.

Although slow and incremental, there have been moves in the past two decades towards introducing market mechanisms into the hitherto command-and-control world of spectrum management. This has led to concerns that while an unfettered market-based approach might increase economic efficiency, it could also militate against achieving other policy goals such as plurality and diversity of the media, or the protection of life.

In practice, although increasing demand for spectrum means we are likely to see more allocation and assignment mechanisms which maximise efficiency, neither model (nor indeed the licence-exempt model, which I don't discuss here for reasons of space) will be a universal answer. Spectrum management is a highly technical, complex and case-dependent subject, and with the ever-present potential for interference it generally calls for change in a measured and staged way.

The sources of policy options and goals

Although the telecommunications arena has seen a shift in the last two decades from direct

1. The views expressed are the author's own and do not necessarily reflect those of the Commission for Communications Regulation. Many thanks to Jim Norton, Samuel Ritchie and Dermot Nolan for comments on the draft, and also to Jim Connolly, Jonathan Evans, Susan Fleming, Catherine O'Brien and Dave Thom for their work on ComReg's spectrum strategy, which informed much of this chapter.

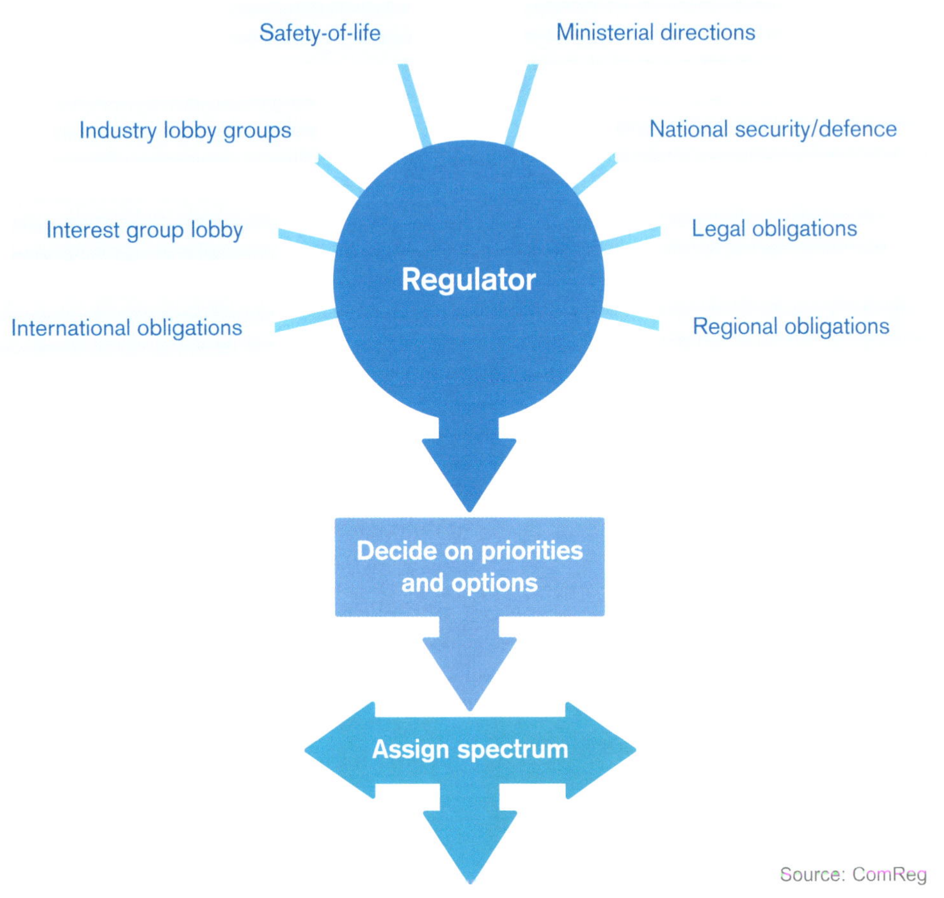

Figure 1: Source of policy goals

government involvement (both in operation and regulation), the spectrum management scene across Europe is more diverse. In some countries you'll find the independent regulators handling spectrum management as either a unique function or in combination with other responsibilities such as telecommunications, posts, media and other network industries. In others, responsibility for spectrum management remains, or is shared, with the relevant Ministry. It's generally true, though, that governments are more involved in spectrum than in the overall telecommunications area, particularly where safety-of-life and national security are concerned.

In implementing policy goals, regulators must balance a number of different, and potentially conflicting, inputs as shown in figure 1.

The International Telecommunications Union (ITU)[2] Radio Regulations, which have the status of an international treaty, contain obligations concerning how spectrum is used for aviation, maritime and satellite services (among others).

Regional obligations within Europe include implementing EU Directives and Decisions and, where appropriate, CEPT Decisions. NATO members have further obligations regarding the use of spectrum. There are other powerful voices to be heard as well: special interest

2. The ITU is a specialised agency of the United Nations.

groups such as the amateur radio fraternity and, in some countries, a very powerful community radio forum. Industry lobby groups, especially where these industries are multi-national and contribute significantly to wealth and job creation, are a powerful force.

On the other hand, it is also vital that the needs of non-commercial interests, such as safety-of-life services, should be met. Historically, national security and the defence forces have been strongly represented in national, regional and international spectrum policy.

How, then, do the needs of these disparate interest groups, and the necessity to meet international obligations, feed into the regulatory structure? Simply, the process varies by country. In Ireland, for instance, any spectrum used by the defence forces is exempt from licensing by the regulator. Also, as part of national IT policy, the Minister for Communications, Marine and Natural Resources has directed the regulator to facilitate consumer access to broadband services through both spectrum and telecommunications policy.

Once all these inputs are considered, it is the unenviable task of the regulator to prioritise them and decide how to assign spectrum in order to achieve policy goals.

Traditional spectrum management

Spectrum has traditionally been managed on a command-and-control basis, with clear priorities to avoid interference and to safeguard national spectrum rights. From the time that Marconi was granted his first British patent in 1896, regulation was not far behind: just seven years later, the Berlin conference of national radio administrations laid down the basis for the International Radiotelegraph Convention (IRC).

This was followed by action to overcome interference problems caused by unregulated radio transmissions, both internally and across borders. The 1927 International Radiotelegraph Conference and, later, the International Telecommunications Union (ITU) allocated separate frequency bands to the various radio services of the time, and put in place arrangements for national licensing.

This need for international co-ordination arose from the nature of spectrum use itself, and the fact that what happens in one frequency range in one location affects not only adjacent locations, but also adjacent frequency bands. In radiocommunications, a signal carrying the desired information (be it voice, data or video) is superimposed on a higher, 'carrier' signal; this allows each service to occupy a different part of the spectrum. The way in which the information signal is superimposed on the carrier is called the 'modulation scheme'.

Older-style modulation schemes superimpose the signal in analogue form on a characteristic of the carrier wave – that is, they cause it to vary continuously. Examples of this are amplitude modulation (AM) and frequency modulation (FM). On the other hand, digital modulation schemes cause a characteristic of the carrier wave to vary discretely, in order to carry a digital signal – a stream of 1s and 0s. Digital modulation schemes use a series of pulses as the carrier: the pulse amplitude or pulse width may be modulated to carry the signal, or the frequency or phase may be modulated discretely, i.e. they may be shifted from one level to another to carry the information.

One of the effects of modulation is to disperse the signal above and below the carrier frequency. The minimum band of frequency assigned around the carrier required for the transfer of the information is referred to as the 'in-band' signal; the 'out-of-band' signal is the unwanted spill-over into neighbouring frequencies caused by practical implementation limits in the transmission chain. Some modulation techniques are cleaner than others in this regard, but the degree to which the signal spreads out on either side of the carrier – its bandwidth – also depends on a number of other factors. So a signal using one carrier frequency can appear, albeit at a lower level, at another frequency which is apparently quite distinct from it in the spectrum. Managing this phenomenon – known as interference – is the most crucial issue for spectrum managers worldwide.

The overarching strategy for managing interference is 'band harmonisation', which can happen both at global and regional[3] levels. At the global level, the usage of the frequency bands, rights to use them and the agreed rules between countries are determined at the World Radiocommunications Conferences (WRC)[4] of the ITU. Each WRC produces a document known as the Final Acts which update the Radio Regulations currently in force; both documents have the legal status of international treaties.

Currently, band harmonisation on the global scale can be relatively technology-neutral (for example, television which includes the PAL, SECAM, NTSC, DVB-T and 8-VSB standards in the same frequency bands). However, it is not service-neutral;[5] the radio spectrum is divided up into bands in which only certain services may be provided. There may be differences between the three regions[6] of the ITU as well as between nations within each region.[7]

The Radio Regulations are intended to allow for the use of a frequency band, while minimising the risk of interference to the users of the band in the territories of other countries. The rules may detail what type of emissions may be used, suitable methods for determining interference, guard bands, protection ratios, notification periods and so on. This overall process is usually referred to as 'co-ordination' and, typically, can take between three months for a terrestrial station to five years for a space station.

A change in usage of a band typically requires an agreement known as a 'Common Position' (CP) from regional organisations, and is then nominated as an agenda item for the next 'competent' WRC. This is a lengthy process and can take a minimum of three to five years before final agreement. So although band harmonisation minimises and quantifies the risk of harmful interference, the necessary notification periods, co-ordination and possible agenda items at WRC make it a time and resource-intensive business. It can also limit the flexibility of the spectrum manager in allowing for new and innovative use of spectrum – some of which may even be currently unused.

3. For example: the Conference of European Postal and Telecommunication Administrations (CEPT) in Europe and the Inter-American Telecommunication Commission (CITEL) are two such regional agencies.

4. These occur between every three to four years.

5. AS, AMSS, BS, BSS, FS, FSS, MS, MSS etc are examples of services defined by the ITU and are: Aeronautical Service, Aeronautical Mobile Satellite Service, Broadcasting Service, Broadcasting Satellite Service, Fixed Service, Fixed Satellite Service, Mobile Service and Mobile Satellite Service respectively.

6. Region 1: Europe and Africa, Region 2: The Americas, and Region 3: Asia

7. Different services used within a band are detailed in the footnotes to the Master International Frequency Register (MIFR) which is Article 5 of the Radio Regulations.

Traditional spectrum management, therefore, is not service-neutral and has only limited scope to be technology-neutral. This has led to two different kinds of inefficiency. Firstly, as more services and applications were developed, they tended to be allocated spectrum at higher and higher frequencies.

But not all frequencies are equal. Below 60 MHz, transmission characteristics are strongly influenced by the way the signal is reflected and/or absorbed by the various layers of the ionosphere, which depends on the time of day, the season and the solar weather linked to the sunspot cycle. Higher frequencies (above 60 GHz) suffer from excessive absorption by water particles in the atmosphere. In addition, the attenuation (weakening) of signal with distance from the transmitter site varies inversely with frequency, so that for a given transmitter power and antenna gain, lower frequencies will 'travel' further than higher ones.

The 'cream' of the frequencies (in that they have attractive propagation characteristics and can be used to provide all services, including mobile) are found between 250 MHz and 2.5 GHz. Naturally, these are also the ones that have already been allocated to existing applications. However, they are not necessarily the most heavily used, and this brings us to the second type of inefficiency. In order to secure frequencies and allow new services to be developed, regulators have effectively had to guess which, among a number of competing services, are most likely to be successful, or more 'deserving' of spectrum – in other words, to pick winners.

The successes (2G mobile phones or GSM, and satellite TV) and failures (ERMES, TFTS[8]) of this approach are well documented and some are discussed later in this chapter. The problem is that recovering unused spectrum can take many years, since current spectrum 'owners' will be reluctant to give it up, even if they're not using it.

The paradox is that, while it is very difficult to find 'clean' spectrum in attractive frequency bands for new applications, studies by Ofcom and the FCC have shown that, even at lower frequencies, existing spectrum bands are not that heavily used. The reasons are many. Avoiding interference with neighbouring systems, in either space or frequency, may require certain 'dead zones' in which frequencies cannot be used.

An example of this is the 'guard band' between television channels, which is designed to stop the signals from one channel spilling over into an adjacent channel, causing interference and degrading the signal. Demand for the service may be intermittent, so that the frequencies are used intensively at certain times and, at others, not at all. Or, the frequency may have been assigned to a service for which no demand materialised, or not to the extent that was forecast.

On the one hand, then, we have a system of frequency management which is good at avoiding interference and safeguarding national rights (for example, for television broadcasting), but which results in under-utilising large chunks of prime spectrum. On the other, we have hugely increasing demand for spectrum, driven largely by growth in mobile services, but also by many other (telecommunications and non-telecommunications) applications. At the same time, both economic and technical developments are likely to change the way in which supply-and-demand are married.

8. Enhanced Radio Messaging System – the pan-European radio paging system; Terrestrial Flight Telephone System

Economic drivers of change in spectrum management

Traditionally, telecommunications applications have been thought of, and regulated, as belonging to separate 'families': fixed, using copper wire, cable or fibre to the home or premises, and aimed primarily at peer-to-peer communications; mobile, using the wireless spectrum, again for one-to-one communications, with lower capacity than fixed links but near-ubiquitous access; and broadcasting, a one-way, one-to-many service over cable, satellite or terrestrial radio.

Now, these separate markets are increasingly converging in terms of the services and content being delivered. The latest generation of mobile phones can receive and display video material and games and, in the near future, will incorporate digital TV and/or audio receivers. A new variant of the European digital TV standard, DVB-H (Handheld), has recently been established, and Digital Audio Broadcasting has evolved into T-DMB (Terrestrial Digital Multimedia Broadcasting), with both standards optimised for delivery of digital content to small handheld devices such as mobile phones. The UMTS standard also includes a broadcast element, MBMS (Multi media Broadcast Multicast Services).

The distinction between fixed and mobile telecommunications is also becoming increasingly blurred. Fixed operators are seeking to add mobile or nomadic aspects to their service (for example, through portable broadband technology such as WiFi and WiMax), just as mobile operators are seeking to add low-cost calling and broadband to their services. This convergence can range from simple single-billing solutions to converged fixed and mobile end-user devices. The Unlicensed Mobile Access (UMA) standard allows end-users, with a single handset, to access voice services over mobile or fixed broadband networks using VoIP. More and more voice traffic is being carried over mobile networks, while the latest generation of fixed wireless technology has a 3G mobile pedigree, enabling non-line-of-sight connection and enabling portable user terminals to be deployed.

Consumers face an increasing choice of communications options, whether fixed, mobile, or broadcast. Demand for mobility is well established, with the huge success of 2G mobile phone technology. The shift in demand from pure voice to voice plus data – and, increasingly, video – means that applications are more and more bandwidth-hungry. From the regulatory point of view, as the number of options increases, so too does the difficulty of predicting how markets and technologies might develop in the future. The regulatory environment needs sufficient flexibility to cope with this uncertainty.

Technological drivers of change in spectrum management

To meet the demand for converged services, wireless technologies are emerging at a faster rate than traditional spectrum management can handle. From the consumer point of view, technology is evolving to meet the trends in demand identified above for greater functionality. However, it would be simplistic to assume that it's the most technologically advanced services, with the highest functionality, that automatically drive the greatest demand; consider the success of the humble SMS, a limited text-only service without even a full alphanumeric keyboard. In general, however, consumer demand has emerged over time for increased access to content, mobility and interactivity.

Advances in digital technology have led to developing more and more complex modulation schemes and a higher degree of intelligence in transceivers. Ultra-wideband (UWB) radio promises the capability of sharing spectrum with other uses and technologies; it 'spreads' the signal over a very wide range – several Gigahertz – rather than requiring exclusive use of a relatively narrow band of frequencies.

Other techniques under development include software-defined radio (SDR); listen before talk (LBT), where the equipment can 'listen' to other users of the spectrum before deciding what frequency and, in some cases, what modulation technique, to use; cognitive radio; dynamic power control (DPC); and spread spectrum techniques.

These techniques can use what are currently 'white spaces' in the spectrum, gaining additional capacity without interfering with current users. Smart antenna technologies use software to improve the ability of antennas to extract a particular, wanted signal from a mass of competing signals.

Paradoxically, while you'd have thought spectrum managers would be pleased with new technologies that are designed to use spectrum more efficiently, they don't sit easily with those managers' methods. Their main priority is the need to ensure efficiency while safeguarding all spectrum users from harmful interference.

This concern has meant that licences traditionally defined the spectrum, equipment and technical parameters of the radio system. Fitting in other technologies which can 'underlay' existing ones (by operating within the noise floor of spectrum allocated to other uses) or 'overlay' them (by using frequencies which are temporarily unused by other users) threatens the exclusivity traditionally enjoyed by licensed users of frequencies.

This exclusivity offers protection, not only against interference, but in many cases against potential competitors. It follows, then, that when you change traditional frequency management techniques to accommodate new technologies, you'll create losers as well as winners.

The move towards market mechanisms in spectrum management: liberalisation and trading

Economists since Ronald Coase (1959) have argued that allocating radio spectrum by administrative fiat makes little sense; establishing a market for spectrum in which owners could buy, sell, subdivide and aggregate spectrum parcels would lead to a much better allocation of the resource.[9] The argument is that markets are more efficient at allocating resources than regulators, as their ability to capture all the information on supply and demand is far greater, and much more rapid, than that of a centralised planner.

A more liberal approach to spectrum management would define rights clearly, preferably in both a technology and service-neutral way, and then assign the rights to individual users via an auction with the provision that they could subsequently be traded openly. Thus the market would decide not only who should have the licence (because the auction would assign it to the most efficient operator who places the highest value on the spectrum), but what services should be provided: if one business model failed, the rights would be bought by another operator with a different model.

9. Faulhaber G. & Farber D.: Spectrum Management: Property Rights, Markets and the Commons

The clear definition of rights and obligations is absolutely crucial in this scenario, because if interference problems arise, the transaction costs of settling them may be prohibitively high, restricting trading and negating any benefit from liberalisation.[10]

Internationally, we now see a trend of relying more strongly on market forces in spectrum management. Some countries have kept control over functions such as spectrum allocation (i.e. deciding what spectrum should be used for what purpose and, sometimes, using what technology), while increasingly relying on market mechanisms such as auctions to assign spectrum (i.e. to decide who gets the rights to use it). Other countries have moved further towards deregulating spectrum management by allowing for market-based allocation of spectrum use. Some of the options available to implement a more liberal spectrum regime are:

- **Leasing and sharing**
  This typically involves a partial transfer of a licensee's rights to spectrum, either for a limited period and/or for a portion of the spectrum encompassed in the licence. Licensees can realise some income from unused spectrum, while market entrants can also get access to spectrum. It also creates a financial incentive for licensees to utilise their spectrum more efficiently. Spectrum leasing arrangements have been authorised in the USA.

- **Change of ownership**
  Change of ownership means transferring licence rights to a third party in a secondary market, without any change of use or reconfiguration. It can increase economic efficiency by reassigning the spectrum to a party that values it most, hence exposing the opportunity cost of spectrum. A number of countries have introduced some form of change of ownership via secondary markets, including Australia, New Zealand, Guatemala, the UK and the USA.

- **Spectrum re-configuration**
  Spectrum re-configuration allows for the aggregation or partitioning of spectrum according to need. For example, a licensee could make efficiency gains by switching to technologies that use less spectrum, and then partition and sell the spectrum that it no longer needed. It also provides greater flexibility, in that a user can only buy what they need. However, this gives rise to a concern that spectrum re-configuration would result in excessive fragmentation.

  According to the ITU, there are three reasons why fragmentation may not be a problem. Firstly, because portioning the spectrum may devalue the asset; secondly, because the cost of carrying out such transactions may be high; and thirdly, because users may hold onto unused spectrum for expansion of their network at some future date. Where this policy has been introduced, for example in Australia and Guatemala, fragmentation does not appear to have become an issue.

- **Change in use**
  Allowing changes in spectrum use provides greater flexibility and exposure to market forces than any of the policies above. As many spectrum decisions as possible are devolved to the market, since market players are in the best position and have the most information to determine what allocations of spectrum would be optimal for meeting consumer interests. In the

10. Cave M. & Webb W.: Designing Property Rights for the Operation of Spectrum Markets, Papers in *Spectrum Trading*, Warwick Business School, Aug. 2003

countries that have introduced spectrum trading to date, the majority have also allowed for some form of change of use.

Social, cultural and policy issues

For many applications, how spectrum is used is more than just an economic issue: it has important social, cultural and political implications. Public sector broadcasters use spectrum that is specifically allocated for broadcasting use under international agreements. They provide services which are generally free at the point of consumption, and are funded by a range of mechanisms including a licence fee and general taxation.

In many cases, these broadcasters do not pay for the spectrum they use; in others they may only pay an administrative fee which does not reflect the value of the spectrum. This issue is highly topical in the light of the recent ITU Regional Radiocommunications Conference, RRC '06, which dealt with the future planning of certain broadcasting bands and the planned switchover to digital terrestrial television (DTT) transmission.

Digital technology is able to transmit the same number of channels using only one-third to one-sixth of the bandwidth. This opens up the possibility of releasing some current 'broadcasting' spectrum for other applications, such as mobile broadcasting or fixed broadband wireless access – an issue commonly referred to as the 'digital dividend'. While some of the dividend may be absorbed by increased broadcasting demands, to transmit more channels or provide High Definition Television (HDTV) services, there is certainly scope for re-thinking other uses of the band.

Other important issues include the widespread availability of broadband, and the availability of spectrum both for essential services and for enhanced public services such as transport management and safety initiatives. In Ireland, extending access to broadband and encouraging take-up are key aspects of government policy, because of its economic and social importance. Wireless technologies are increasingly emerging as attractive alternatives to cover rural and remote areas, where upgrading existing infrastructure can be expensive.

Future requirements for government services such as defence, security and public safety may need to be considered. This includes upgrading existing analogue mobile communication services, which could lead to releasing existing spectrum in the UHF bands for civil use, and potential requirements for broadband communications such as wireless CCTV.

Radio spectrum can play a significant role in the transport sector, enabling cost-effective traffic monitoring and toll collection, and in distributing real-time information to public transport users. For example, you'll find real-time information on local bus services in a number of European cities, using a combination of vehicle tracking technology and either a private or public mobile radio network to relay the information to bus stops.

Tensions between a market-based approach and the achievement of policy goals

With the introduction of market mechanisms into spectrum management, there has been much debate about the best way to achieve policy goals other than maximising economic efficiency and consumer welfare. In the past, spectrum was usually gifted to public service broadcasters, defence forces and emergency services.

Today, the climate has changed: as the demand for bandwidth-hungry applications such as data, mobile TV and audio has increased, there has been increasing pressure on other traditional users of spectrum to justify their privileged access, or at least to prove that they are using it efficiently. The debate on digital switchover has focused minds on perhaps a wider issue: how to balance market mechanisms with the need to ensure that spectrum is available for services of general economic or social interest, while ensuring efficient utilisation.

The latter point is important, particularly if spectrum is free to certain organisations: as demand for a free good is potentially infinite, spectrum could be used wastefully in the absence of incentives to use it efficiently. In this context, market mechanisms may have a role to play, at least in identifying the opportunity cost of the spectrum.

While the debate has an unfortunate tendency to portray spectrum management models as either the unfettered operation of market forces or the dead hand of command-and-control, the reality is that spectrum management is much more technical, complicated and case-dependent than either model might imply.

As previously explained, not all spectrum is of equal value for all applications, and frequencies cannot be easily contained within a particular area. A signal from one country can spill over borders into another. This is particularly crucial in the case of broadcasting services, where transmitting stations use high sites and high power, and can cause interference at a range of several hundred kilometres – hence the greater need for international co-ordination. Recognising this, where change is being introduced, this is usually done on a measured and staged basis.

Economic efficiency is an important responsibility for spectrum managers, but they must also consider the tools available to them to achieve other objectives. A range of policy approaches is possible, ranging from pure free-market to pure command-and-control – but, most likely, involving a combination.

One pragmatic argument for gifting spectrum to free-to-air broadcasters was that a fee would result in either an increase in the licence fee or a direct subvention from government. But there's a drawback here: this approach gives no incentive for efficiency, with the result that large tracts of valuable spectrum are underused when modern technology could provide both broadcasting and other services which consumers want.

A pure free-market approach would hold that spectrum is an input just like any other: broadcasters are expected to pay for land, office buildings and intellectual property, so why not spectrum? The theoretical benefit of this approach is that it ensures spectrum goes to the most efficient user and (provided that change of use is allowed) to the most efficient use.

Thus, the services delivered would be those for which the user is most willing to pay. On the other hand, this rationale does not sit easily with the public-service model of most public sector broadcasting, since it raises the risk that the broadcaster might not secure the required spectrum. The European Commission, in its 2005 communication on *A Marketbased Approach to Spectrum Management in the European Union*[11], showed a commendable willingness to think the unthinkable. It stated:

"It may also be necessary to reassess the assumption that broadcasting as a public

11. Communication from the Commission to the Council, the European Parliament and the European Economic and Social Committee and the Committee of the Regions: A market-based approach to spectrum management in the European Union, COM (2005)400 final

interest automatically requires terrestrial spectrum, since coverage obligations increasingly can be fulfilled by means other than terrestrial wireless transmission, given the growing reality of convergence and multiple platforms."

This, however, would be a step too far for most policymakers.

A happy medium, which seeks to recognise current realities but to create incentives for efficiency, might be to 'grandfather' existing spectrum rights to public sector organisations, while making these rights subject to re-packaging and trading. In this case, inefficiencies inherent in the current system would persist for some time – but the hope would be that they would erode over time, as the public sector recognised the opportunity costs of holding on to spectrum it did not need, and gradually released it to the market.

While in theory this would provide a 'glidepath' towards a more efficient system, it would essentially depend on the public sector organisation becoming a spectrum management organisation in addition to its statutory functions. You could argue that any such body is unlikely to be as efficient as a private sector body specialising in spectrum management, and the incentive is likely to be diluted by a perception that this issue is peripheral to the body's main function. On the other hand, if the organisation actually becomes a successful seller-on of unwanted spectrum, there may be resistance to perceived windfall gains from a national resource.

A second intermediate approach might be to require broadcasters to pay for spectrum, but at a subsidised rate, or use a form of incentive pricing to encourage them to use efficient technology. This requires rather more intervention on the part of the regulator, in terms of setting technical efficiency criteria and estimating the value of spectrum, but may still be useful in encouraging efficiency.

The extent to which spectrum utilisation is related to the 'essential' or 'public service' element of the service would also need to be defined. In other words, not all aspects of a service may be equally deserving of free or reduced-cost spectrum. Broadcasters may be obliged to pay for fixed links they use to link their broadcasting sites together, and emergency services may use commercial mobile spectrum as well as that set aside for uses related to safety-of-life.

There is probably no single 'right' approach: striking the balance between safeguarding public services and encouraging efficiency will vary according to the public service concerned, how desirable the frequency band is for other services and the preferred frequency management model. In general, though, while there are drawbacks to a system of grandfathering rights which allows trading and spectrum subdivision, it probably represents the best and most achievable form of trade-off.

International harmonisation

As a valuable common resource, there is greater pressure to bring spectrum allocation policy under the ambit of the European Commission to ensure a harmonised approach to spectrum allocation across Europe. Member States now need to balance European policy against national policy goals in order to avoid a 'one-size-fits-all' approach that may leave valuable spectrum lying fallow. For example, the drive to establish a harmonised band for mobile satellite services across Europe may result in spectrum remaining unused in smaller countries where no such market exists.

Probably the greatest success of harmonisation was the spread of GSM[12] as the *de-facto* world digital mobile phone standard. This started in the late eighties in ETSI, as the successor to the then-growing analogue mobile phone service TACS.[13] What originally started as a prestige niche product became, by the forces of harmonisation, standardisation and the sheer size of the market, the ubiquitous world standard in just ten years.

An example of a failed standard would be the European ERMES[14] paging standard, which resulted from around three years' harmonisation work in CEPT and around the same again in standardisation in ETSI during the early 1990s. The resulting product was initially a minor success; however its market share was first strongly diminished by the ubiquitous nature of GSM, and then effectively wiped out by the advent of SMS[15] messaging.

The most successful atypical product and usage which came about in a standard harmonised band is the IEEE 802.11 (WiFi) family of standards, using the 2.4 GHz band. During the 1990s, this band was harmonised in Europe by CEPT for a low power radio LAN product. The European standard product (HiperLAN) was commercially unsuccessful; however the frequency it used was the same as a product standard developed by the IEEE[16]. This led to the rival American standard subsuming the European one, and it is now used in the same harmonised band.

While harmonisation and standardisation can lead to successful products such as GSM, and can simplify spectrum management once the system is in operation, they are highly labour-intensive in the planning stages of a new or novel system. They can also limit the flexibility with which the spectrum can be re-allocated should the product fail.

Moreover, successful harmonisation relies on sufficient market demand to ensure spectrum is efficiently used in terms of channel occupancy.[17] In smaller countries, the demand for a product may simply not justify the resources needed by the spectrum management agency to take part in the harmonisation and standardisation processes in any great detail. This is how compulsory harmonisation can lead to a less-than-optimal occupancy of large chunks of spectrum, and small countries having to follow in the wake of the larger states' progress.

The current proposals for a greater use of market mechanisms in spectrum management have led to an interesting debate about the interaction between harmonisation and flexibility. There is a fear that, if change of use is allowed, bands which are currently harmonised on a pan-European basis could become fragmented; economies of scale in both equipment and service markets could disappear; and the net result could be to make the use of spectrum less, rather than more, efficient.

The European Commission, in its *Communication on Spectrum Policy Priorities for*

12. Global System for Mobile Communications
13. Total Access Communication System
14. Enhanced Radio Messaging System
15. Short Message Service
16. Institute of Electrical and Electronics Engineers of America
17. This is, in general, the time that any transmitter spends on a particular RF channel; however, it can also be thought of as the number of users that any channel can serve.

Digital Switchover[18], has stated that "a part of the spectrum dividend should be earmarked for harmonisation", adding in a footnote that the term 'harmonisation' should be understood in the wider context of enabling the deployment of pan-European services, including the common adoption of flexible approaches for spectrum management such as spectrum trading and unlicensed use.

The footnote is an interesting addition, since the paper on a market-based approach seems to argue against the approach of earmarking spectrum for particular uses. It is arguable that, if all Member States introduced flexible use of spectrum and trading, the market itself would tend to harmonise spectrum where this would be beneficial. If consumers in different Member States ascribed similar values to the same spectrum – and, in particular, if they valued 'roaming' – then, over time, particular spectrum bands would gravitate towards their most efficient use.

Harmonisation in this sense would lead to a better outcome for consumers than harmonisation by regulatory fiat, since the services would be determined by the market rather than determined in advance, with all the usual difficulties involved in 'picking winners'.

Given the diversity of approaches among Member States, however, such an approach is likely to take many years to develop; the Commission's plan to use part of the digital dividend for pan-European services may be the only realistic way to achieve harmonisation in the medium term. On the other hand, as previously mentioned, the harmonised approach may not be best for all countries.

Perhaps the best approach is to think of spectrum harmonisation initially in terms of harmonisation of bands, in the sense that a uniform technical design for rights of use (to cover, for example, channel bandwidth, maximum EIRP, block edge masks, guard bands and out-of-block emissions) is applied internationally. Further harmonisation could be applied to methods of spectrum management within bands so that, for example, common bands are allocated to licence-exempt services; are set aside for public service purposes (subject to measures to ensure efficiency, as described above); or are subject to market-based rules such as flexibility and trading.

The Commission initiated a major study on spectrum trading[19], which estimated that the net gain from the introduction of spectrum trading combined with flexible rights of use would amount to 8-9 bn per year. It further estimated that, if one country decided not to reform its spectrum management policy by adopting such measures, this would create costs for others. Conversely, when one country decided to join the reform project, this would yield an additional benefit to the others. 60-70 per cent of the benefits would be attributable to national reforms, while 30-40 per cent would accrue as a result of making the transition in all the EU countries.

A common approach to spectrum management would open up the possibility of market-driven harmonisation, while leaving the possibility for individual member states

18. Communication from the Commission to the Council, the European Parliament, the European Economic and Social Committee and the Committee of the Regions: EU spectrum policy priorities for the digital switchover in the context of the upcoming ITU Regional Radiocommunication Conference 2006 (RRC-06), COM (2005) 461 final

19. "Conditions and options in introducing secondary trading of radio spectrum in the European Community", Analysis, DotEcon and Hogan and Hartson

to find their own solutions to national issues such as broadband availability, without having to reserve spectrum for applications which may never develop.

A combination of band harmonisation (which in many respects might not differ too much from current methods of frequency planning, although licences would be technology and service-neutral) together with harmonisation of the method of spectrum management, might deliver both the technical efficiency of traditional methods with the economic efficiency of markets, to the benefit of consumers.

An economic evaluation of spectrum allocation policy[1]

Thomas W. Hazlett

Professor of Law & Economics, George Mason University

The standard spectrum allocation approach posits three alternative models: 'exclusive use,' 'commons,' and 'command and control'. The regulator's role is specified as selecting between these models, case-by-case. This framework, however, yields the anti-consumer outcomes widely associated with traditional regulation. The analysis of property rights found in the law and economics literature offers clarification. It demonstrates that the conventional approach mistakenly conflates ***access regimes****, rules coordinating wireless usage, with* ***property regimes****, high-level rules determining who is initially authorised to make such choices. The debate over spectrum allocation policy will gain coherence when, rather than seeking to impose preferred spectrum access regimes, public policy enables a competitive market process to discover them. This does not foreclose the option for the Government to provide particular access regimes determined to be under-supplied by private spectrum owners, but moves such decisions into a better informed environment, where opportunity costs are visible. To enable this efficiency-enhancing outcome, it is necessary for spectrum regulators to move away from case-by-case (or block-by-block) administrative allocations to a general regime of exclusive property rights.*

Introduction

In summarising conventional thinking about the regulator's role in radio spectrum allocation, the US Federal Communications Commission issued its Spectrum Policy Task Force Report in November 2002[2]. It defined the regulator's spectrum allocation task as selecting among three alternative models, which it defined as follows:

- **The 'exclusive use' model**. A licensing model in which a licensee has exclusive and transferable flexible use rights for specified spectrum within a defined geographic area, with flexible use rights that are governed primarily by technical rules to protect spectrum users against interference.
- **The 'commons' model**. This allows unlimited numbers of unlicensed users to share frequencies, with usage rights that are governed by technical standards or etiquettes but with no right to protection from interference.
- **The 'command and control' model**. The traditional process of spectrum management in the United States, currently used for most

1. This paper is adapted from Thomas W. Hazlett, *The Spectrum Allocation Debate: An Analysis*, IEEE Internet Computing (Sept/Oct 2006), 52-58; www.computer.org/internet

2. SPTFR," http://www.fcc.gov/sptf/reports.html

spectrum within the Commission's jurisdiction, in which allowable spectrum uses are limited based on regulatory judgements.[3]

The Report's conventional policy prescription is that each of the models has its appropriate place, and that regulators should examine frequency use possibilities case-by-case. While the Government should generally move away from rigid rules, circumstances in each wireless application will be determinative.

This paper attempts to clarify the anti consumer elements in this policy choice framework, which confuses *access regimes* (how resources are employed by consumers and producers) with *property regimes* (how control over organising choices is defined). While 'exclusive use' and 'commons' are recommended as liberal alternatives to traditional regulation, the bureaucratic selection process that continues to impose regimes, block-by-block, *constitutes* 'command-and-control'. Some forms of wireless market organisation are chosen, and others excluded, not by competitive markets but via administrative allocation.

This approach is revealed by considering the law and economics literature on the general evolution of property rights. Before considering this alternative analytical framework, however, I'll look at some consequences of the conventional approach.

The conventional trichotomous framework

The standard spectrum allocation framework fundamentally mis-characterises the choices confronting the regulator.

First, the exclusion uniquely associated with 'exclusive use' is just as essential to 'licence-exempt' bands or 'command-and-control'. In each instance, government effectively prohibits certain wireless activities so that others are more productive. Moreover, assigning exclusive rights is not associated with 'exclusive use' but with a method for determining how rules will enable multiple users to access the radio spectrum resource.

Indeed, 'exclusive use' licences awarded to cellular operators, for example, permit intense 'spectrum sharing' in cellular bands, where millions of rival users (i.e. subscribers) use airwaves and network infrastructure under rules designed by competing wireless carriers. Conversely, unlicensed rules impose limits on users that, for instance, exclude the use of CDMA spread spectrum devices. In either case, exclusions are employed to facilitate 'spectrum sharing'. The essential difference lies in the nexus of control: which parties are ceded authority to impose rules and, so, what incentives will govern the evaluation of economic trade-offs.

Second, the policy selection structure is incorrectly formulated. Basic property rights – what may be called *high-level property regime* choices – do not uniquely determine the organisation of the market. Assigning basic control over a band of airwaves begins, not ends, the crafting of an access regime: a set of rules that will allow end-users to appropriate resource rights. To conflate the initial assignment of property rights with the ultimate set of rules governing consumer use of wireless devices leads to fundamental misunderstanding.

Most pointedly, it suggests that when a certain type of spectrum access – a means of organising spectrum sharing – is determined by regulators to be socially beneficial, the state should then impose that organisational

3. SPTFR, p. 5 (Word version)

structure. The approach is demonstrated in the claim that the Government should set aside unlicensed bands, in the same way that it adds social value by allocating land to public parks:

"A mechanism based on markets, such as an exclusive use model, will be most efficient in most cases. However, government may also wish to promote the important efficiency and innovation benefits of a spectrum commons by allocating spectrum bands for shared use, much as it allocates land to public parks."[4]

In fact, government does not generally allocate land, the principal resource used to create public parks, but rather defines general ownership interests. Such private property rights then enable government to create public parks. The means by which state agencies acquire land rights (first appropriation, purchase, condemnation or gifting) are of secondary importance here. What is primary is that the land market actively benefits economic decision-making even when the ultimate access regime is a 'commons'.

First, general property rules eliminate barriers to productive use of assets; society need not wait for the state to set resource access rules, parcel-by-parcel. This actively advances competition in services and fosters innovation. This is of crucial importance in spectrum allocation, where regulatory barriers to entry have long deterred competitive entrants and new technologies.[5]

Second, the asset market provides valuable information on the opportunity cost of alternatives. We can see how much additional acreage would cost New York City, should it seek to expand Central Park, or how much revenue it could generate by selling a slice of it.[6] Without private property rights in land, these values are concealed, and the government agency supplying the amenities of a public park lacks the necessary inputs for efficient decision-making. This is the central shortcoming we encounter in planned economies: the so-called 'socialist calculation problem'.

Eliminating entry barriers imposed by government spectrum allocation permits alternative technologies and business plans to be tested in the marketplace. It is this discovery process, where costs and benefits are internalised by responsible economic agents (i.e. asset owners), that reveals efficient means of organising wireless markets.

Less generous spectrum ownership rights also permit transactions, but limit their scope and, therefore, truncate the market's discovery process. For example, the power limits imposed by government in unlicensed bands enable some useful forms of co-ordination to occur, but not others. Crucially, they force fundamental market organising decisions (setting power limits, technology standards and so on) into the political/regulatory arena, sacrificing information that would be gleaned from marketplace competition.

The cost of any one particular market design is revealed by transactions between resource owners. Trial and error is fundamental, as innovative ways to please consumers inevitably emerge from uncertainty. The distribution of exclusive ownership rights enables a diversity that encompasses

4. SPTFR, p. 38

5. See, e.g., Hazlett (2001), and the section entitled *Silence of the Entrants*

6. New York City routinely sells 'slices' of Central Park by franchising private concessions. It permits restaurant owners, e.g., to enjoy exclusive rights for a fee. These revenues help offset the cost of maintaining the public park. The ability to reasonably evaluate trade-offs under a regime of exclusive ownership rights drives efficiencies across a wide swath of activities, supporting provision of public goods.

technologies, services, business models and co-operative structures. The property rights structure does not preclude a 'commons' but rationalises it.

Property rules: the law and economics approach

In 1967, Harold Demsetz pondered the question: "Why do we see exclusive ownership of some resources, but not others?"

He offered a simple model wherein property rights are costly to define and enforce, and so come into being when benefits exceed their expense. He reasoned that it was efficient for native tribes in Canada to treat beaver habitat as open to all when demand for skins was quite limited, and then likewise efficient for exclusive rights in beaver hunting grounds to emerge when the European fur trade was introduced, increasing the benefits associated with conservation and ownership.

This approach has widely influenced scholars seeking to understand how rules governing economic activity evolve. Four basic property regimes have been defined:

- Open Access, which allows exploitation without limit;
- State Property, which limits use via rules crafted by government officials;
- Common Property, which limits use via rules crafted by a group of owners; and
- Private Property, which limits use via rules crafted by a single owner.[7]

In general, individuals behave differently under the incentives of alternative property regimes. Economic agents also realign incentive structures by reconfiguring rights. So, private property evolves into common property when a corporation is created with the private assets of common shareholders, and state property emerges from common property when a government agency purchases land from a corporation.

The optimal property rights regime permits beneficial co-ordination of economic activity. The 'tragedy of the commons', better thought of as a 'tragedy of open access', results when users who could benefit from conserving resources are stymied. This is often framed as a transaction costs issue, or one of ill-defined property rights.

With spectrum, government policy states that it aims to 'minimise interference'. In fact, potential conflicts are a by-product of any wireless activity; efficient rules maximise the total value of wireless applications rather than minimise disruptions. Efficiency demands that each increment of interference (the damage to third-party communications) takes place where the benefits of the activity causing the conflict exceed the costs imposed.

The spectrum regime in the UK, the US and virtually all other nations is best described as state property, also known as 'administrative allocation'. A government agency is charged with defining spectrum use rights for various parties. Examples include when 3G wireless licences are assigned by auction, traditional broadcasting licences are authorized or when unlicensed bands are set aside for users of government-approved equipment. A further reconfiguration of use rights then evolves.
So even though in WiFi hotspots, for instance, non-exclusive spectrum rights (and equipment standards) are specified by the state, the marketplace often creates its own rights – such

7. See, generally, Lueck and Miceli (2006)

as when access to wireless local area networks (WLANs) is denied to non-subscribers or non-employees (in enterprise WLANs).

In US cell-phone networks, the state generally awards relatively broad, exclusive airwave rights, approximating *de facto* private spectrum property. Depending on the country, operators have latitude to select services, technologies, network architectures and price schedules. This discretion is then employed to re-assign spectrum use rights to third party vendors (such as handset makers), application or content providers (say, to transmit photographs, offer online services or deliver music downloads), rival networks (such as Blackberry) and, of course, to subscribers. In the latter case, each call made by a mobile subscriber can be seen as a transaction reassigning 'exclusive use' rights from the licensee to the consumer – the primary economic function of the mobile network.

It is not ironic, but an implication of transactional efficiencies created by clearly-defined private rights, that this results in massive spectrum-sharing. Millions of parties enjoy simultaneous access to given bandwidth, facilitated by the investment in significant sunk capital provided by the network.

This infrastructure, a complementary input to spectrum in the provision of wireless services, is a key component in productive deployment of spectrum. The absence of such assets may constitute a 'tragedy of the commons' whose magnitude dominates social losses due to 'interference'.[8]

Unlicensed bands are not 'commons' but are administratively allocated. The basic organisational rules are not crafted by joint owners, but by government regulators. The mechanism used to co-ordinate spectrum use is equipment regulation, most generically consisting of power limits. These are exclusionary devices intended to separate users geographically. This co-ordinates rival wireless users, realising social benefits, such as when cordless phones or WiFi routers are used in licence-exempt bands. But the social costs of such co-ordination are not limited to the administrative expense of imposing rules, as is sometimes asserted. Indeed, the principal costs are likely to be accounted for by the opportunities sacrificed when alternative access rules – say, rules that would result if exclusive ownership rights were issued to a spectrum owner – are extinguished.

In an actual commons, owners set access rules and then internalise costs and benefits from resource use. In unlicensed spectrum, rules are imposed to determine basic resource appropriation issues by an outside party – the regulatory agency. Resource costs or benefits are internalised only as they are communicated politically. Our experience with 'public interest' spectrum regulation underscores that these decisions characteristically squander rich possibilities for productive use of airwaves.[9]

Spectrum has been usefully exploited under some allocations imposed via this framework, but this fails to remedy the anti-consumer consequences of administrative allocation. The existence of some benefits does not ascertain the existence of net benefits. It is well known, for instance, that the traditional licensing regime governing TV broadcasting has produced socially-useful services. Americans buy some 25 million TV sets annually, investing about

8. 'Common interest tragedies' result in inefficient under-use or over-use of a resource. See Fennell (2004), Hazlett (2005).

9. This was Ronald Coase's (1959) intuition regarding the spectrum allocation system. Further research has supported this view. See Hazlett (2001).

$8 billion[10] – far more than they do in, say, WiFi routers or modems.[11]

The TV band is nonetheless a textbook example of socially wasteful under-utilisation (over-allocation) of radio spectrum. Bandwidth with very high value in alternative employment is set aside by government planners for an over-the-air delivery platform that nearly 90 per cent of US households actually pay to bypass via cable or satellite. Given that the incremental cost of moving to 100 per cent cable and satellite distribution for broadcast TV content is approximately just $3 billion, while the social value of the 402 MHz (now walled off for broadcast TV services), is likely to exceed $2 trillion,[12] the state property regime for radio spectrum dissipates (vast) net social value. The willingness of consumers to purchase millions of devices that can use these frequencies does not mitigate this anti-consumer outcome.

Maximising the social value of wireless

The FCC believes that "The Commission's rules for unlicensed transmitters have been a tremendous success"[13] and that the policy implication is that it should allocate more spectrum for unlicensed use. The logic mimics the view that, given that there are 2.7 TV sets per US household,[14] the Government should allocate more spectrum to the TV band.

The regulatory approach to incremental unlicensed allocations lacks the multi-layered analysis needed.

The first layer identifies the appropriate margin. The productive use of one set of frequencies does not imply gains from additional bandwidth. South Korea is known to have the highest concentration of WiFi hotspots in the world, but uses only the 2.4 GHz ISM band. Regulators there have yet to authorise either the 1997 U-NII or 2003 US allocations for unlicensed devices in the 5 GHz band, some 555 MHz in aggregate. This implies that, no matter how popular the inframarginal allocations may be, incremental unlicensed bandwidth may not be productive. The limited economic activity observed in the US in the unlicensed 5 GHz or PCS bands (30 MHz in the 2 GHz band allocated in the mid-1990s) is consistent with this view.[15]

A second stage identifies the property rights alternatives. Wireless activities are not uniquely associated with particular regimes. Unlicensed devices such as WiFi routers or cordless phones can use licensed frequencies; a cellular phone service can be supplied using licensed bands (as it typically is) or unlicensed frequencies (as it is for certain links created via multi-mode handsets).

Assuming, *arguendo*, that additional bandwidth for a spectrum 'commons' is desirable, it is still necessary to devise a path to that end. One mechanism is to impose, by regulatory fiat, unlicensed access rules. That is the implication of the SPTFR's tripartite decision tree. But this approach fails to incorporate secondary rights

10. Eric A. Taub, *The Big Picture On Digital TV: It's Still Fuzzy*, NY TIMES (Sept. 12, 2002); http://tech2.nytimes.com/mem/technology/techreview.html?_r=1&res=9801E0D61531F931A2575AC0A9649C8B63&oref=slogin The revenue statistic assumes average U.S. sales prices are equivalent to international prices, where 2004 sales of $48 billion accounted for 150 million sets. Tekla S. Perry, Digital TV's 100-Meter Dash, IEEE SPECTRUM (June 2005) http://www.spectrum.ieee.org/jun05/1222

11. Total worldwide wireless LAN equipment sales total about $3 billion. See Hazlett (2006)

12. See Hazlett (2001b), Hazlett & Muñoz (2004)

13. FCC (2002), par. 6

14. Leichtman Research Notes, 1Q 2005; http://www.leichtmanresearch.com/research/notes06_2005.pdf

15. In contrast, note the extremely high valuation implicitly placed on additional cellular (liberal rights) spectrum by consumers, as estimated using an international dataset in Hazlett & Muñoz (2004).

reassignments. For instance, by assigning spectrum ownership, 'commons' could be created if profitable opportunities were to be discovered. Importantly, competitive market forces would search, test and reveal a variety of alternatives, gauging their value. As FCC spectrum analysts Evan Kwerel and John Williams write:

"Future expansion of dedicated spectrum for unlicensed use could be obtained through negotiation between the manufactures of such devices and spectrum licensees... Competition between licensees would ensure that fees reflect the opportunity cost of the spectrum. Alternatively, manufacturers of low power devices might form a bidding consortium to acquire additional spectrum in our auction. If there is a continued desire as a matter of public policy to provide spectrum for such devices on a 'free' basis, the FCC itself might purchase the spectrum in the auction, essentially reducing overall proceeds to the Treasury. This would have the advantage of making the opportunity cost of such allocations more explicit."[16]

What the FCC calls a 'private commons' is not simply a theoretical possibility. Cellular networks organise complex spectrum sharing, equipping customers with handsets, frequency access rights, and wireless network infrastructure in exchange for subscription fees. Some access is, at the margin, 'free' (for example, within-plan, on-net, or off-peak minutes, as well as text messaging), and customers use the network without discrimination.

However, just because the forms of co-ordination differ from the manner in which unlicensed bands operate does not imply that a wider use of the unlicensed regime should be mandated. Quite the contrary. To the extent that cellular networks are liberally regulated and competitive, the market adopts organisational modes serving consumer interests. Included here, most notably, is the demand by customers to use extensive fixed network infrastructure. The transactions costs of providing wide-area network investments under unlicensed rules are relatively high, depriving consumers of desirable options.[17]

The third layer considers the spectrum allocation process as a whole. What are the implications of allocating spectrum in the current tripartite case-by-case vs. a liberal ownership regime? The consensus of economists is that the latter eliminates the barriers to competition posed by the current system, permitting rational economic decisions to be made.[18] This does not reflect a preference for one type of market organisation over another; economists are agnostic as to what type of network architecture, or non-network wireless devices, are optimal under various circumstances. As Kwerel & Williams offer, in their argument for a 'Big Bang' to endow an active spectrum market:

"One possible arrangement would be for a licensee or group of licensees covering a particular band throughout the US to charge manufacturers a fee for the right to produce and market devices to operate in that band.

16. Kwerel & Williams (2002, p. 31; footnote omitted)

17. This does not imply that non-network wireless devices, such as those employed in WLANs, would not be widely deployed were more liberal spectrum allocations made under exclusive rights. By expanding the supply and lowering the cost of access, it is predictable that additional applications, business models, and technologies would be offered. It is also observed that numerous telecommunications networks bundle their 'exclusive rights' network services with devices that access unlicensed bands, as seen in mobile phones using Bluetooth headsets or digital subscriber line modems that embed wireless routers using 2.4 GHz spectrum.

18. See Rosston & Hazlett (2001); Cave (2002); Hahn & Wallsten (2006)

Such contracts could provide different grades of access for different fees, thus providing for a wider range of uses than are possible under the current rules."[19]

The passage suggests two critical considerations. First, there are myriad ways to co-ordinate wireless use, even assuming an 'unlicensed' structure. The trial and error of the market is a reliable mechanism for testing alternatives.

Second, it is not efficient for the state to designate market structures. To assume that only by government regulation can a 'commons' be provided is to mistake an access regime for a property regime. And to assume that forcing a particular set of unlicensed rules on spectrum users creates efficiency is to ignore the underlying characteristic of administrative allocation: that the state lacks the information or incentives to effectively evaluate the trade-offs between rival alternatives.

The pro-consumer policy framework dispenses with the state's case-by-case rule makings. Instead, exclusive rights to spectrum should be generally distributed to owners – some of whom, no doubt, will be public agencies. The unlicensed bands which host wireless devices, such as the 902-928 MHz and 2.4-2.4835 GHz bands, would be fairly certain to remain, as now, *de facto* state property.

Indeed, the reallocation of unlicensed bands is difficult due to the transaction costs that are generated by unlicensed rules. The transactional efficiency of ownership rights lies in the fact that a responsible economic party is enabled to co-ordinate spectrum sharing. This is widely misunderstood. 'Licensing costs' are characterised as an expense associated with the award of exclusive rights,[20] but these are actually a by-product of the regulatory system and are entirely remedied by more efficient *government policy*. The *marketplace* costs of trading exclusive rights are quite modest in the wireless phone market, where over five trillion minutes of use are purchased by global customers each year.

The tragedy of the anti-market

Liberalising spectrum policy empowers competitive spectrum owners to experiment with alternative network architectures or access models. It also permits governments to evaluate, with economic data gleaned from actual transactions, whether particular forms of spectrum access need to be subsidised. Just as a public park is best provided when the costs of the park are known and competitive land-use alternatives permitted, exclusive spectrum ownership facilitates efficiency.

Some regulatory regimes are liberalising. Countries such as Australia, New Zealand, Guatemala and El Salvador instituted sweeping reforms in the 1980s and 1990s, radically advancing the distribution of exclusive spectrum rights.[21] The United Kingdom has recently adopted a framework for simultaneously eliminating licence restrictions and expanding the bandwidth allocated to these liberal licences, with the aim of allocating about 70 per cent of spectrum under 3 GHz in this way by 2010.

Yet, most countries continue to allocate spectrum case-by-case. Since December 2002, for instance, the FCC in the US has been enmeshed in rule-making to regulate TV band 'white space'. The policy would impose rules

19. Ibid

20. Carter et al. (2003)

21. For a description of these reforms, see Hazlett (2004)

determining how unlicensed devices and broadcast stations co-exist. An alternative means of organising shared use would be to issue TV band overlay rights giving new licensees full flexibility subject to interference protection for incumbents. This would delegate the creation of an access regime to competitive owners.

Sen. Larry Pressler (R-SD), then Chairman of the Senate Commerce Committee, proposed this solution in May 1996.[22] It was opposed by a wide range of political interests, including TV broadcasters and, today, regulators simply ignore the licensed overlay as an alternative regime choice.

It was much the same in the 1990s, when the FCC allocated 30 MHz for unlicensed PCS (U-PCS). Apple Computer then argued that wireless LANs were in need of additional spectrum. Without considering the option that Apple acquire spectrum rights and configure them for LAN services, the FCC mandated a spectrum set-aside, imposed standards (including listen-before-talk) and set power limits. Yet, virtually no use has been made of the bands. Meanwhile, adjacent licensed PCS bands host intense traffic. The 30 MHz squandered on U-PCS would provide billions of dollars in annual consumer surplus gains if PCS operators were able to acquire them.[23]

Apple suffers no penalty from the misallocation, a lack of accountability that may be called a 'tragedy of the anti-market'. The future is unknowable, and the value of rival technologies, each with its own optimal mode for organising spectrum access, is uncertain. What is certain, however, is that society loses the benefits of market competition when it socialises risky deployments of radio spectrum.

22. The author assisted in the preparation of this proposal. See Hazlett (2001)
23. See Hazlett & Muñoz (2004)

References

Brito, Jerry. 2006. *The Spectrum Commons in Theory and Practice*, Stanford Technology Law Review (forthcoming)

Carter, Kenneth R., Ahmed Lahjouji, and Neil McNeil. 2003. *Unlicensed and Unshackled: A Joint OSP-OET White Paper on Unlicensed Devices and Their Regulatory Issues*, Federal Communications Commission Office of Strategic Planning and Policy Analysis, Working Paper No. 39 (May)

Coase, Ronald H. 1959. *The Federal Communications Commission*, 2 Journal of Law & Economics 1 (Oct.)

Coase, Ronald H. 1960. *The Problem of Social Cost*, 3 Journal of Law & Economics 1 (Oct.)

Demsetz, Harold. 1967. *Toward a Theory of Property Rights*, 57 American Economic Review: Papers and Proceedings 347 (May)

Demsetz, Harold. 2003. *Ownership and the Externality Problem*, Chapter Eleven in Anderson and McChesney, eds., Property Rights: Cooperation, Conflict, and Law 282 (Princeton, NJ: Princeton University Press).

Federal Communications Commission (FCC). 2002. *In the Matter of Additional Spectrum for Unlicensed Devices Below 900 MHz and in the 3 GHz Band*, Notice of Inquiry, ET Docket No. 02-380 (Rel. Dec. 20)

Fennell, Lee Ann. 2004. *Common Interest Tragedies*, 98 Northwestern University Law Review 907 (Feb.)

Hazlett, Thomas W. 1990. *The Rationality of U.S. Regulation of the Broadcast Spectrum*, 33 Journal of Law & Economics 133 (April)

Hazlett, Thomas W. 2001. *The Wireless Craze, the Spectrum Auction Faux Pas, the Unlimited Bandwidth Myth, and the Punchline to Ronald's Coase's 'Big Joke': An Essay on Airwave Allocation Policy*, 14 Harv. J. on Law & Technology 335 (April)

Hazlett, Thomas W. 2001b. The *U.S. Digital TV Transition: Time to Toss the Negroponte Switch*, AEI-Brookings Joint Center for Regulatory Studies Working Paper 01-15 (Nov.)

Hazlett, Thomas W. 2005. *Spectrum Tragedies*, 22 Yale Journal on Regulation 242 (Summer).

Hazlett, Thomas W. 2006. *'Barbed Wireless' and the Vertical Structure of Property Rights*, unpublished manuscript

Lueck, Dean, and Thomas J. Miceli. 2006. *Property Rights and Property Law*, A. Mitchell Polinsky and Steven Shavell, eds., Handbook of Law and Economics (forthcoming)

section 5

Global areas of focus

25

25
25

The genie is out of the bottle

Charles Leadbeater

Author, *We-think: the power of mass creativity*

Nick Jaffe has finished for the night. He puts his earphones to one side, closes the lid on his laptop and sits back in his chair, his nightly communication to his worldwide audience complete. He is not quite sure where they are, nor how many there are, but he knows they're out there because he can see his nightly sessions – a mixture of music, chat and ranting – being downloaded.

Nick, of course, is not broadcasting. He is a podcaster and just 13. His studio is his bedroom. For him, taking part in mainstream radio would be like going to a restaurant with his parents wearing a suit and tie: it's not clear why you would do it.

Nick produces media in a way that someone 20 years his senior could only have dreamed of and he consumes media completely differently as well.

He rarely watches television or listens to the radio. He acquires most of what we wants to watch from the internet, from sites such as YouTube. He most likes short films and comedy sketches which no one over the age of 15 will have heard of. He carries most of what he wants to watch in his iPod, his device of choice.

He gets the content from various aggregators of podcast materials and through recommendations from his mates. Nick's media life is pretty much a seamless cycle of production and contribution, reviewing and sharing, watching and listening.

Nick is announcing the arrival of a new media environment, one that will disrupt and reconfigure the world of mass-produced, industrialised media. In a sense, we are moving from an era in which information and media was produced for the masses, to one in which it will also be produced *by* the masses.

The corollary is that we are moving from an economy of mass production to one in which innovation and creativity also become mass activities, rather than being confined to an elite of journalists and broadcasters; special people working in special places.

The audience (at least a large chunk of it) that used to simply watch and listen, passively, now want – and is able – to take part and have their say. They will no longer just sit slumped on their sofas; they can take to the stage themselves, become part of the action, at least some of the time, in a way they want. If the last 50 years have been about the creation of cultures, organisations and infrastructures for mass-media consumption, the next 50 will be about mass-media participation.

Mini-media has never looked bigger

Let me sketch, necessarily briefly and baldly, the main features of this emerging world of production by the masses, before going on to assess the benefits it could bring, the challenges it poses to incumbent industrial media production and the issues that it raises for regulators.

The means of media production are becoming increasingly widely distributed. The iPod generation do not just want to watch and listen wherever they are; increasingly, they also want to create and contribute. More than 1bn people worldwide – and rising – have the capacity now to become mini-media producers.

Some may be professionals freelancing in their own time, but most are a rising breed of Pro-Am producers: amateurs like Nick Jaffe, who do it because they love to rather than because they're paid to, yet who operate to high standards of production. They want to do it well, judged by the standards of the communities they operate in.

Those mini-media producers are linked not just by infrastructure but also by shared platforms and commons, such as MySpace, the web site aggregator, Second Life, the mass player immersive computer game and Wikipedia, the online encyclopaedia, which allow them to exchange, share, combine and review information, cultural and media production.

Industrial-era information producers can only really work if the people contributing are employees; journalists, for example – professionals with the skills to use the very expensive equipment required to make media content.

In this world the producers were employees, who worked 9 till 5, five days a week. Viewers and listeners did not get much of a look-in. But the new media communities, the likes of Wikipedia and Second Life, allow a sliding scale of contribution: people can either contribute a lot (most edits on Wikipedia are made by a thousand or so core contributors) or a little.

These new media communities can only extend the range of potential contributors by adopting a modular design for their products.

Wikipedia is broken down into tens of thousands of particular articles. Linux, the open software programme, is broken into many thousands of smaller modules. This means people can contribute to just the piece that interests them, where their skills are relevant, without having to be involved in the organisation as a whole. As long as the modules fit together, like Lego bricks, then it doesn't matter.

Participative media is encouraging new kinds of personal freedom – individual activities like Nick Jaffe's – only because it is also creating new kinds of social production; more or less tightly-knit forms of collaboration to create, aggregate and distribute content.

These collaboratives – like Wikipedia and Linux – are organised without being organisations. Wikipedia, which now has more than 1m articles in English, 400,000 in German, and more than 1,000 articles in over 90 other languages, has just a handful of full-time employees. It does not have a big head office, corporate awaydays, an HR department, induction programmes for new staff or any of the other paraphernalia of modern corporate life.

Indeed in many respects, seen through the lens of the profit-driven market and the hierarchical corporation, these collaborations seem quite outlandish. They scramble all the categories we rely on to order the industrial world: consumers can become producers at least part of the time; demand can create its own supply so long as the means of production is in the hands of the users; the amateur, long a marginalised and fringe figure, is playing a new role alongside (and sometimes in competition with) the professional; and leisure, once just a time for relaxation, is now also for activities that look very much like satisfying work, requiring the application of skill, effort and learning.

These collaboratives bring together many distributed and decentralised contributors – but they are not like markets. Generally, they are not organised by price signals, yet supply seems to respond to demand.

Indeed, many of the participants are not directly motivated by financial gain. To the extent that these Pro-Am media producers create markets, often they are niche. Production by the masses creates millions of markets with a few consumers, whereas hit-driven industrial era production was designed to create a few products – songs, TV shows – with millions of consumers.

And these markets behave much more like conversations; they encourage an interplay between producer and consumer, not just an exchange of goods and money. People who want to be participants and players want different things from people who just want to be couch potatoes. Players in any game want equipment to play with, a pitch on which to play, people to play with and some shared rules to play by.

They do not necessarily want a service, they want tools so they can do it themselves and a platform on which they can participate with others. Couch potatoes want good service: anytime-anywhere media, at the flick of a button. Commons-based production offers something much more fundamental and potentially radical: tools that allow for mass participation.

The content and tools are often held in more or less loose forms of common ownership. People draw on, and give back to, the resources of the commons, whether that is by sharing content for a computer game, contributing to the source code of a software programme or by pooling information, as with Wikipedia.

Commons-based media production is not a free-for-all, and nor is it a bazaar. The likes of Wikipedia and Slashdot, the geek discussion site, have succeeded only because these communities are structured to take decisions in reliable ways. They are not just self-organising crowds or swarms. They are organised without being an organisation; they make decisions without having their kinds of hierarchy of industrial-era organisations. In the case of Wikipedia these governance arrangements are both formal and informal.

As its founder Jimmy Wales likes to put it, Wikipedia relies on: anarchy (anyone can do whatever they want); democracy (contentious decisions are voted on); meritocracy (the best ideas should win out, judged against the ideal of the neutral, informative encyclopaedia); aristocracy (people who have been around in the Wikipedia community longer tend to get listened to more); and finally monarchy (in the form of Wales himself, if all else fails). These new forms of social production can operate at all scales of operation: from the global to the intensely local and small-scale.

By this stage, many people will be feeling a queasy sense of déjà vu from the superficial parallels between this story and the hype that surrounded the dot.com boom.

Isn't this new way of organising ourselves to produce, edit, distribute and engage with information and cultural production at best just a marginal pastime, on the fringes of the media world or, worse, a mere passing fad? Is this new way of organising going to prove durable, and will it spread from where it started in computer games, open source software, music file sharing and the world of wikis, into the mainstream of how organisations work?

Co-operative and mutual forms of organisation are not new. Swiss villages have long organised their pastures and woods as commons. But elsewhere agricultural commons were extinguished long ago in favour of private property and investment to boost productivity.

Savers pioneered mutual forms of organisation in the 19th century. But in the last few years many of these mutuals have turned themselves into traditional shareholder-owned companies. The history of capitalism is littered with heroic but failed attempts to create co-operative factories. Why should commons-based media production, Wikipedia-style, be any different?

Commons-based media: here to stay

There are several reasons for thinking this is more than a passing fad scurrying across the margins of the economy.

- First, these new collaboratives are emerging in information, software, entertainment, culture and media, which are among the fastest-growing sectors of the developed economies. These sectors are not marginal but central to modern economies and indeed modern life.
- Second, while these new patterns of organisation are still emerging, the larger ones – and there are plenty that stay very small – nevertheless seem to have some powerful and durable features. They are driven by cheaper, more distributed technologies. They speak to values of individuality: they allow people more scope to be the author of their life, to express what matters to them and makes them distinctive.
- Third, they do not depend – as earlier efforts at collaborative self-organisation did – on people buying into alternative or altruistic values. These collaboratives grow because they do what it says on the tin. If you want to buy anything in the world the best place to go is probably the eBay trading platform. People do not use Linux open source software just because they do not like Microsoft; they do so because it works. YouTube is not just an interesting experiment; it allows film producers to find audiences, and audiences to find content they want. Wikipedia is not perfect but it is a good starting point to find the answer to many questions.
- Fourth, these social and collaborative ways of organising have powerful economic benefits in terms of competition: they are very low-cost compared with traditional media. On any day more people turn to Wikipedia than the online versions of major US newspapers. Online games that mobilise the contributions of thousands of players get a very low-cost workforce of co-developers. These collaboratives often provide the most potent competition to incumbents and their established high cost models. Imagine a start-up coming up with an alternative to Microsoft's PowerPoint programme. No venture capitalist in their right mind would back it, no matter how good it might be. Why take on Microsoft? Competition to powerful incumbents will often not come from the market but from groups of amateurs who will carry on

innovating even when there is no money to be made. That is one reason to imagine a future in which our choices as media producers and consumers will be a mix of those from large global corporations with professionals and branded products and low-cost global collaboratives of amateurs and small producers. The power of the corporation will be contested not by other corporations but by a completely different social model of production, with wholly different economics.

- Fifth, these new collaborative forms of organising seem to have hit on a new recipe for innovating en masse and at scale. Like most organisational innovations this involves combining some already well-known ingredients – peer review, modular product design, team-based working – but in a new way. Peer review is an established technique among academics and scientists: eBay, YouTube and Slashdot are taking it into new areas, at scale. Linux and Wikipedia have found ways to create complex, information-rich goods and services, at scale, by mobilising and combining the ideas of many contributors.

- Sixth and finally, these new collaboratives are finding ways to respond to their own weaknesses. Wikipedia is still not as accurate as Encyclopaedia Britannica, which itself is not as accurate as most people thought. One objection from traditional media stems from the cacophony of the material available on vast sites like YouTube and MySpace. How do people find their way to what they might want? Don't they need a navigator? Isn't that a role that only skilled professionals – such as journalists – can play? Well perhaps, but it also turns out that the best of these communities work by allowing people to flag and recommend what might be relevant. Content is self-sifted by the participants. A similar but slightly different objection is about quality. Many of these collaboratives work only because they have low or non-existent barriers to entry: it is very easy to take part and contribute. But how can these very open, easy to access self-organising sites be trusted, if there is no one looking after quality? Don't we need gatekeepers – professionals like journalists and regulators – to inspect and assess for quality? Again these collaboratives seem to be evolving their own, distinctive solutions to these questions, relying on self-help and peer review to trial, debate, test and sift good ideas from bad. For example, on 15 November 2004 Robert McHenry, a former editor–in-chief at the Encyclopaedia Britannica, published an article mocking Wikipedia as the 'faith-based encyclopaedia' highlighting in particular an article on Alexander Hamilton. Hamilton's biographers disagree on whether he was born in 1755 or 1757. Wikipedia glossed over this debate fixing it at 1755. McHenry argued this showed it could not be trusted in the way a professionally-produced encyclopaedia could. But within hours of McHenry's article being published the reference was corrected and over the following days all dates and references in the article were checked. Within a week Wikipedia had a version which was clean and correct. Ironically, McHenry's critique had triggered precisely the collaborative self-correction mechanisms which should make Wikipedia so robust over the long run. Rapid feedback, peer review, many people looking for problems and providing solutions, provides a recipe for rapid improvement and high quality.

It is still early days but it seems likely that these highly social and distributed forms of media production, sharing and consumption will be durable. The dot.com boom was fuelled by venture capital money looking for the next big thing. Too often, it just found a better way to get pet food to people in a hurry. What we are witnessing now is a wave of social innovation which is fuelled by a mixture of cheap technology, amateur passion, simple economics, individual expression and loose collaboration.

The lack of money, at least at the outset, is part of what makes this wave so powerful and durable. In media and culture, lack of resources is as likely to produce radical innovation as a well-funded corporation.

Social media production: three key benefits

All well and good, the sceptic might reply, but even so why does all this matter? Or to put the question in a slightly different way: is this just a way for kids to display pictures of themselves on the internet and for bands to market their MP3 files – or could it bring larger, more significant benefits to society?

The answer, tentatively, is yes it could. The dot.com boom offered faster pet food. Social media production offers improvements to the quality of our societies in terms of fundamental values: freedom, democracy and equality.

First, this wave of social innovation is good for personal autonomy. How we consume information, and what news and views we can access, is fundamental to how we see the world and make decisions. The kinds of cultural activities we engage with have a huge bearing on who we think we are, where we come from, the story we tell about ourselves. Seen in these lights, traditional mass-media suffers from several limitations, which social production addresses.

Criticisms of traditional mass-media stem from its economics. The high capital costs of creating systems for creating and distributing content – employing hundreds of journalists and others in expensive office space in London, buying printing presses and building studios and so on – means media organisations need to find big markets to attract advertising, or win support for public subsidy.

That may result, however, in the number and range of views and voices traditional media can air being too limited: there isn't enough time, money and capacity to reflect the diversity of what people have to say. New, more distributed forms of production and sharing allow many more views to be garnered from many more sources, often from those that would not pass the tests of return on investment or public value. A related critique is that high capital costs put media organisations in the hands of only a few people – corporate owners or state-appointed executives. Concentration of power is a bad thing if it suppresses diversity and debate.

As yet, the more distributed forms of production have escaped this problem: Jimmy Wales does not decide what happens in Wikipedia in the way that Rupert Murdoch can influence his editors and empire. If distributed media allows for more diversity of expression then it should also create a more open, contested culture. It should be possible to see any issue – and so our own role in it – from several different vantage points. That in turn should help to make us more self-aware, reflexive and critical.

Think of media as if it were a sport. Many of the people who watch football also play it. Their participation as amateurs enhances their enjoyment as consumers: they are two aspects of the same interest. The same will be true for media in future.

For most of the last century in media we've had only the equivalent of a small number of Premiership teams to watch. In future, alongside the media equivalent of the Premiership, we need to develop the media version of Hackney Marshes where scores of amateur football teams play.

Social production allows us all, even if only in small ways, to become participants rather than recipients. That means, at the risk of sounding pompous, we have new opportunities to learn how to be free, through what we choose to make and contribute to, rather than through what we choose to buy and consume.

Industrial production robs people of a sense of agency. The cliché is that you are a cog in a machine. In the last two decades the way some people work in large corporations has shifted towards more self-management and self-organisation.

But the underlying story is the same: for most people work is a necessity, a way to make a living but not a way to express yourself. In the industrial era that was something you could do only at leisure or through consumption. Social production is changing all that: it recreates the possibility that people can find a sense of autonomy through work, albeit not work for a corporation.

People who produce, for free, new computer games, encyclopaedia articles, software, music and films are finding ways to express themselves through voluntary labour. That is why social production offers the possibility of a deeper sense of freedom based on participation and not just consumption; taking part, not just consuming. It also allows a wider range of motivations to come to the fore.

Most people have diverse motivations. They are neither purely altruistic, completely self-interested or obsessed by the power and status conferred by hierarchy. Non-market, social forms of production allow people to do things because of their passions, interests and skills, rather than because of the financial rewards or because they are told to by their boss.

The second reason why this wave of innovation is important is that social production of media should be good for democracy and the public sphere. Television and mass-media have provided the information backbone to our public life: that is where issues are debated, and politicians and others appeal for our attention and occasionally our votes.

As a tool for democracy, mass-media has been criticised on several grounds. Concentration of ownership gives undue weight to the views of a few. The passive send-and-receive broadcast model means that people are treated as targets for well-honed messages rather than as citizens and participants. The need to reach large audiences with slickly-produced commercials requires money and that in turn creates opportunities for corruption in party funding. Politics is turned into a spectacle, part of the entertainment business, searching for an audience: George Galloway's appearance on Celebrity Big Brother is merely the most recent and infamous example of this trend.

Social production of media would create an alternative backbone, one based on the idea of the commons: a shared platform for debate and ideas.

The media commons would make it far easier for people to have their say, to voice their views and to get organised: it promotes basic democratic values such as self-organisation, free association and self-regulation. Modern political parties, themselves creatures of the industrial era, now find themselves constantly outflanked by social movements and campaigns, often born by self-organising networks. These new forms of political engagement are not based on send-and-receive models of communications; they are more like vast rolling conversations.

We have already seen in the US and the UK that bloggers and campaigners can provide an important new check on the power of traditional media, forcing newspapers and news channels to pick up stories they have ignored or dropped.

The rise of blogs-slash-political campaigns such as the Daily Kos, which emerged out of the ruins of the Howard Dean campaign, show that social media can get organised and have an impact in its own right.

The third power behind this wave of innovation is that, in the long-run, commons-based media should be good for equality and global development.

At first glance it is far from clear why there should be any connection between media, poverty and equality. Why should people who need clean water, food and HIV drugs be remotely concerned with how middle class kids in the developed world share their MP3 files? About 25,000 people a day die from diseases caused by a lack of clean water. Set against that challenge, the debate over the merits of social media versus traditional media seems beside the point. But at Yochai Benkler puts it in The *Wealth of Networks*:

"Information, knowledge and culture are core inputs into human welfare. Agricultural knowledge and biological innovation are central to food security. Medical innovation and access to its fruits are central to living a long and healthy life. Literacy and education are central to individual growth, to democratic self-governance, and to economic capabilities. Economic growth itself is crucially dependent upon innovation and information. For all these reasons information policy has become a critical element of development policy and the question of how societies attain and distribute human welfare and well-being. Access to knowledge has become central to human development."

Stacked up against that challenge, how does traditional media fare compared with social media production? Proprietary systems for owning and controlling knowledge limit its flow and direct it to where people can pay. That is why so much pharmaceutical research is devoted to diseases of the rich and corpulent and so little to diseases of the poor.

In most scientific and cultural fields one person's output becomes another person's inspiration or input. If proprietary controls – such as patents and copyrights – put up the price of inputs, then it will price out of the market some innovators who cannot afford to pay the fee to license access to the knowledge. The alternative to proprietary systems for spreading knowledge and ideas has been international versions of traditional public service broadcasters, often state-funded and, at times, politically motivated.

Once again social media production – we might call it 'barefoot' media – offers some distinct advantages.

As Amartya Sen has argued, good government depends on democracy, and democracy depends on the free flow of information.

To the extent that social media production is less easy to control than traditional, concentrated broadcast media, then authoritarian regimes have fewer options to keep their populations in ignorance.

Commons-based media is relatively low cost and so more readily applicable to problems of the poor. It does not rely on employing high cost, professional journalists or researchers. Open and collaborative models encourage self-help and self-reliance.

Open source software, for example, gives people tools to help themselves rather than relying on the services of a corporation, from whom the software was bought. Not many people in the developing world can afford to buy the Encyclopaedia Britannica. But if you have access to a computer and a modem, you have access to Wikipedia. Indeed in some African states educators in cities are now downloading Wikipedia onto CDs to distribute to villages which have computers but no phone lines.

Wikipedia's aim is to become the Red Cross of information, to put a free encyclopaedia in the hands of everyone on the planet, in their own language. Wikipedia's example is just the leading edge of what could become a new global information commons, stretching from information and media into culture and science.

A combination of social and technological innovation is making possible new forms of collective organisation, for how we create, share and distribute information, media and cultural products.

These new ways of organising ourselves tend to be highly distributed, collaborative, participative and very low-cost. The new forms of structured self-organisation – witnessed now across fields from software and computer games to music and basic information sharing – could bring our societies very large benefits in terms of competition, efficiency and innovation, freedom, democracy and social justice.

But they also pose a significant challenge to all institutions – not just media organisations – that have relied on high barriers to entry and professional control of knowledge and information. Doctors, teachers and journalists, the organisations that employ them and the places they work are all being changed by this trend from centralised to more distributed activity; mass markets to niche; broadcast communication to conversation; consumption to participation; passive to interactive. This is a world in which, as Benkler puts it:

"All the means of producing and exchanging information and culture are placed in the hands of hundreds of millions, and eventually billions, of people around the world, available for them to work with, not only when they are functioning in the market to keep body and soul together, but also, and with equal efficacy, when they are functioning in society and alone, trying to give meaning to their lives as individuals and as social beings."

The continued rise of social media production will not necessarily compete with, still less displace, traditional media corporations. Indeed they could complement one another and many corporations will see opportunities in creating their own versions of social media – witness News Corporation's purchase of MySpace.

The computer games industry shows that proprietary ownership of the core game can be combined with massive subsequent player development: Electronic Arts increasingly provides not just games but platforms and tools for communities to develop games. Second Life, the highly immersive game created by Linden Labs, takes this one step further and allows players to create the environment together.

Large computer companies such as IBM are very successful in making money from the open source Linux operating system by selling related services, rather than software.

The point as far as the media is concerned is that the whole domain of media production has expanded: the range of possible contributors and distributors has widened. All sorts of interesting hybrids, collaborations and complements are likely to emerge from the interaction between traditional, industrial-era media and the new commons-based systems of social media production.

It does not have to mean war. But it could because these emerging models present a huge challenge to the established incumbent models of Hollywood, the music business and broadcasters, both public and commercial.

The media industries have relied on high capital costs for creating and distributing content for their competitive advantage. In the old days they could see the competition coming from a long way off because it needed a lot of money and equipment. But over the last decade all that has changed.

It is becoming harder and harder to spot where the competition is coming from in a world in which a 20 year-old college dropout can write a file sharing programme on a borrowed laptop which eventually upends an entire industry's distribution and business model: Napster. We now live in a world where any newspaper reader can also become a commentator and publisher. Where bands can create a following online without a recording studio or a record deal.

In this world, it's not surprising the incumbents have sought out new ways to shore up their position. High capital costs no longer provide a sufficiently high barrier to entry. So instead, over the past two decades, there has been a massive expansion in the coverage of intellectual property, copyrights and patents, to make new forms of social production too costly or too risky.

This extension of intellectual property is presented as merely protecting creators against theft. But one motive is protectionist in the economic sense: a rearguard action to protect an incumbent business model against disruptive, low-cost competition. All of this will make it much harder for consumers to become producers and participants; to cut, paste, add, amend, share.

Some rules of thumb for regulators

If we are now living through a contest between traditional, industrial-era models and emerging, social models, what are the implications for media regulation? Perhaps the best way to answer that is to propose some fairly broad and bold rules of thumb that regulators should follow.

More than ever it matters to be clear about our ultimate goals: why we value media in terms of extending freedom, deepening democracy and improving social justice. Different forms of ownership and organisation have to be judged against what they do by these yardsticks, not just in the UK but globally.

Given the enormous potential contribution social media production can make, the basic question all regulators have to ask is whether enough space is being left for these new collaborative forms of activity to emerge, or whether their habitat is being scorched before they have even got going?

The most important thing for regulators to avoid is interventions which unwittingly or deliberately sustain incumbents while quashing scope for disruptive new entrants to emerge.

Extensions to copyrights, patents and other forms of intellectual property may well reward current rights holders without encouraging them to innovate more, and simultaneously make it harder for new collaborative models to emerge, which might in turn be the basis for new commercial and business models.

Media regulation itself is largely a product of the era of industrial mass-produced media. Regulators were the ultimate arbiters of taste and decency and bulwarks against over-powerful concentration of ownership. But in the new landscape the best alternative to over-powerful big media might not be regulation but increased competition from low-cost social media production.

The overall aim of regulation should be to get the optimum mix of social and commercial, open and closed, proprietary and shared ownership media models. Britain needs to be the best place in the world for media, of all kinds. That means being at the leading edge of encouraging the open and social models now emerging.

The future of public service media in this more participative age will thus be vital: the BBC, created as a broadcasting service should become a platform for mass participation, the centrepiece of Britain's cultural and media commons.

Wikipedia and its ilk present as stark a challenge for the BBC as well as commercial broadcasters: if Wikipedia can provide such excellent online information with just a handful of employees why does the BBC need all that space out at White City? Why did the BBC never come up with anything as radical, disruptive and global in its reach?

It is difficult to know how much law and regulation will be able to do, to resist the trends now at work.

No matter how much Hollywood, the recording industry and broadcasters put into lobbying, they will not curtail the trend towards cheaper, more powerful and distributed technology; they will not curtail people's tendencies to want to have their say, share what they find interesting and find what they need at lowest cost. Law might slow and perhaps divert some of these forces – but it cannot dam up the source.

Modern societies have developed in the context of mass-media and industrial information production, which have shaped our view of where ideas come from, how debate takes place, who can be a media producer and who is merely a sofa-borne consumer. A genuine shift is underway, from production for the masses to production by the masses, which will mean, as Benkler puts it:

"Information and communications are core elements of autonomy and public, political discourse and decision-making. Communication is the basic unit of social existence. Culture and knowledge, broadly conceived, form the basic frame of reference through which we come to understand ourselves and others in the world…the basic components of human development depend on information and innovation and how we disseminate its implications."

The rise of social, commons-based media production allows us to imagine how we could reorganise ourselves, encouraging greater freedom, democracy and possibly justice, while also promoting innovation and efficiency.

Nick Jaffe is at work in his bedroom. So are millions of others. The genie is out of the bottle. It is not going back in.

This article is based on a chapter in Charles Leadbeater's book *We-Think: the power of mass creativity*, which is available in draft, online for download and comment, at www.charlesleadbeater.net or www.wethinkthebook.net. A wiki version of the text will be available at Wikimedia.com.

The book will be published, incorporating comments from this open review, in Summer 2007 by Profile. This chapter draws on the arguments in Yochai Benkler's book, The Wealth of Networks, Yale University Press, 2006.

Processes and institutions: new perspectives on policymaking and regulatory authorities

Philip Booth

Editorial and Programme Director, Institute of Economic Affairs
Professor of Insurance and Risk Management, Cass Business School, City University

Introduction

It's often the case that technical economic work on regulation focuses on equilibrium models of market competition and how regulatory action can, in theory, move a market closer to its theoretical welfare-maximising position.

In this chapter we consider the implications of viewing competition as a process rather than as an equilibrium end-state. We also consider the implications for regulation of the potential failure of regulatory institutions.

We conclude that regulatory structures should, in the first place, remove impediments to the process of competition. Secondly, the actions of regulators should be constrained and their domain clearly defined. This conclusion is a further justification for pursuing a 'light touch' regime combined with mechanisms for nurturing institutional competition. As such, the analysis is complementary to that undertaken by, for example, Armstrong (2005) and by Ofcom itself.

The market failure justification for regulation

One traditional economic approach to regulation is to cite particular ways in which markets fail to maximise welfare, and to propose addressing those 'failings' through regulation. Examples of such market failure include:

- the absence of perfect competition (e.g. natural monopoly);
- the presence of social costs and benefits;
- missing markets, such as incomplete markets in insurance; and
- the presence of public goods.

Many regulatory bodies in the UK have adopted the market failure approach to regulation. This means the development of regulation can involve a process by which the regulatory body identifies market failures and then develops instruments focused on 'correcting' them.

For example, where perfect competition is absent, price regulation is often used to try to re-create the price structure that would exist in a perfectly competitive market. If social costs and benefits exist, then taxes, subsidies or regulations are often used to attempt to equalise private and social costs. Public goods can be produced by a government, or by a corporation on its behalf, and made available to all, financed by general taxation or by a general tax. Sometimes this approach will involve conducting a cost-benefit analysis of regulation (often known as a regulatory impact assessment (RIA)) before deciding to take action.

This approach can have advantages. The most obvious is that it imposes a discipline on regulators to demonstrate that there is actually a problem to be addressed by regulation. It can also provide regulators with a rigorous framework for determining the appropriate instrument to use in attempting to address the problem.

For example, the appropriate instrument to address a problem of social costs from a particular economic activity might be a tax, whereas incomplete markets in information may be better addressed by the compulsory provision of information.

Rigorous economic justification may well lead to better-directed regulation than a seat-of-the-pants approach with administrative or political discretion being exercised by politicians and bureaucrats. A further advantage of making the justification for regulation transparent, using an economic framework, is that it facilitates rational debate, backed up by empirical evidence about the nature and extent of regulation.

Problems with the market failure approach

Despite its advantages, the market failure approach does have serious shortcomings. Indeed, they're so acute that we cannot be confident that any act of regulation will lead to an improvement compared with the 'failed' market.

It has two main sets of problems. The first is that it ignores important insights into the nature of competition. The second is that it does not take into account the imperfect nature of regulatory institutions.

Insights from Austrian models of competition

The textbook model of perfect competition is a situation where there is perfect knowledge, and where identical products are sold at a price equal to marginal cost. This leads, when combined with other assumptions, to all opportunities for welfare maximisation being exploited. But it should be obvious that a perfectly competitive market cannot exist. If we had perfect competition prevailing, there would be no innovations or product differentiation. If consumers or producers were to discover new knowledge, either it would have to be shared immediately with all others in the market or the state of perfect competition would come to an end. New knowledge and innovations occur continually in real life markets.

So competition should be understood as being the *process* by which consumers and producers seek new knowledge to enable the production of new goods or existing goods at lower cost, thus enhancing welfare. If perfect competition were to exist, the *process* of competition would have come to an end.

Crucially, though, it is the process of competition that enhances welfare, not the achievement of some theoretical equilibrium end-point. As Kirzner (1997) points out, the process of competition ensures that incentives exist for greater welfare through the correction of entrepreneurial error (finding lower cost methods of production, finding products that satisfy consumers better, and so on). If regulators are to exist, they best serve the market by removing inhibitions on the process of competition[1] rather than by trying to recreate the hypothetical result that arises from so-called perfect competition.[2]

You could argue that the perfect competition model is a useful reference point to help us understand markets and their outcomes. As it is a model, its assumptions do not have to be valid for it to produce reasonable predictions.

There is some truth in this. But, if the absence of perfect competition is to be used as a justification for forms of regulation that undermine the process of competition, there will be serious consequences for consumer welfare. Kirzner puts it this way:

"Now the mere failure of a theoretical picture to replicate with precision all features of the reality it seeks to explain, is not necessarily fatal for the usefulness of that theoretical picture. But mainstream theory filters out of the picture those aspects of reality which are at the core of an adequate explanation for market phenomena." (Kirzner, 1997, p 30)

In other words, if the regulators' reference model, when judging regulatory intervention on market failure grounds, involves the features of a so-called perfect market, the regulator is using the wrong model against which to justify intervention.

There's a second problem, too. The absence of perfect competition means that there are some undiscovered opportunities for enhancing consumer welfare.

But what are they? We cannot know because we require the process of competition to discover them. A regulator cannot know what the undiscovered opportunities for welfare enhancement in an imperfect market actually are.[3]

This problem is neatly summed up by Hayek (1982, Volume III, p 70) "If the factual requirements of 'perfect' competition are absent, it is not possible to make firms act 'as if' it existed."

Writing specifically about broadcasting, Peacock (2004) suggests:

"A static approach with competitive equilibrium as the norm ignores the important fact that it is the very existence of disequilibrium which indicates to entrepreneurs that there may be opportunities for entrepreneurs to make more than normal long-run profits, encouraging them to use process and product innovations to improve such prospects. In the context of broadcasting, this could include ways of eroding government influence in the market by legal means, such as through new methods of transmission." (Peacock, 2004, p 40)

Peacock cites the example of 'rediffusion' which the Nobel Laureate Ronald Coase had shown was heavily restricted by government intervention, which also limited the development of the cable market more generally.

Insights of public choice economics

Public choice economics also has some uncomfortable messages for regulators.

The most important premise of public choice economics is straightforward: we should not assume that people will behave in one way in the political arena and in a different way in the economic arena.

In the economic arena we generally recognise that agents act in their own best interests and that they have imperfect knowledge.[4] In the political sphere agents will have those

1. There are plenty of inhibitions on the process of competition within the broadcasting market.

2. See also Kirzner (1992) for a fuller exposition of these ideas.

3. If the outcome of perfect markets could be known by regulators, detailed central planning of economies would work.

characteristics too. That is not to say that all agents in the political sphere will act only in their own best interest: altruism is possible in both the political and economic arenas. However, it is prudent to adopt a working assumption of the pursuit of self-interest.

There are a number of implications that arise from combining the assumption of the self-interested participant in the political process with our understanding of various administrative aspects of the regulatory process:[5]

- Bureaucrats cannot 'correct' market failure, even if they wished to do so, because they lack the information to know what the outcome of the market process would have been had the 'failure' not existed.
- Bureaucrats will act in their own best interests, taking courses of action that will lead to promotion and advancement. They are likely to wish to avoid scandal, thus becoming risk-averse, so they may regulate to reduce risks to a greater extent than consumers desire. They will also wish to increase the size of their regulatory bureau.
- Electors in general have no interest in being perfectly informed about political issues because the probability of an individual's vote impacting on the result of an election is infinitesimal.
- Because of this, there are information asymmetries between regulatory bureaux and those to whom they are ultimately accountable: electors. Thus electors are at a relative disadvantage when assessing the merits of proposed regulations.
- Where the benefits of government action are concentrated among particular voter groups, or institutions or companies, such groups have an incentive to lobby for increased regulatory protection. Where the cost of such regulation is dispersed among voters, the losers will have no incentive to lobby to oppose increased regulation because the expected cost of lobbying to the individual voter will be large relative to the expected benefit.
- Politicians will, other things equal, respond to the preferences of the 'median voter' rather than act to create regulatory institutions that might address genuine problems of market failure.

The features described above tend to bias political institutions in favour of a greater level of regulation than will lead to welfare maximising solutions. It will also bias political institutions in favour of forms of intervention that benefit concentrated interest groups.

Public choice economics does not always lead to specific and strong conclusions about the design of regulatory systems. Instead it leads in the direction of some important cautions, as well as towards the view that some forms of institution are likely to give rise to better results than others. The main caution can be outlined as follows:

"One [area of policy into which public choice economics has been integrated] is simply a lack of enthusiasm for government as a solution to problems. The view that government is the automatic perfect solution to innumerable problems no longer exists.

4. Indeed it is, of course, the problem of imperfect knowledge that leads to information asymmetries that regulation is often designed to address.

5. See Tullock et al (2000), reprinted with revisions in the US as Tullock et al (2002), for a clear and full discussion of these issues; see also the original texts on rent seeking and public choice economics such as Tullock (1967) and Buchanan and Tullock (1962). Tullock et al (2000, 2002) contains chapters specifically on telecommunications and the internet.

Not very long ago, the simple proof that the economy did not function perfectly was regarded as an adequate reason for governmental action. Today, we start from the knowledge that the government also does not function perfectly and then make a selection between two imperfect devices" (Tullock et al, 2000, pp 11-12).

Tullock suggests that we should try to assess the costs and benefits for each of two imperfect approaches when thinking about regulatory problems.

It will not be enough, though, to suggest that this conclusion should lead us to a view that we can rely on RIAs to assess whether to regulate in particular instances. One of the problems highlighted by public choice economics is the 'information asymmetry' that exists between regulators and those to whom they're answerable. We cannot expect the general public or Members of Parliament to equip themselves with the information to judge the costs and benefits of regulatory actions and to appraise RIAs.

More generally, not regulating at all might well be better than creating an institutional structure that will be able to expand its power relatively unchecked by those to whom it is nominally accountable. In general, there is little incentive for those to whom a regulator is accountable to hold regulators to account,[6] yet there are strong incentives for regulators to over-regulate.

Although it is difficult to make precise and detailed regulatory prescriptions from public choice economics, we can make inferences about the form of institutional structures that might be most appropriate:

- government agencies should work over the smallest area possible (e.g. at the State rather than Federal-level in the USA, and country rather than Union-level in the EU);
- where possible, there should be competition between regulators;[7]
- regulators should have very clear and specific objectives;[8] and
- regulators should be constructed in such a way that they are relatively impervious to being manipulated by interest groups – particularly by concentrated producer interests who have a strong incentive to lobby.

These conclusions do not conflict with our earlier deduction that regulators should work to remove inhibitions on competition rather than correct so-called market failure.

UK regulators and the market failure approach

In this section we consider briefly some approaches to regulation that have been proposed by the financial regulator, the Financial Services Authority (FSA), and Ofcom, both by way of contrast with each other and also to compare with the conclusions of the analysis above.

The FSA has made explicit statements about their approach to regulation – though their statements have not always been consistent,

6. There is even less incentive for voters to hold politicians to account for the regulatory structures they set up, given the dispersed nature of regulatory costs.

7. This is possible in financial regulation where, for example, stock exchanges can compete for business on the basis of their systems of regulation or, on a more limited basis, through the mutual recognition approach to regulation in the EU.

8. Ofcom's objectives are reasonably clear and limited; on the other hand, the FSA's objectives are extremely general and offer a large amount of discretion to the regulator.

as shown by the following quotes from FSA documents:

"In meeting our objectives in a manner consistent with the principles of good regulation, we have adopted a regulatory approach based on correcting market failure... There are, however, numerous cases where unregulated financial markets will not achieve the best outcome due to some form of market failure, making action on our part **necessary**." (FSA, 2003 – my emphasis)

On the other hand, in a speech in 2004 (see FSA, 2005, p 12), the FSA's Chairman Callum McCarthy said:

"In the FSA's work, a principle we have enunciated…is that regulatory action should only be taken when there is market failure … The strong – and to me correct – test goes beyond that: there must be both market failure and the prospect that intervention will provide a net benefit…Identification of a market failure should not lead to the assumption that regulatory failure is less likely or less costly. It is an open and empirical question, which needs analysis on a case by case basis."

Clearly this quote is incompatible with the earlier FSA sentiment that market failure renders regulatory action *"necessary"*. It also suggests that the FSA believes that the appropriate approach to regulation is a market failure analysis followed by an RIA, to assess the costs and benefits of regulation.

Nevertheless, the approach has weaknesses. Our understanding of the process of competition, and the economic processes that exist in political structures, leads us to conclude that we should adopt institutions and norms for regulation that limit discretion and apply principles, rather than allowing regulators *carte blanche* to pursue the 'correction' of so-called market failure, even where tempered by an RIA.

The legal framework for the FSA also falls short of the public choice principles enunciated above. Two of the four main statutory objectives of the FSA are: "maintaining public confidence in the financial system" and "securing appropriate degrees of protection for consumers". These give enormous amounts of discretion to the FSA.[9]

Ofcom's regulatory principles do seem to recognise the need for institutions not to be judge and jury in determining whether they expand their regulatory role (see Ofcom, 2005, p 9).

Ofcom states that it will operate with a bias against intervention, but with a willingness to intervene where *required* (bold in original). And, perhaps most importantly, it states that it will regulate where there is a specific statutory duty to work towards a public policy goal that markets alone cannot achieve. Thus Ofcom is suggesting that it is happy to be restrained.

We can illustrate how this restraint might work by giving two examples of where Ofcom, hypothetically, might consider intervention. In the first example, let's assume that there is a statutory duty to broadcast news in Urdu. If the market did not provide that service, Ofcom would use regulation. This is not using discretion – merely discharging pre-defined statutory duties.

To take another example, spectrum scarcity might lead to higher-than-perfectly-competitive

9.It is also notable that promoting competition, or removing inhibitions to competition, was excluded from the FSA's objectives.

prices for television advertising slots. But the implication of the Ofcom statement is that regulation would not be used to 'rectify' the absence of perfect competition in that regard: the market is providing advertising, despite its imperfections.

In a sense, it is difficult to justify any regulation if one accepts the Austrian analysis of the process of competition.[10] However, it can be said that the model that Ofcom states it uses is more compatible with that model of competition than, for example, the approach to regulation of the FSA.

Ofcom also states, (Ofcom, 2005 para 2.6) that "we believe that the benefits for citizens and consumers are potentially largest where markets are open, new entrants can compete against incumbents, investment is encouraged and innovation flourishes". This would seem to be a statement in favour of ensuring that the 'process of competition' should be allowed to take place, rather than a statement in favour of artificially recreating a hypothetical, and perfectly competitive, market outcome.

The arguments applied to public service broadcasting

When spectrum is scarce, there is concern that a free broadcasting market will be dominated to too great an extent by the preferences of the median viewer.

The traditional argument is that if, for example, there are two broadcasters, then both will move towards satisfying the tastes of the median viewer in order to capture the greatest share of the market. There will be many types of programming that may not take place under this arrangement, even though they may be desired by a substantial proportion of the population.

Furthermore, broadcasting in an era of limited spectrum has public good properties – it is very difficult to charge for individual programmes. In such a market, all broadcasting may be 'underprovided' and certain types of broadcasting, including programmes of educational value that may be attractive only to a minority of people, may well not be provided at all: those who are willing to pay for such services cannot communicate their preferences effectively.

There are various potential solutions to this public good problem. These include regulating broadcasters that are given licences for the scarce spectrum and the provision of free-to-air public service broadcasting (PSB) (see Ofcom, 2004b, chapter 3).

The current methods of finance of public service broadcasting are combined with a regime in which entry to the market is restricted and in which funding for public service broadcasting is institutionalised in a small number of broadcasting organisations – most notably, the BBC. There are serious constraints here on the process of competition.

Developments in satellite, digital and internet technology remove the impediments to competition caused by spectrum scarcity. Yet the mechanism of funding the BBC, through a licence that has to be purchased to watch any television at all, regardless of whether you choose BBC services or not, couldn't have been better designed to bring to the fore the problems of regulation discussed above.

10. Aside, of course, from regulation generated by the market process itself and basic legal frameworks that are required for the operation of markets.

The Corporation is both subject to regulation and is the recipient of a large amount of public funding; it is a concentrated producer interest;[11] and it produces programmes free at the point of use in a way that effectively prohibits competition. Any competitor providing a similar product would have to charge for it.[12]

New technological developments undermine the main justification for intervention to support public service broadcasting. These issues were discussed in Ofcom (2004a,b), but had been discussed much earlier by Alan Peacock in CFBBC (1986).[13]

Public service broadcasting no longer has obvious public good properties. Nevertheless, some believe that it has externalities attached to it, as it helps to build a cohesive society and helps to inform and educate (see Ofcom 2004a,b). As Ofcom (2004b) puts it, "There is a sustainable rationale for PSB. The Market will continue to under-provide programming that we value as citizens and as a society."

We do not debate the merits of this argument; we simply note the justification for regulatory and governmental support for PSB and ask how it can best be supported, given our discussion of the nature of competition, institutions and public choice economics.

The externality argument does not provide a rationale for concentrating PSB into a single producer. If there are external social benefits from the production of a particular product, the appropriate mechanism of intervention is the provision of a subsidy to equalise social costs and benefits.

It should be noted that the arguments we have used against the market failure doctrine also lead us to the conclusion that governments and regulators cannot determine the optimal extent of any such subsidy. Consequently, this chapter does not argue that such intervention is better than no intervention at all. It merely argues that, if intervention is to take place, then the appropriate form of intervention is to subsidise PSB on a fully competitive basis.

In fact, the economic justification for regulatory encouragement for PSB in the digital age is very similar to that of subsidising arts and blue-sky research in universities. In these latter cases, there is competition between different forms of institution that can benefit from the subsidies and grants provided by the Arts Council and Research Councils. Furthermore, because the purposes for which subsidies are given are defined widely – and, in the case of Research Councils[14] and the Arts Council, many forms of institution are

11. The BBC is very effective in promoting its interests to the public at large, to regulators and government (as private producer interests frequently are too). For example, the report *Building Public Value* (BBC, 2004) promotes the BBC's own case very well. On p 6 it even suggests that the 'right broadcast might save my life'. This report, and others like it, is produced at the expense of the licence payer. The BBC as a concentrated interest group has an incentive to use money in this way. The dispersed interest group of licence-fee payers cannot articulate their interests easily. Furthermore, they suffer from 'rational ignorance' – there is no point becoming informed about the issues and campaigning for a different system of finance for public service broadcasting because the costs to an individual of campaigning far outweigh the expected benefits which will be widely dispersed.

12. Unless another producer's product is so superior that the excess of its market value over the alternative provided by the BBC covers its cost of production, a competitor broadcaster will not be able to produce content profitably. If it were able to produce the content profitably under these conditions, it would not be public service broadcasting!

13. See also Elstein (2005)

14. It can certainly be argued that the concentration of funding in Research Councils has institutionalised particular approaches to research. However, two points could be made here. Firstly, Research Councils go far beyond subsidising blue-sky research and therefore have become larger and more important than they would be if they only provided grants on a competitive basis for blue-sky research. Secondly, the situation would surely be worse if all the funding for blue-sky research were to be concentrated in a few government-sponsored institutions.

allowed to submit applications for subsidy – new ideas from a wide variety of competing groups can be financed.

These methods of subsidising the provision of arts and education allow, to a greater extent than is the case in public sector broadcasting, the competitive processes to continue. Furthermore, they disperse the potential interest groups who benefit from regulatory action, making regulatory capture less likely.

A 'Public Service Broadcasting Council' could distribute money on a competitive basis and would only need to finance public service adaptations to programmes. Ofcom (2004b) considered this approach and decided against it. In making their judgement, it is clear that they looked at the advantages of greater competition for PSB funds,[15] but they did not look at the constraints that public choice economics suggests should be placed on institutional discretion and the concentration of power within institutions.

Competition between regulators?

Assuming that there is some role for regulation within the broadcasting market, it is reasonable to ask whether the process of competition should apply to regulators. For example, at the EU level, should we look to have a single regulator or should regulators operate at member-country level? We should be careful in applying arguments used to analyse the functioning of markets to institutions that operate outside markets. Nevertheless, there are some lessons we can draw.

There are three main forms of EU regulation that could be adopted. We could have a single pan-EU regulator; we could operate a system of mutual recognition whereby a market participant can choose to locate in any EU country, regulated by the regulator in the home country, but then trade in all other EU countries; or we could operate a system of completely independent national regulators, but with trans-national agreements to facilitate free trade, consistency of regulation and so on.

There would seem to be two advantages of the mutual recognition approach. Firstly, a large number of regulators would be free to innovate and learn from each other so that good practice could be copied. Secondly, there are incentives for regulators to provide a set of regulatory services that is compatible with the flourishing of business. Regulators could lose business because regulated companies could move to a different regulator while continuing to operate in all EU countries.

We cannot know in advance the 'optimal' set of regulations and it is possible that separate national regulators with mutual recognition may help us discover this better than a single regulator at the EU level.

There are also public choice justifications for a mutual recognition approach. Regulators would be constrained from over-regulating because, if they over-regulated, they would lose regulated companies to alternative regulators. Furthermore, the smaller the domain over which regulators have jurisdiction, the more accountable they are likely to be to the group of citizens they are supposed to be serving. EU political institutions are particularly unaccountable to the people they serve.

15. Indeed, the benefits of institutional competition were taken very seriously with an external working paper being published on the subject. Ofcom proposed, instead, the financing of a 'Public Service Publisher' that would commission and publish public service broadcasting and make it widely available. This would provide some competition for the BBC.

Prima facie, it would seem that mutual recognition may work best in areas such as insurance and investment services, where regulation is designed to provide a package of services to consumers (for example, a safe market in which to transact shares or the provision of information to purchasers of financial products). Consumers can then choose the jurisdiction under which the financial services they use are regulated.[16]

While the above arguments have merit, they are not necessarily conclusive in the case of broadcasting regulation. If national control of regulation (but with mutual recognition) is preferred, then why not allow Parish Councils to set up broadcasting regulators?

It is conceivable in broadcasting and telecoms that there is a genuine EU public interest. For example, with regard to competition regulation, regulation is likely to be more liberal if the market is defined at the EU, rather than national, level.

Similarly, in order to facilitate competition, one may wish to prevent individual EU governments from placing restrictions on companies based in another country from having access to their own. However, it could be argued that these objectives should fall under the remit of competition policy and trade policy rather than broadcasting and telecoms regulation.

We can only repeat that, while we can learn some lessons for regulation from our analysis of competition, the parallels are very imperfect.

Conclusion

Our understanding of competition as a process in which institutions evolve and compete to satisfy consumer demand, combined with our understanding of how government institutions behave in practice, informed by public choice economics, leads us to this conclusion: namely, that regulatory discretion should be minimised and that the main role of regulators should be to facilitate the *process* of competition.

If public service broadcasting is to be subsidised because it has externality benefits, such subsidies should be transparent and dispersed across a wide range of bodies free to compete.

Some lessons have also been drawn for regulation at the EU level. However, we have added a caution here. A competitive market economy facilitates the discovery of new institutions, new products, new ways of production and new services that enhance welfare. Regulation operates outside the domain of the market and regulators are not subject to the same conditions as firms operating within a market economy.

16. Indeed, it is a pity that the Financial Services Action Plan is being rolled out with the harmonisation of so much regulation at the EU level in a system that is perfectly amenable to mutual recognition.

References

Armstrong M. (2005), Public Service Broadcasting, *Fiscal Studies*, Volume 26, Number 3, pp 281-300

BBC (2004), *Building Public Value: renewing the BBC for a digital world*, BBC, London, UK

Buchanan J. M., (1962), *The Calculus of Consent: Logical Foundations of a Constitutional Democracy*, University of Michigan Press, Ann Abor, US

CFBBC (1986), Report of the Committee on the Financing of the BBC, Cmmd.9824, London: HMSO (the Peacock Report)

Elstein D., (2005), Public Service Broadcasting in the Digital Age, *Economic Affairs*, Volume 25, number 4, pp 68-70

FSA (2003), Reasonable Expectations: regulation in a non-zero failure world, The Financial Services Authority, London, UK

FSA (2005), *International Regulatory Outlook*, Financial Services Authority, London, UK

Hayek F. A. (1982), *Law, Legislation and Liberty*, Routledge & Kegan Paul, London, UK

Kirzner I. M., (1997), *How Markets Work: Disequilibrium, Entrepreneurship and Discovery*, IEA Hobart Paper 133, Institute of Economic Affairs, London, UK

Kirzner, I., (1992), *The Meaning of the Market Process: essays in the development of Modern Austrian Economics*, Routledge, London, UK

Ofcom (2004a), *Ofcom Review of Public Service Television Broadcasting Phase 1: Is Television Special?*, Office of Communications, London, UK

Ofcom (2004b), *Ofcom Review of Public Service Television Broadcasting Phase 2: Meeting the Digital Challenge*, Office of Communications, London, UK

Ofcom (2005), *Ofcom's Annual Plan 2005/06*, Office of Communications, London, UK

Peacock A., (2004), *Public Service Broadcasting Without the BBC?*, IEA Occasional Paper 133, Institute of Economic Affairs, London, UK

Tullock G., (1967), The Welfare Costs of Tariffs, Monopolies and Thefts, *Western Economic Journal*, Volume 5, June, pp 224-232

Tullock G., Seldon A., Brady G. L. (2000), *Government: Whose Obedient Servant?* A Primer in Public Choice, IEA Readings 51, Institute of Economic Affairs, London, UK

Tullock G., Seldon A., Brady G. L. (2002), *Government Failure: A Primer in Public Choice*, The Cato Institute, Washington DC, USA

The changing nature of regulation: policy, process and accountability within the European Union

Graham Mather

President of the European Policy Forum

One of the most striking elements of the contemporary debate on economic regulation in the EU is the frequency with which similar issues surface in different sectors. Telecoms, financial services and energy often provide instructive comparative examples.

When Viviane Reding raised the idea of a European regulator, she did so in a rather subtle and interesting way. She said that "the most effective way to achieve a real level playing field for telecom operators across the EU would, of course, be to create an independent European telecoms regulator that would work together with national regulators in a system similar to the European system of central banks". The new European body would be "a light European agency, independent from the Commission and from national governments", and it could "ensure by guidelines and, if necessary, instructions that EU rules are applied consistently in all Member States".

Mrs Reding's idea has not attracted much support across the EU. Yet it does provide an opportunity to reflect on complex interrelationships between national regulatory and competition authorities, their counterparts in other Member States, the networks which have arisen to accommodate them, and the European Commission itself. The parallels between sectors are significant in economic and as well as institutional terms.

In financial services, for example, the idea of a single regulator for the European securities and financial markets has been mooted on a number of occasions. And in energy, one of the immediate reactions to the realisation that gas supplies to Ukraine might be restricted was the suggestion from the EU's energy commissioner, Andreas Piebalgs, that a single EU energy regulator might be necessary to tackle the problem.[1]

Yet in neither case has a single regulator emerged. In securities markets the system designed by Baron Lamfalussy has spawned three pan-European committees which bring together existing regulators from across the Member States. CESR, the Committee of European Securities Regulators, is the longest established. CEBS, the Committee of European Banking Supervisors, based in London, is developing well. And CEIOPS, the Committee of European Insurance and

1. Energy ministers rejected the idea at their meeting on 14 March 2006.

Occupational Pensions Advisors, is developing its work.

In energy, the Council of European Energy Regulators and the European Regulators Group for Electricity and Gas provide an effective means of informing and guiding EU and member state policy development. CEER recently, for example, highlighted the problems of independence of national regulators, noting that:

"In some countries, energy regulators continue to share authority with other actors, i.e. ministries enjoy power to approve, reject or amend the energy regulatory authority's decisions, or they have regulatory competencies in their own right. Furthermore, in some countries there is more than one energy regulator, at national and regional levels. The more agencies that are involved in decisions, the more likely that inconsistent regulation between Member States is likely to arise and persist. The CEER supports the Commission's assessment that 'as a minimum, regulators should have clear powers of the issues covered in the Directives and with independence from industry and government, especially where the state retains a large shareholding in energy businesses."[2]

The network model has also proved effective in competition policy. The European Competition Network has proved a useful parallel network to the work of DG Competition, helping to ensure the effective and consistent application of Articles 81 and 82 of the Treaty. National competition authorities, working together in the ECN, thereby provide something of a check-and-balance on the operation of the competition functions of the European Commission, and effective staff work means that lead authorities can emerge to handle issues and cases affecting a group of Member States.

A network of European telecom regulators therefore parallels almost exactly the networks which have been set up to look at securities regulation, banking supervision, insurance and occupational pensions supervision, energy regulation and the interaction of the competition authorities themselves.

The European Regulators' Group (ERG), for telecoms chaired from January 2006 and for one year by Kip Meek of Ofcom, is therefore typical of a general approach in that it operates under a framework established by the European Commission and in parallel with the Commission's own operations in telecom regulation.

The advantages of regulatory networks

How do these regulatory networks operate in practice? The advantage is that they can provide an exchange of intelligence which is based on informality and flexibility; they can bring regulators across the EU up to speed with best practice, quickly and cheaply; they can provide informal but effective sounding-boards for exchanges of ideas and data, intelligence and future plans; and they can, as Commissioner McCreevy said in another context, come early to the party with approaches based on subsidiarity and close connection with regulated entities.

These are sound advantages which also facilitate peer group review and the fast adoption of regulatory techniques which have proved effective. There is no real downside –

2. CEER Regulatory Benchmark Report 2005

the debate on stronger control from the Commission can always resume if the networked regulators run into difficulties.

When the European Policy Forum convened a conference in Brussels to look at telecom regulation in the run up to i2010, some 14 telecom regulators and 13 competition authorities took part. The agreement reached between Ofcom and BT to open up access to its wholesale networks generated particular interest from across the EU. Many regulators in Member States said that they thought this approach, based on competition law but operating by a voluntary agreement and with tight benchmarks and constraints, had useful lessons for adopting elsewhere.

Similarly the ERG is an important force for subsidiarity and diversity. At its May 2006 meeting in Vienna, for example, it high-lighted several points of divergence with the Commission's proposals on mobile roaming services in the single market. Drawing on their experience as national regulators, ERG members warned that the then Commission's proposals on roaming could have a negative material impact on many consumers and also on industry, with a threat of monthly subscription rates on roaming or the withdrawal of roaming services altogether.

In terms of remedies, the ERG has identified concrete examples of best practice to allow the development of a common position among national regulatory authorities (NRAs). It has launched an experimental approach on collaboration between NRAs in various markets, and on specific best practices, and this is producing progress reports on each of the experimental initiatives. In addition to work on competition issues, the ERG receives reports from the Commission on current cases and publishes detailed documentation of much of its work on its website.

The approach – developing co-operation across Europe, focusing on best practice, sharing information and ideas but lacking any top-down authority or control – does seem to fit very well with concepts of a decentralised yet constructive and co-operative approach to the opening of European markets.

ERG activity adds to, rather than takes away, from knowledge and effectiveness of the Commission. It's difficult to see how a stronger central regulator, with the role of the NRAs reduced to that of the Euro-zone central banks where all key decisions are taken by the European Central Bank (ECB), would offer significant advantages.

The risk would be a temptation to introduce single solutions and make shortcuts to the solution of market problems. Where abuse of a dominant position in a member state occurs by an incumbent operator, or where there is cartel activity, Europe already has its EU-level and national level competition authorities wit the ability to work together to take action.

Another drawback of a single regulator is the fact that national regulators know their own markets better, and each one is different from the next. Subsidiarity to be able to adapt remedies to national circumstances is very important here, as a one-size-fits-all solution is unlikely to work.

Advancing the cause of competition

Under the Lisbon process, opening markets to competition from suppliers domiciled in other Member States is of key importance to the Union's economic objectives. Yet progress has been disappointing.

The original technique by which market opening was to be secured – the Directive has – proved an inadequate instrument to secure a significant opening of both pan-European financial services and energy markets. Entrenched economic interests have exercised a powerful grip on some member state governments. Neither energy nor financial sectors have faced the revolutionary technological changes which have helped to open up telecoms and IT.

In these two areas, Brussels has therefore shifted to more direct and aggressive techniques. Sector inquiries have been opened by DG Competition on three fronts: the European gas and electricity markets, the supply of business insurance and retail banking. Dawn raids have taken place on the offices of a number of incumbent European energy suppliers and many expect that cases for the abuse of dominant position will be brought with the prospect of extremely large fines and possible structural remedies.

In energy markets, as noted above, there are also signs that the European Commission is seeking to insist on national regulators being independent from ministries in Member States. It now sees this approach as more promising than a centralisation of regulatory control. In a similar way, it is looking at whether European action could help Member States to unblock planning and other systems which currently impede the speedy introduction of new energy infrastructure, storage and interconnection facilities.

The problem is one of protectionism of national champions, the European Commission already has ultimate legal power to take national regulators or Member States to the European Court of Justice – an effective solution to this problem, as has been recently shown in Spain and France.

In Spain, one example is the case of Endesa and the takeover by the German company E.ON. The European Commission rejected the case, made by the Spanish authorities, for the application of the public interest clause in the Merger Regulation (Article 21.4) by which Member States can impose additional regulation, on the grounds that the measures are disproportionate.

Looking beyond price

These changes in approach represent a recognition that some of the economic issues are changing in relative significance. I sense a shift from an absolute focus on price competition and the ability of customers to switch suppliers, towards a greater focus on security of supply and quality of service.

It is becoming obvious that while policymakers will continue to seek to open up competition, drive down prices and facilitate easy customer switching from one supplier to another, their eyes have now been caught by two significant and potentially competing objectives: the need to secure adequate investment and supply, and the need to secure continuous service of satisfactory quality.

These competing strands can be seen in the European Commission's white paper on energy, published in 2006, which shifts the emphasis towards developing trans-European energy infrastructure, reserves of storage capacity and external diplomatic and security initiatives – all focused on strengthening the reliability and assurance of energy imports. While price competition remains important to consumers and a significant part of the Lisbon strategy, the new broader focus is concentrating on supply rather than price.

We can also see a rather similar phenomenon in telecoms: as mobile telephony and broadband services become part of the daily fabric of life, and price competition makes them almost generic in nature, interest is shifting to new areas such as the development of next generation networks, the building of infrastructure capable of carrying super-fast and super-wide traffic, and the value-added experiences which will result.

As the supply of basic connectivity becomes almost generic and falls further in price, telecom customers will be looking in particular at the quality and reliability of their experience. Certain large suppliers in both cable and fixed telephony have had seriously unsatisfactory customer service records. Some new entrants have allowed their marketing to suggest reliability that they cannot deliver. Providers with an interest in making their services 'sticky' have been slow to facilitate migration to competitors, and have made it difficult to unsubscribe. And all suppliers have a temptation to bundle services together in a way which may make it difficult to compare prices for particular services, or indeed to change supplier.

When should regulation hold back?

It is clear from this discussion that EU regulatory design is beset by a number of contradictions and dialectics – the most profound of which is the tension between the need for large-scale investment and the maintenance of effective competition and focused regulatory action. Crucial to the debate on how to foster investment in new generation telecoms networks is the issue of regulatory forbearance: the idea that the relevant regulatory authority should refrain from regulating a specific market for a fixed period of time in order to encourage investment in new technology and services.

The German proposal regarding the established incumbent Deutsche Telekom has helped to clarify the issues and unite other national regulators and the European Commission in shared concern. A period of regulatory forbearance, possibly for three years, has been suggested in order to provide an incentive for the established incumbent, Deutsche Telekom, to develop a national broadband network which would partly rely on VDSL technology.

As the European Policy Forum recently argued, such an initiative would preclude rivals from competing with Deutsche Telekom, which would gain first-mover advantage across a wide range of telecom services. Furthermore, where Deutsche Telekom competes for global business against rivals such as BT or Telecom Italia, it would be placed at an unfair advantage in that it could shut out rivals from providing bit access services in Germany; yet its competitors, such as BT, would be required to open its own broadband networks to Deutsche Telekom.

More generally, there is a conceptual problem with regulatory forbearance. Rather than establishing a clear position in which a sector or function is, or is not, subject to regulation for reasons which align with economic and competition rules, it creates a kind of regulatory limbo in which regulatory inaction *pro tem* may, or may not, be followed by a return of the regulator.

Instead, it would make sense to build the new regulatory regime for next generation communications technologies on three foundations.

- Where effective competition is established, the abolition of *ex ante* regulation for all consumer services, in order to encourage a suitable environment for the development of new forms of innovation and competition across these cutting-edge technologies;
- Refocusing *ex ante* regulation on clearly perceived infrastructure bottlenecks. The onus should be on solutions that are based on competition law and voluntary initiatives in the case of replicable facilities and services.
- An urgent need to pay attention to rapidly emerging markets with the onus on encouraging investment in next generation core and access networks. In this context, there should be a priority on promoting equitable voluntary access arrangements, as demonstrated in the legally-binding undertakings made by BT on the access infrastructure it operates within the UK.

Impact assessments: a key to quality

One of the most important means of developing and improving the quality of European regulation, and implementing the better regulation agenda, is the technique of impact assessment. Ofcom, with other UK regulators, has moved quickly to build these into its systems, and UK best-practice developed by what is now the Better Regulation Executive has played a significant part in influencing an impressive EU model.

The European Commission's most recent guidelines on impact assessment now match world best-practice and, together with strengthened activity in the Council Secretariat and the European Parliament, hold out the prospect that future EU legislation and regulation will be submitted to comprehensive and systematic impact assessment. The EU's Better Regulation agenda remains of high policy significance with senior Conservatives wholly committed to its development.

However, as my colleague Frank Vibert and I have argued in recent work on this subject for the City of London Corporation,[3] it would be unwise to become self-congratulatory about the successful introduction of the new techniques. Although they now aim to align legislation and regulation at European national levels with regulatory best-practice and the needs of businesses, financial institutions and civil society, this has taken some years to achieve and the quality of impact assessments remains patchy.

To some degree, the better regulation initiatives, and impact assessment systems in particular, are counter-intuitive. They provide a brake from the dynamic to legislate and regulate which is a powerful phenomenon in modern societies. Legislators have traditionally seen the production of statutes as their core activity. Regulators exercise their functions primarily through the promulgation of regulatory measures. A system which poses constraints on this process – whether by requiring a more thorough assessment of risks, a more careful quantification of costs and benefits, an exploration of unanticipated consequences or a wider quantification of relevant factors – needs to be durable to survive.

3. *Evaluating Better Regulation*, City of London Corporation, October 2006

At one level, this durability is achieved by buy-in among those operating the systems. If legislators and regulators believe that impact assessments will help them to produce better-crafted, more successful and more acceptable measures, then the prospects for the system are positive.

On the other hand, the impulse to legislate often generates demands for speedy action in the aftermath of adverse public events, economic problems, national crisis or developments that generate widespread public concern. Even the slowing down of the process because of the need to produce an impact assessment is, in those circumstances, likely to be unwelcome.

In the same way, the existence of standard methodologies for assessing costs and benefits makes impact assessments contingent and contestable. You can already see signs that non-governmental bodies are commissioning their own impact assessments where they believe that the official version may in some respects be inadequate, or may fail sufficiently to bring into account factors which they deem especially relevant. This makes a strong debate likely around impact assessments, and although this is a healthy development it may generate demands for the process to be truncated or indeed by-passed.

While impact assessments, by their very nature, require thorough technical work it is impossible wholly to separate the economic and other elements which they examine from the stuff of politics. Strong political pressures may seek to overcome the more technical analysis which impact assessments generate and explicitly or implicitly, to override them.

Realising the benefits of impact assessments

All these factors make it desirable to establish systems which will help to ensure that the benefits of the impact assessment system are entrenched among European institutions, national governments and sector regulators. There are two requirements.

The first is a 'compliance' role to ensure that there are means of scrutinising and auditing the operation of impact assessment systems to ensure that these assessments have been carried out where they ought to have been; and at a suitable level of quality; and that their conclusions have been properly fed into the legislative, regulatory or policy-making processes.

The second requirement is for a 'lesson-learning' function to draw wider conclusions from the measures which have been introduced following the deployment of impact assessments. Because impact assessments provide quantification of objectives, techniques, costs and benefits they introduce, for the first time, a framework against which legislation and regulation can be evaluated and reviewed, yielding useful lessons for the future.

At one level, this lesson-learning function allows the position down the line, after the introduction of legislation or regulation, to be compared with the objects of the measures themselves. Using the impact assessment figures as benchmarks of intentions, it is possible to attempt a more informed review of whether policy aims have been achieved; of whether costs and benefits have been broadly in line with expectations or significantly adrift from them; and whether unanticipated effects or developments have intervened to change the outcomes from those anticipated or desired.

This evaluation or audit role is being developed at Ofcom, notably in its Consumer Panel's development of a toolkit to assess whether consumer interests have been effectively handled in policy initiatives. The technique is capable of being very widely adopted in regulatory systems across the sectors, and it meshes well with two further areas which are at large in the EU regulatory debate.

While the EU is rapidly developing its processes for regulatory co-operation and networks, and for the impact assessment of legislation and regulation, it has found it more difficult to simplify the *acquis communautaire* of existing directives and regulations. Indeed the Commissioner responsible for this area, Gunter Verheugen, has gone on record suggesting that Member States are not enthusiastic about this process, which could result in a need to unravel some of their own legislation in force.

Once again, we are hitting a problem of counter-intuitive measures with legislators being invited to undo some of their work rather than to pile ever-more measures on top of an unchanged settlement. Audit techniques can assist here, and the proposals we have put forward to introduce systematic audit and review of the way impact assessments have worked can also be used to revisit the legislative *acquis*. For if Europe cannot meet the ambitious simplification targets it has set itself to roll back redundant legislation and regulation, it will fail to meet its Lisbon aspirations and lose a valuable opportunity to improve its global competitive position.

It would also miss the opportunity to demonstrate an emergent and important new form of accountability. The European institutions are, by their nature, preoccupied with complex technical economic and regulatory issues. Regulators work hard at national level to communicate effectively with their stakeholders, but find it more difficult to be visible and accountable when working with their 24 European counterparts – or interacting with the Commission, Council and Parliament, all of whom have struggled to achieve effective dialogue with, and accountability to, citizens and consumers.

Even where Europe tries to break out of old legal forms into new results-focused ambitions such as the Lisbon Process, using co-operation and flexible approaches, the task can be a hard one. Few Europeans stopped in the street would have an immediate instinctive grip on the benefits of this 'open method of co-ordination'. As the protracted discussions with the European parliament over the Lamfalussy committees have found it can be extremely difficult to achieve effective regulatory co-operation married to parliamentary callback and review.

Yet there are signs that regulatory networks are beginning to achieve this. Because they are based on national regulatory authorities, they promote subsidiarity. Because they are not top-down led, they encourage lesson-learning and peer review. Because they have strong existing stakeholders, and focused media attention, they bring with them existing and effective accountability systems.

As new audit and review techniques are developed, they lend themselves to external scrutiny, including that of national parliaments. Any review of the way in which regulation is changing across the sectors in the Europe of 25 must be optimistic about the durability and effectiveness of the systems which are emerging.

The convergence continuum model: a framework for analysing regulatory reform in Asia-Pacific

Steve Burdon
Visiting professor at University of Technology, Sydney

Introduction

During the last five years, Asia-Pacific has improved from third to first place in the delivery of new technologies among the regions of Americas, Europe and Asia-Pacific. As this dramatic shift has taken place, significant changes have occurred in government policy and regulation of Technology, Media and Telecommunications (TMTs). The more technically-advanced countries have progressed from duopoly competition to a multicompetitive situation and are now on the threshold of convergence competition. In this chapter, the issues underlying these changes are examined and a conceptual model is suggested for exploring different scenarios for future TMT competition and policy.

Importance of TMTs in Asia-Pacific

The importance of TMTs and their transformational impacts on the world, and on the Asia-Pacific region in particular, have been underscored by initiatives of both the World Bank and Asian Development Bank in rating Telecommunications and Technology investments as Category 1 priorities for investment funds.

TMT sectors create direct and indirect benefits in the countries where they are located, delivering economic and social development. The growth of these industries results directly in new, highly-skilled jobs and wealth. The extent of these direct benefits depends on how large the sectors are in relation to the economy, and their rate of growth. For many developed and emerging economies in Asia-Pacific this is significant, but for a number of countries direct benefits have a limited impact. For these countries, what's even more important is the indirect impact; the transformation in the way individuals, businesses and other parts of the society work, communicate and interact.

Particularly interesting are the benefits of TMTs on productivity, as TMT diffusion levels rise across all countries. For developed countries, the impacts can be significant as an Australian Government forecast indicated: "On the basis of quite conservative assumptions…(ICT) will remain the main driver of productivity growth over the next 20 years."[1]

There has been global acknowledgement that beyond their role as a facilitator of

1. *Forecasting Productivity Growth: 2004 to 2024* (2006) Australian Government – Department of Communications, Information and the Arts, March, p.1

communication, TMTs have been a development enabler with significant macroeconomic impact, and a tool to achieve internationally-agreed goals and development objectives.[2] Alleviating poverty, hunger and sickness in developing nations are part of these goals. The appropriate policy and regulatory settings for TMTs can deliver innovative applications in government, commerce, education and many other areas, with a profound impact on people's lives.

In 2004, Asia-Pacific laid claim to the highest share of investment in TMTs. This reflected the leading position of several Asia-Pacific economies in the newer TMT sectors, including 3G where Asia-Pacific has 48.6% of the world's subscribers, the Americas 44.5% and Europe 5.5%. In the case of broadband, Asia-Pacific has 41% of the world's subscribers, the Americas 30.5% and Europe 27.7%.

Major changes are taking place in China where its telecommunication investment alone stood at $US27bn in 2004, almost 15% of the world's total. Europe and the Americas made up 33 and 23% respectively, with 4% invested in Africa, and 3% in Oceania.[3] This is a remarkable change which sees Asia-Pacific as the world leader for new technology subscribers in an increasing number of services.

The role of regulation policy

Regulation policy plays a vital role as an enabler of information societies. Where policy settings and regulatory reforms are contextually appropriate, we can directly link them to the success of new communications platforms, innovative business models and new services.

There is an argument that says that once public assets are sold and government ownership has been relinquished, government should withdraw from direct intervention in the industry as a matter of course. However, the historical evidence of telecommunications and media 'liberalisation' regimes directly contradicts this notion. Indeed, the fact that TMT services remain high-profile and emotive issues which are critically important to electoral constituents mean that you can be sure that government will continually intervene in industry developments. For example:

- Timed local calls have historically been a major political issue, and governments in Singapore and Australia ignored the economic rationalist argument in legislating them out of consideration.
- Broadcasting is heavily regulated in all countries, both in terms of content and number of operators. Televised content and access to local programmes and culture are sensitive electoral issues. France is usually seen as the leading advocate for maintaining high levels of local cultural content. Within Asia-Pacific, Australia and Hong Kong both have legislation which supports their policy objective of increasing, or at least maintaining, local content airtime.

Profile of Asia-Pacific region

Make-up of Asia-Pacific

It is difficult to define Asia-Pacific as an homogeneous region, or to find a common denominator. Indeed, many debate whether the Indian sub-continent, Japan and Oceania are even part of the geographic region. For the purpose of this chapter, they are included in a geographic definition. Segmentation is possible on the basis of developing,

2. United Nations World Summit on the Information Society (2005) Tunis, November
3. *Trends in Telecommunication Reform 2006* (2006) International Telecommunications Union, February

emerging and developed economies – and, rather uniquely among the generic regions of the world, Asia-Pacific has a significant number of countries in each of these categories. These differences are highlighted further when you consider the aspects of culture, values and government regimes.

It is perhaps in terms of economic trade and time-zone commonalities that the countries can be grouped. Intra-Asian regional trade in 2005 was $US1.4 trillion and is forecast to grow to $US1.9 trillion by 2008.[4] This makes trade within Asia one of the biggest markets in the world, with the largest growth rate.

TMT policy and regulation in Asia-Pacific

There's a significant variation in the state and range of the individual countries' national regulatory authorities (NRAs). In a few countries, liberalisation in their tele-communications industries has yet to occur, while for many others bringing broadband to the masses continues to be a major challenge for NRAs. A number of countries are currently working through their options.

- Thailand's telecommunications sector will be fully liberalised in 2006. At the time of writing, some of the ongoing issues include establishing an independent regulatory body and privatising the state monopoly operator.
- The Malaysian Communications and Multimedia Commission was established in 2001 with oversight of economic and technical regulation, consumer protection and content regulation of TMTs.[5]
- In South Korea, the broadcasting and telecommunications industries still have separate regulators, despite being commercially and competitively advanced in convergent services.
- Australia's competitive landscape features bundled triple-play service offerings,[6] and vertical and horizontal integration of its two largest communications providers. Its communications and broadcasting regulators were merged in 2005, but competition policy regulation rests with a separate authority, the Australian Competition and Consumer Commission (ACCC).

The diversity apparent in Asia-Pacific countries suggests that TMT regulatory reform and development will proceed very much based on the decisions of individual countries. To some extent, the differences will be tempered by the close intra-regional economic ties. Poorer countries are likely to adopt or emulate those elements of the policy and regulatory structures of the wealthier countries that could improve their own economic position.

Key issues affecting policy and approach

National interest

Both democratic and authoritarian government regimes of the Asia-Pacific region have traditionally emphasised five-year strategic plans with stated objectives. In contrast, the Westminster-style democracies of the USA and Australia have shied away from such concepts – perhaps owing to the political ramifications linked to ambiguous success.

4. *The Physical Internet* (2006), The Economist, June 17

5. Malaysian Communications and Multimedia Commission (2006), http://www.cmc.gov.my/about_us/roles.asp (accessed: 17 August)

6. In Australia several companies already offer discounts for a 'triple-play' bundle of Pay TV, broadband and voice.

The OFTA objectives are:

• Enable HK to be recognised as world class telecoms centre for doing business
• Ensure HK has available high quality telecoms services at competitive prices
• Ensure HK has high benchmark performance versus world's best practice in four categories:*
  • Regulatory framework
  • The development and effectiveness of competition
  • Consumer benefits from competition
  • Impact on the investment behaviour of operators

*Source: OFTA website, April 2006

Table 1: Hong Kong's proactive objective-based regulation

As digital convergence continues to increase global trade and intensify competition through deconstruction and reconstruction of business models, many of the Asia-Pacific countries appear to have woven TMT regulatory objectives within the fabric of their national strategic plans:

- In Japan, the MITI/METE has taken a proactive approach to industry development, and the appointment of senior government bureaucrats into the senior management positions of the telecommunications entities NTT and KDD has precipitated the review of national interest issues.
- In Hong Kong and, to a lesser extent Singapore, high living standards have accrued from their respective positions as hubs for global and regional trade and, increasingly, tourism. Hong Kong's regulatory framework stands out for its proactive, results-focused management of competition (see table 1).

Socio-cultural norms and the political process

In discussions about media content and broadcasting, all countries have social policies supported by content regulation covering culture and sexual behaviour, political parties and religion. (Table 2 provides an example.) Broadcasting content and licensing are especially constrained in virtually all countries. Values, language, ethnic, religious and cultural differences are perhaps more evident in Asia-Pacific than most regions, and content judged as being offensive can vary widely from country to country.

It's not surprising, then, that throughout Asia-Pacific, rules and regulations vary significantly and the global media conglomerates of the west have a different freedom in Australia and New Zealand relative to Hong Kong, India and China. In a comparison of the context in which television regulation was shaped and developed in the countries of Hong Kong, Japan and South Korea, the governments played a central role in television broadcasting

7. Kwak, K, (1999) *The Role of the State in the Regulation of Television Broadcasting in South Korea*, Media International Australia incorporating Culture & Policy, Australian Key Centre for Cultural and Media Policy, No. 92, Brisbane, pp.65-79

Censorship is practised to:
• Preserve our traditional Asian values such as importance of the family, respect for elders and filial piety
• Maintain racial harmony and religious tolerance amongst our people
• Protect our social fabric and ensure social cohesion
• Protect children and young persons from corruption by undesirable materials
Source: Singapore Government MICA website, August 2006

Table 2: Content regulation by Singapore's Media Development Authority

and could exercise control through the basic regulatory channels. Further, political factors exerted a greater influence than economic and technological change on the nature and practice of government control over television broadcasting.[7] In Western societies, religion is separated from the government but the legislation supports their Judeo-Christian value systems.[8]

Digital convergence is elevating the discussion of different value systems and approaches on a global basis. In the short-term, the value systems of individual governments will be maintained while any excesses viewed as "contrary to the general population interest" are likely to be increasingly threatened. Over the longer-term, it is likely that increased regulation will be imposed on the basis of national interests. It is believed that Asian-Pacific influence on net policy and governance is sure to grow in coming decades.[9]

Telecommunication and media providers' profitability

Broadly speaking, policy should encourage new infrastructure investment and new content development. These two objectives pose different challenges for policy makers and NRAs as they make decisions on implementing competition policy and interconnect conditions.

Telecommunications carriers are required to build significant infrastructure to deliver the new services of 3G mobile and broadband.
As the cycle of development shifts to multiple competitors, the question of access to the new networks is critical. The financial returns for utilities (this could include telecommunication networks) are generally in the order of 10 per cent but the price/earnings ratios for incumbent carriers has historically been closer to 14 per cent (see figure 1).

An analysis of the PE ratios indicates that the markets expect the future profitability of telecommunications carriers to move increasingly closer to that of utility organisations. Practical examples of this shift include the restructuring of NTT in 1999 into three separate entities including one for networks, and moves within Australia for unincorporated joint ventures between carriers to share the infrastructure of 3G networks.

8. For example, News Corporation will accept the status quo in Hong Kong and not criticise the political parties in China but will argue vehemently and try and exert its media power to influence the Government of the day in Australia and New Zealand in order to achieve its business ends.

9. Horovitz, R (2005) *Media Licensing, Convergence and Globalization*, Conference: Activism: Re-drawing the Boundaries of Activism in a New Media Environment, Budapest, Hungary, 14-15 October

Figure 1: Telecommunications carriers forecast price-earnings ratios for 2007

A classic conundrum for regulators has been striking the right balance between a fair financial return while also creating a competitive market for services. The argument advanced by incumbent operators is that they should be allowed to retain all the benefits of their investment before they could be expected to invest further in expanding high-speed, last-mile infrastructure.

Past history, however, provides little encouragement for that view. In the early stages of broadband deployment, incumbent operators hesitated to introduce broadband services. This may have stemmed mostly from fear of cannibalising revenues from established services, such as leased lines and dial-up internet access – particularly in countries where there was very little inter-modal competition from cable companies or other facilities providers. Countries with the most extensive broadband penetration have generally achieved this through strong competition – either from a healthy cable television industry, such as in Canada, or from new market entrants who have capitalised on easy access to facilities through strict unbundling requirements, as in Japan and South Korea.

Where regulatory decisions are poor or delayed they can slow-up the availability of new infrastructure and services for consumers. Some argue that the relatively low penetration rate of broadband services in Australia and New Zealand can be laid at the door of untimely regulatory decisions. One solution might be for governments to quantify future policy objectives for all industry parties and publish international benchmarking of comparative services and pricing.

But whatever approach governments take to regulators, one thing is critical if industry is to make significant investments with medium-term paybacks: namely, certainty of approach. When the New Zealand Government unexpectedly declared open access to Telecom New Zealand's (TCNZ) local loop,[10] the TCNZ share price experienced within a week one of the most dramatic drops ever seen in the telecommunications industry. Although the Government obviously believed such action was necessary, but it was not a good outcome for the stakeholders..

The convergence continuum model – a four-phase conceptual framework

Development of the convergence continuum model

An analysis of TMT regulatory trends in Asia-Pacific countries indicates that a customised approach to regulation is required for each country, appropriate to their specific political, cultural and economic positions. In other words, there is no one-size-fits-all, best-practice regulatory model that can be universally applied to optimise the competing requirements and tensions of government, industry players, consumers and national social and economic interests.

The convergence continuum model is a conceptual framework proposed for understanding the impact of digital convergence on policy and regulation by using a four-phase approach to TMT competition development (see figure 2).

Once the first phase of a country's development has passed and universal access to basic services is available, there is significant evidence that promoting competition results in a quicker uptake of technology and more competitive prices. Interestingly, China created

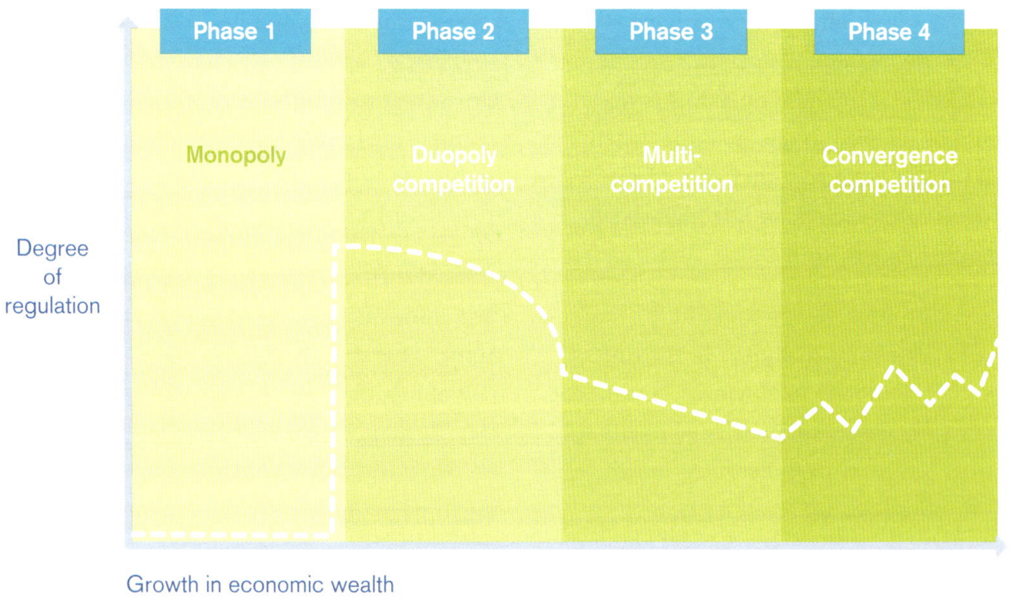

Source: Professor Steve Burdon (2006) Criteria Research and Analysis Pty Limited

Figure 2: The convergence continuum model

10. *New Zealand: Telecom NZ takes a pounding*, (2006) AsiaMedia News Daily, 5 May http://www.asiamedia.ucla.edu/article.asp?parentid=44894 (accessed: 17 August 2006)

two government-owned operators to compete with each other to achieve these benefits.

As expected, incumbent operators fight to defend their natural monopoly while powerful media conglomerates engage in trying to create new monopolies – or at least to minimise competition, by owning or controlling the new convergent services of content and carriage wherever they can.

Key to these strategies are two things: the level of competition authorised by countries (based on the number of competitive players), and the types of services opened up to competition.

Phase 1 – monopoly provision

The initial provision of basic network infrastructure requires a significant investment without an established customer revenue stream. Traditionally, the governments of developing countries have made these investments through government-owned entities. Equally, this approach has also been deemed appropriate by developed countries as well. For example, the Singapore Government was anxious to develop a broadband economy at a very early stage, so it initiated its own network structure and invited tele-communications operators to invest.

During the monopoly phase, governments can be innovative in terms of obtaining finance for the large investments involved. In Vietnam, an interesting BOOT arrangement[11] was initiated using hard-currency international settlements to finance the building of an international telecommunications network. And since the late 1980s in Thailand, the private sector has been encouraged to participate in Thailand's telecommunications market by entering into joint ventures with the relevant state enterprises, or by obtaining expensive concessions from one of the state enterprises in the form of 'build-transfer-operate' contracts as opposed to the normal 'build-operate-transfer'.[12]

Phase 2 – duopoly competition

As countries satisfy the basic service provision requirements of Phase 1, they progress to a telecommunications liberalisation regime with limited competition – usually a duopoly. Invariably this coincides with a shift in status from developing economy to emerging economy, as illustrated by Thailand and Indonesia.

During this phase, the policy objective is to tip the competitive balance towards neutrality and gently erode the incumbent's in-built competitive advantage. Key issues of inter-connect, pricing, universal service number portability and so forth are regulated to enable the newcomer to build a viable position. Furthermore, retail price caps are often instituted as a method of curbing the natural advantages of the incumbent.

In relation to market entry, additional carrier licences are often auctioned or sold via a

11. Under a BOOT contract, a private sector company or organisation agrees with a government agency to build, own and operate a piece of infrastructure, for example, a road, hospital, prison or water filtration plant. After a defined time period the asset is transferred to the government agency, which does not have to fund the initial capital cost of the asset or facility. Instead, the private sector company agrees to build, own and operate the infrastructure for which it earns a commercial rate of return that includes recovering the capital and operating costs over the period of the contract.

12. Presently, private firms build and operate concession networks over a specified period of time and share their revenue with the TOT, CAT or PTD. Ownership in such networks is transferred to the state enterprises immediately after completion of construction and installation as Thai law does not allow private ownership of telecommunications infrastructure.

Phase 1	Phase 2	Phase 3	Phase 4
Bhutan	Afghanistan	China	Australia
Brunei	Bangladesh	India	Hong Kong
Fiji	Indonesia	Malaysia	Japan
Laos	Mongolia	South Korea	
Myanmar	New Zealand		
North Korea	Philippines		
Papua New Guinea	Singapore		
Thailand	Sri Lanka		
Timor-Leste	Taiwan		
	Vietnam		

Source: Professor Steve Burdon (2006) Criteria Research and Analysis Pty Limited

Table 3: Distribution of Asia-Pacific countries on the convergence continuum model

combination of auction and 'beauty contest'.[13] The latter concept is a means of delivering other perceived benefits. In Australia, for example, the second licence-holder was required to roll out a national infrastructure plan with specific technology requirements. The approach that's taken for awarding or auctioning licences is a critical policy issue. Europe's NRAs used a free-market approach, reaping huge financial windfalls from auctioning 3G wireless licences. However the winning operators were so financially-strapped as a consequence that there were long delays in network building; eventually, a few of the operators even surrendered their licences.

In contrast, most of the Asia-Pacific NRAs took a more measured approach – albeit after seeing the impacts of the free-market approach in Europe. The Australian and Japanese Governments adopted the view that a one-off large fee would be less useful to their policy objectives than an approach of smaller or incremental fees accompanied by actual network build and commercial service launches.

As the incumbent operator becomes less dominant, and the playing field levels, regulatory controls over services such as international calls and bundling of services should, at least in theory, be rolled back.

13. Traditionally three methodologies have been used for licensing:

1. Auction: traditionally seen as the most efficient mechanism for putting an economic value on spectrum and allocating it to the firm that uses it most efficiently.
2. Free licences: providing the licences free of charge to the existing telecom players, for example where 3G spectrum allocation can be viewed as an extension of 2G spectrum allocations considering that the operators are allowed to provide services with a technologically neutral licence. To avoid hoarding of spectrum, a minimum spectrum charge can be taken for the period until the service provider rolls out services.
3. Beauty contest: the government invites applications which are then rated according to some pre-set criteria. Licences are allocated to those who the government believes best meet the stated requirements.

Phase 3 – multi-competition

The regulatory shift from Phase 2 of duopoly competition to Phase 3 – a multi-competitive situation – is based on the judgement of whether adequate efficient competition has been achieved in the marketplace, as shown by industry profitability and community benefits.

This situation is likely to evolve as three scenarios are developed:

- firstly, the regulatory mechanisms for achieving a more level playing field have been achieved by interconnect conditions, universal service and other regulatory action;
- secondly, the more advanced networks and technology platforms become, the more likely it is that newcomers will develop a network advantage and/or possibility of creating a niche market; and
- thirdly, that as the demand for services grows, new competitors can build sustainable market share without major detriment to the existing players.

China and India's emerging economies appear to have arrived at this multi-competitive Phase 3 relatively quickly, perhaps learning from the experiences of Western countries regarding the development of regulatory practices. In addition, low-cost technologies exist today that facilitate access, enabling developing countries to 'leapfrog' over older, more expensive technology and, indeed, avoid legacy issues.

It is interesting to note that the FCC in the USA was the first regulator to introduce competition; at that time, this was in the most advanced communications market. It could be argued that other NRAs such as Ofcom have now developed more advanced regulatory models and the FCC's decision to allow competing technologies for mobiles resulted in a comparatively low penetration of this important technology. This is not meant as a criticism of FCC policy; it merely suggests that initiating regulators do not always make the best decisions and, with the benefit of hindsight, fast followers can capitalise on the initiators' experiences.

Some commentators believe that the endpoint of the multi-competition phase will be minimal, or no, regulation. There is little evidence to support this view, as new network structures and services can give rise to unanticipated issues requiring regulatory intervention or proactive action.

Phase 4 – convergence competition

Most TMT industry players in Phase 3 are already positioning themselves to gain competitive advantage from convergent network opportunities, or from opportunities created by owning both content and carriage.

Phase 4, on the other hand, is characterised by bundled triple-play service offerings and converged TMT NRAs. However, countries with broadband networks and significant numbers of internet and mobile users are already beginning to face a new set of issues that come with convergence. VoIP services are already achieving a double-digit share of international traffic in a number of Asia-Pacific countries. These new activities are disrupting broadcasting and media markets as new web-based services are developing.

Recently, there has also been rapid growth in the number of media players, initially through a proliferation of free-to-air stations and the multi-channel availability of pay-TV stations. New media on the internet has also added to the convergence dynamic.

Types of convergence	Examples
Platforms	Fixed to mobile
Services	TV over broadband
Devices	Mobile TV
Industries	Consolidation within segments

Source: Professor Steve Burdon (2006) Criteria Research and Analysis Pty Limited

Table 4: Four generic types of convergence

Where there is separate regulation of broadcasting and telecommunications, a number of conflicting approaches may occur, such as content being regulated for broadcasting but not for telecommunications. Some countries such as Australia and Hong Kong have already moved to address this by combining telecommunications and media NRAs.

Several commentators in recent times have written about the emerging shifts of convergence and these can be best reviewed as four generic types (see table 4).

New regulatory challenges in the age of convergence

Countries with advanced TMTs are facing a fresh set of dynamics around convergence, heralding a new paradigm which is likely to require a significantly different approach.

Forecast impact of convergence

Precisely what impact convergence will have on consumers in society is still a matter of some debate. However, a recent survey found that telecommunications executives are in little doubt that the impact will be significant

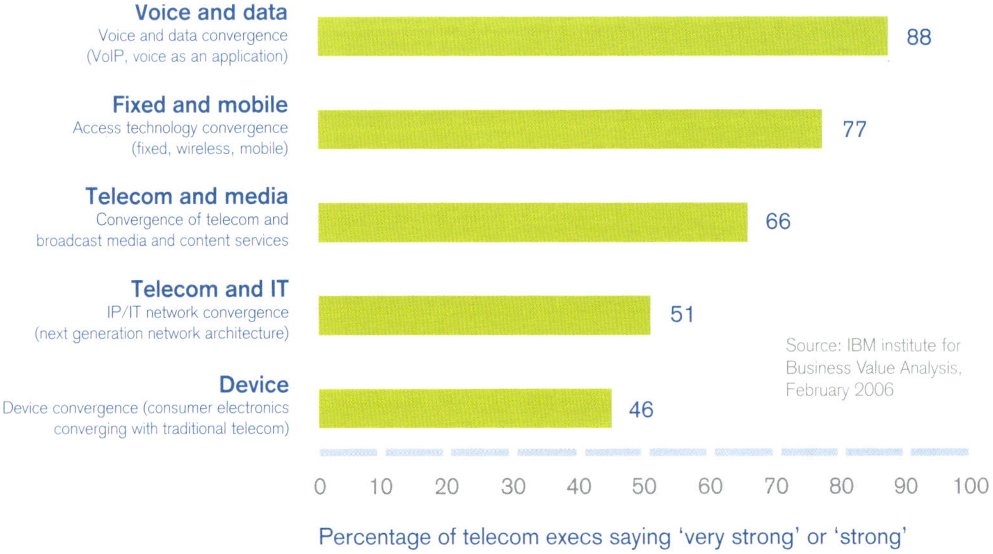

Figure 3: Impact of different types of convergence over the next three years

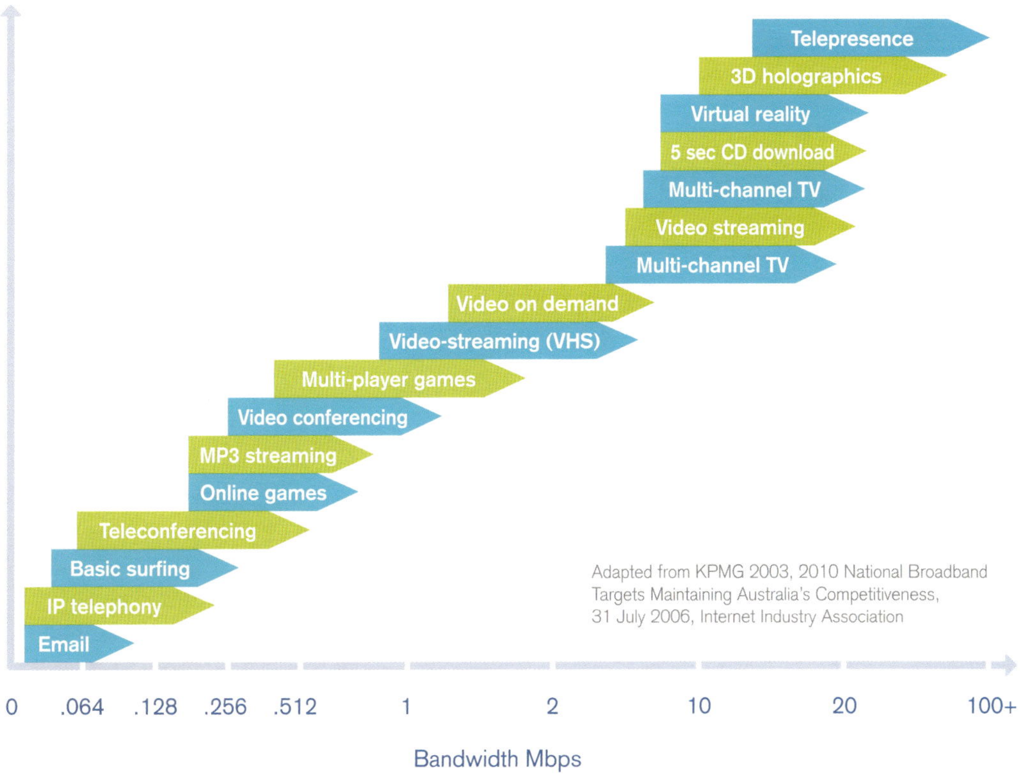

Figure 4: Indicative application bandwidth demands

over the next three years, across the generic types of convergence (see figure 3).

Development of new convergence services

Although the inter-relationship of the four generic types of convergence is still evolving, there appears to be one essential gating factor – namely broadband availability – for a whole suite of new convergence services in the entertainment sector. This, coupled with broadband's impact on the health and education sectors, is one of the reasons why many Asia-Pacific countries have become intent on fostering its accessibility and affordability for widespread adoption. Figure 4 shows a suite of new services that can be developed as bandwidth increases from 2 to 100 Mbps.

Regulatory challenges for convergence services

As market boundaries shift, organisations in one market may be confronted with a new type of competitor or a changed market situation. Consequently, the traditional views on market definitions that NRAs have relied on in the past could become redundant, with a new complexity arising in the scope and difficulty of regulatory oversight between separate broadcasting and telecommunications services.

As a case in point, in July 2006 South Korea's broadcasting regulator and its Ministry of Information and Communications ended a lengthy debate over jurisdiction issues, after agreeing to jointly operate web-based television services (IPTV). The two bodies

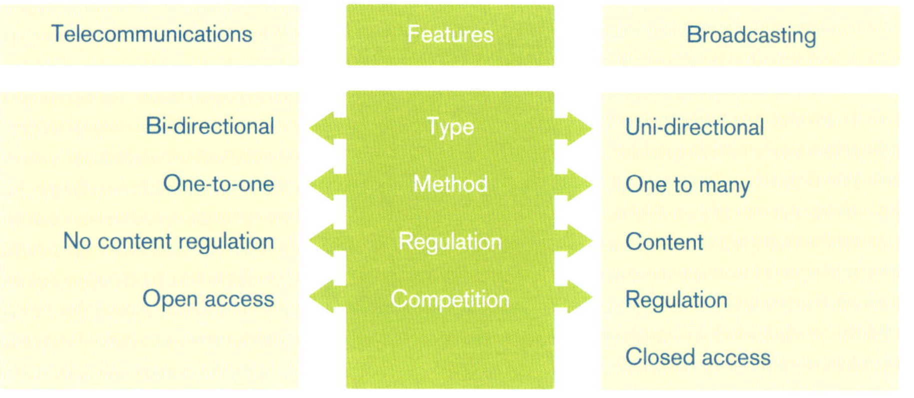

Source: Professor Steve Burdon (2006) Criteria Research and Analysis Pty Limited

Figure 5: Historic differences in key features of telecommunications and broadcasting

were in dispute over whether IPTV should be operated within the boundary of the multimedia or telecommunications sectors.[14]

Some idea of the difficulties involved can be seen in figure 5 which summarises the differences between telecommunications and broadcasting communications.

Emerging convergence factors of new media, internet and advertising

Audience fragmentation, web distribution and consumer-generated content are all forces contributing to a new media state of play, disordering the traditionally neat distinctions between audiences, content creators and advertisers. As M. Walsh writes in *The New Biz of Showbiz:*

"Content is created by consumers within a network of connections and exchanged seamlessly. Even traditional entertainment products like movies and music are annotated with user reviews, remixed or simply exchanged across peer to peer networks."[15]

Content that was once fixed on distinct platforms – TV, radio, print, cinemas – is becoming unbundled and accessible across the web. Some of this is due to piracy but, increasingly, consumers are also accessing legal music, movie and TV show programming provided by the original content owners through Apple iTunes or their own websites. The internet has evolved into a cost-effective and fluid distribution system for hitherto exclusive content, eroding the influence of the content programmers, governments and regulators who have become accustomed to making decisions about the availability and timing of content on behalf of audiences.

This networked model of media has far-reaching implications stretching way beyond merely exchanging entertainment content. It also impacts on what gets found and consumed, and creates new advertising models. Traditional broadcast and print advertising is easily and heavily regulated for values and appropriateness in keeping with the national population norms.

14. Choung, A. (2006) KOREA: *Regulators clear path for Web-TV*, Korea Times, 19 July, http://www.asiamedia.ucla.edu/article.asp?parentid=49415 (accessed: 16 August 2006)

15. Walsh, M. (2006) *The New Biz of Showbiz*, http://www.thefourthestate.com/2006/06/the_new_busines.html#more (accessed: 15 August 2006)

On the new networked media, however, the strategies of both regulators and advertisers need a fundamental re-think as young people are clearly spending more of their media consumption time on the web, or playing games. In 2005, 21 per cent of print readers used the web as their primary news source, with blogs consumption becoming de rigueur.[16]

Unlike the traditional media advertising options, there is no predictable model for engaging with social media, when individual consumer opinions are literally the DNA of most web consumption decisions today. Amazon's book recommendation service is based on the consumption patterns of like-minded book buyers. Google's page-rank algorithm works off the density of other people's webpages linking to a particular piece of content, while bands on MySpace survive by the number of friends that they manage to sign up and recruit to their own website. For other companies, blogs have become the latest weapon in the war for consumer attention.

On a globally ubiquitous medium which observes no geographic borders, this imbalance between consumer-generated content and promotion, and commercial advertising, poses regulatory challenges for maintaining the values and cultural norms of individual countries. For any given NRA, overseas-generated content remains outside its jurisdiction but this has not deterred efforts of an increasing number of countries - both democratic and non-democratic - to "curb the perceived lawlessness of the medium".[17]

Now, some commercial advertisers are themselves becoming victims of unlawful practices such as 'click fraud', where people (or computer systems) deliberately click on the online advertisements of competitors, driving up their costs since they pay for every click or visit to their sites.[18]

Some regulation is only to be expected as the medium matures, and the most notable to date has been censorship initiatives. Varying degrees of internet content filtering have been applied in many Asia-Pacific countries including Singapore, Myanmar, Vietnam and China. Furthermore, China has been observed as having the most extensive, wide-reaching and effective legal and technological systems in the world for internet censorship and surveillance.[19]

Three different regulatory approaches for convergence

As competition becomes viable and market forces exert greater discipline on operators' behaviour, it may be wiser for regulators to aim to reduce the regulatory burden on all operators - incumbents and new entrants alike. However, regulators must strike a balance: on the one hand, they must not give incumbents too much latitude to obstruct competition while, on the other, they must avoid imposing so many restrictions that they stifle investment.

There are broadly three regulatory approaches that governments can take.

16. *A Fifth of online users who read newspapers now rely primarily on web editions* (2006) Neilsen/Net Ratings, http://www.nielsennetratings.com/pr/pr_050616.pdf (accessed: 15 August 2006)

17. The OpenNet Initiative (2006)
http://www.opennetinitiative.net/modules.php?op=modload&name=Sections&file=index&req=viewarticle&artid=1, (accessed: 16 August 2006)

18. *Click! Mouse fraud bytes web advertisers*, (2006) Australian Financial Review, Tuesday 15 August, p1,32

19. Internet Filtering in China in 2004-2005: A Country Study (undated) The OpenNet Initiative, http://www.opennetinitiative.net/studies/china/ (accessed: 17 August 2006)

1. They can proactively-manage competition, guided by clearly-set benchmarks for the achievement of national interests. This option would see the NRAs under instruction to proactively regulate, declare services, set prices and generally lay the groundwork for industry development, with guidance from benchmarked progress (see Figure 1 for Hong Kong's approach).

2. Use a reactive 'light-touch' approach to the development of competition and services in consultation with industry and consumers, granting concessions as necessary to persuade industry to provide services and develop infrastructure that would be in national interests.

3. Take a market-led approach by allowing the natural evolution of technological and other developments in the industry eco-systems, only dealing with occasional issues escalated to the NRA which cannot be resolved among the industry players themselves. Under this approach, the industry self-regulates extensively, with NRAs exercising regulatory forbearance and intervening only in the event of market failure.

Implicit in the review of each of these options are considerations about the shifts in balance between *ex ante* regulation, and *ex post* law in managing competition and dealing with anti-competitive conduct.[20]

The proactive-managed approach is already common in Asia-Pacific, and more member countries may follow suit in the future. Its advantages are that it provides for regulatory certainty and, when judiciously exercised, will be a more effective instrument to balance industry profit-maximising goals with national interests.

Summary

The Asia-Pacific region currently leads the world in share-of-investment in TMTs and advanced services such as broadband and 3G penetration, moving from third to first over the Americas and Europe in a matter of five years.

The region is extremely diverse, and for the purposes of this paper a four-phase convergence continuum model was developed as a tool to analyse the current state of play in relation to TMTs, and to postulate future competition and regulatory developments. Country groupings were devised based on the state of competition in the TMT sector and the respective TMT regulatory frameworks. Subjective judgements were then made about the positioning of each country within the convergence continuum model.

At the time of writing, Australia, Japan and Hong Kong were in the position where a single provider could offer bundled voice, data and video services (triple-play), although not necessarily on the same network, and their respective governments had moved to merge their communications and content regulators into one body.

What has emerged is that the Asia-Pacific region is at the cusp of a further paradigm shift as its leading economies move inexorably towards convergence competition. For NRAs that are not already consolidated, one of the key challenges identified includes inter-regulator jurisdictional issues that arise as the

20. Competition law is traditionally ex post in the sense that matters are brought before the appropriate authorities when abuses have been committed or are about to be committed. On the other hand, telecommunications regulations are mostly *ex ante* in nature with proactive market intervention based on specific legislative provisions. *Ex post* regulation is said to be more flexible and less interventionist, leaving problems to be sorted out in the market place until the point where abuses of general rules are committed. However its processes may be too slow in a fast developing communications environment. Cases have to be processed by the appropriate authorities, and the experience is that it may take years to reach a final decision.

old distinctions between broadcasters, telecommunications operators and media players are dissolving.

And as new web-based, globally-available media continues to proliferate with growing broadband penetration, some governments perceive their national values, sensitivities and sensibilities as being under siege. They look to find ways of controlling foreign-generated content, and this state of play is still evolving. Where, historically, technology and telecommunications innovations have led convergence developments, it would appear that new media may only now be doing so.

As we've seen, governments have three possible regulatory approaches open to them, ranging from proactively managing the development of competition; to a light-touch, guide-and-encourage strategy; to a market-led, minimal regulatory structure, with intervention by NRAs only in the event of market failure.

Looking to the near-future, it is likely that a growing number of governments in Asia-Pacific will take the first option. The proactive management approach provides the highest level of regulatory certainty for all players, and has better prospects of optimising outcomes for all stakeholders. Historically, it has achieved results in a number of European and Asia-Pacific jurisdictions, and is probably best suited for managing convergence competition in an era where market dynamics are in a state of major change.

As the Asia-Pacific economies take the lead in an increasing number of TMT sectors, the prime initiators in world regulation – traditionally, the USA, Scandinavia and the UK – could find themselves being challenged. If the current rate of improvement in TMT continues, the old 19th century adage for ambitious Europeans may soon need a revised 21st century version: "Go east, young men and women!"

Acknowledgement

I'd like to thank Kay Ong, Principal of AnalyzeThiz Pty Limited and Senior Researcher at the University of Technology Sydney for her ideas, contribution and editorial assistance in producing this research paper.

Possibilities for deregulation: a case study of Hong Kong

M. H. Au

Director-General of Telecommunications,
Office of the Telecommunications Authority (OFTA), Hong Kong

Introduction

Since 2004, Hong Kong has embarked on a course of deregulation. Some of the *ex ante* regulations introduced at the beginning of liberalisation have been scaled back as competition in the market has become more effective. This paper focuses on economic regulation on interconnection and access. It examines the deregulation steps that have been taken, the considerations behind them, and discusses the opportunities for further deregulation in the environment of fixed mobile convergence and next generation networks.

Some background on Hong Kong

Hong Kong is a Special Administrative Region of the People's Republic of China. The land area is about 1,100 square kilometres; the population is 6.9 million; and there are around 2.2 million households. The Gross Domestic Product in 2005 was about US$25,000 per capita, with the services sector accounting for nearly 90 per cent of that figure. Hong Kong positions itself as an international centre for finance, trading, logistics and tourism.

Hong Kong's economic development has been underpinned by an efficient telecommunications infrastructure and high quality telecommunications services at competitive prices. The market is very competitive and penetration rates are high. Some key statistics of the telecommunications industry in Hong Kong are given in table 1.

Ex ante regulation

Hong Kong has adopted market-driven policies in the telecommunications sector, with all sectors of the Hong Kong telecommunications market having been opened to competition since January 2003. The number of operators is determined by the market, except where physical constraints (e.g. spectrum availability) limit them. All investment in the telecommunications industry is market-driven and there is no government subsidy. The industry is fully privatised and there is no foreign ownership restriction for any type of telecommunications operator, whether facilities-based or services-based. The Office of the Telecommunications Authority (OFTA) regulates the industry under the Telecommunications Ordinance.

It was in 1995 that the local (domestic) fixed network market started a progressive opening-up to competition when three new entrants were licensed to compete with the incumbent. In 2000, licences to new entrants were awarded for wireless technologies (Local Multipoint Distribution Systems, LMDS). The cable

Number of telephone lines	3.8 million
Telephone line density	54% by population
Number of broadband connections	1.65 million
Broadband penetration	23% by population 65% by households
Number of domestic fixed network operators	10 (6 active in retail market)
Number of mobile customers	Post-paid – 4.7 million Pre-paid – 3.8 million Overall – 8.5 million
Mobile service penetration	Overall 123%
2.5G/3G customers as a percentage of all mobile customers	22%
Number of mobile network operators	5 (4 with 3G licences)
Annual revenue from telecommunications services	US$6.3 billion (2004)
Annual capital expenditure in telecommunications sector	US$ 1 billion (2004)

Table 1: Key telecommunications statistics in Hong Kong

television operator was also licensed to operate a broadband internet access service over its hybrid fibre coaxial cable network, effectively forming the fifth wireline local fixed network. Since January 2003, the local fixed network market has been fully liberalised.

Like other markets which develop from a monopoly, the incumbent operator remained dominant in the early days of market liberalisation. It was therefore necessary to introduce *ex ante* regulation while the market was not fully effective to protect consumer interest and to foster the development of competition. It was intended that this regulation should act as a surrogate for market forces and could be withdrawn when the market became effective.

The regulation is contained in a combination of legislation, the Telecommunications Ordinance and the conditions of licences issued under it. One example was that the retail prices of the incumbent operator, including discounts, were subject to *ex ante* approval by OFTA in order to prevent anti-competitive pricing while the incumbent operator retained its dominance in the market. In contrast, the new entrants were only required to publish the ceiling of their prices. This regulation was terminated from January 2005 after considering the status of competition in the market.

Regulation on interconnection and access is by far the most important element of the *ex ante* regulation. The obligation of network operators or 'carriers' to interconnect is contained in the conditions of the licences issued to them. Under these conditions, the licensee is obliged to interconnect with other networks and services and the interconnection charges should be based on the reasonable relevant cost attributable to the interconnection.

Under the Telecommunications Ordinance, OFTA may determine the terms and conditions of interconnection, and issue guidelines on it. Before intervening to make an interconnection determination, OFTA has to consider a number of factors: these include government policy objectives for the telecommunications industry; consumer interests; encouraging efficient investment in telecommunications infrastructure; and the nature and extent of competition among the parties to the interconnection and other relevant factors. That an operator has 'significant market power' or 'dominance' is not explicitly mentioned in the legislation as a factor to be considered by OFTA, but as for all economic regulation, OFTA has to consider if there is any market failure in deciding whether to intervene.

Therefore, these two market characteristics would be relevant factors in the consideration. In Hong Kong, there is also no specific provision in the legislation on 'unbundling of network components', but the definition of 'interconnection' has been extended to include "access to, or interconnection with, any element of a telecommunications network on an unbundled basis at any point that is technically feasible". For this reason, there are two types of interconnection under the Hong Kong legislation: 'Type I interconnection' and 'Type II interconnection'.

Type I is the commonly-known type of interconnection defined in other regulatory regimes such as the EU regulatory framework for electronic services. It refers to interconnection between networks and services such that users connected to one network may communicate with, or gain access to, other users or services connected to other networks.

'Type II interconnection' refers to interconnection by one operator to the customer access networks (e.g. local loops) of another, to enable it to reach its customers. This achieves the same objectives as 'unbundling of local loops' in other jurisdictions. New operators were allowed to interconnect to the copper-based local loops at the telephone exchanges of the incumbent operator, so that they could start offering services to their customers before their own customer access networks were rolled out.

Type II interconnection

Apart from legal treatment, there are a number of features which distinguish the Type II interconnection arrangements in Hong Kong from 'unbundling of local loops' in other countries.

Firstly, Hong Kong has not unbundled network elements other than the local loops. The new entrants are expected to build their own concentrators, backhauls, switches and trunks so that Type II interconnection would facilitate the emergence of some facilities-based competition. Secondly, the Type II interconnection applied only to the copper local loops which were constructed by the incumbent during the monopoly period and has never been extended to newly-built optical fibre cables or wireless access networks. Thirdly, there was a deliberate attempt to scale back Type II interconnection at the stage of further liberalisation in the local fixed network market in the year 2000, as post-2000 entrants are no longer entitled to Type II interconnection mandated by regulation at the exchanges of the incumbent.

This type of interconnection has been effective in opening up the narrowband fixed network market to competition. At the end of 2004, about 11 per cent of local telephone line users were served by the new entrants through Type II. This, combined with self-built customer access networks connecting about 19 per cent of local telephone line users, enabled competition to develop in the local fixed network market. The market share of the incumbent declined steadily to the 70 per cent level. However, the existence of Type II interconnection policy has not discouraged roll-out of self-built customer access networks by some new market entrants. This has something to do with the economics for self-built networks in Hong Kong.

Although the total land area is over 1,000 square kilometres, the terrain is hilly and therefore the population is concentrated into much smaller usable land areas. As a result, the population density in the urban areas can reach 50,000 people per square kilometre. Most of them live in multi-storey apartment buildings, with one building in the private sector typically housing 200 units or more. In contrast, a public housing estate can accommodate thousands of families.

The costs of rolling out optical fibres to these buildings can be shared by a large number of potential customers. Estimates say that 80 per cent of the 2.2 million households in Hong Kong are accommodated in around 8,000 buildings and that it is technically feasible, and economically viable, for the new entrants to lay optical fibres to reach those buildings as an alternative to the incumbent's copper local loops. The legislation facilitates the rollout of self-built customer access networks by giving local fixed operators a statutory right of access to the communal areas of apartment buildings (i.e. areas other than those for exclusive occupation by the residents) to install cables and equipment to reach their customers inside the buildings.

OFTA also adopts a number of active measures to facilitate network rollout. These include co-ordinating road-opening work (to help expedite the issue of excavation permits by the relevant authorities); promoting the provision of adequate equipment rooms and cabling ducts inside buildings; liaising with building managers and property developers (to overcome building access problems); and launching programmes to educate the public (about the merits of facilitating access of multiple networks into their buildings).

Another factor is that the new entrants who came into the market after 2000 have no alternative but to self-build their customer access networks. Even for one of the pre-2000 entrants that was entitled to Type II interconnection, it has instead adopted a policy of self-building – probably because of the perceived advantages of having access networks owned by itself; better control of lead time to serve customers; product differentiation; and the potential to develop higher-performance services for the future.

As a result, facilities-based competition has emerged in the local fixed network market. Apart from 98 per cent coverage of homes by the Digital Subscriber Line-enabled (DSL) local loops of the incumbent operator, over 90 per cent of homes are within the reach of the cable modem service of the cable television network. 71 per cent of homes are within the coverage of the customer access network of at least one other new operator, and 43 per cent are within the coverage of the access networks of at least two other new operators.

The new entrants' customer access networks are typically using fibre-to-the-building technology, with optical fibres laid in underground ducts to a telecommunications equipment room, located on the ground floor or basement of a building. The final connections to the customers' premises in the building are through Category 5 copper wires laid in the vertical risers and horizontal ducts inside the building.

As a result of facilities-based competition, prices for broadband have dropped to very competitive levels compared with narrowband access, and have stimulated demand. Broadband penetration of households has reached 65 per cent, and operators have taken advantage of this high uptake to offer IP telephony and IPTV services over the broadband connections. 'Triple-play' services (i.e. services comprising voice telephony, internet access and pay television) are offered by four fixed network operators and their affiliates. New operators have captured 48 per cent of the broadband market.

This level of competition in the market has prompted Hong Kong to review the regulation of Type II interconnection that was introduced in the initial period of market liberalisation. In the short-term, Type II offered more choices to residents in multi-storey buildings, and increased competition. However, it is also viewed as a low-risk approach of connecting those customers. Its existence could dilute the incentive for investment in alternative customer access networks which have the potential to offer more advanced services, rather than relying on the DSL-enabled local loops.

The benefit of Type II has to be weighed against the cost of slowing down the rollout of more modern infrastructure. In view of the expanding coverage of the customer access networks of the new operators, it has been decided that continuing with mandatory Type II inter-connection indefinitely can no longer be justified. The Government decided in 2004 to end the mandatory application at the telephone exchanges by the end of June 2008, with the exception of cases where the copper local loops of the incumbent operator are still 'essential facilities'. For buildings already connected by more than one customer access network, the withdrawal of Type II interconnection for access to the buildings from the telephone exchanges will be sooner: three years after the buildings received these networks.

Most customers in Hong Kong live in multi-storey buildings. With the optical fibres connecting the buildings terminated in the telecommunications rooms on the ground floor or basement of the buildings, the customers on the upper floors have to be connected through in-building wiring systems normally based on copper cables. These cables are routed through in-building risers and ducting systems.

Owing to space limitations, and difficulties in installing additional cables leading to customers' premises once the buildings have been occupied, the in-building wiring systems can become 'bottlenecks' in some cases. For this reason, Type II interconnection to the wiring systems from the equipment rooms or distribution boxes inside the buildings will be continued. Normally, operators negotiate with each other for mutual access to the in-building wiring systems under their control, with OFTA intervening only if these negotiations fail. So far, OFTA has mediated in inter-operator disputes, but no determinations have been requested.

Type I interconnection

This regulation is essentially to achieve the policy objective of 'any-to-any' connectivity. Every user connected to a telecommunications network can communicate with any other user, or access any service, whether or not the other user or service is connected to the same network or another network. Like other regulators, OFTA has implemented *ex ante* regulation on Type I interconnection. The extent of regulatory attention has depended on whether the interconnection arrangement is symmetric or asymmetric.

Symmetric arrangements

Symmetric Type I interconnection applies between fixed networks for communications between users. The originating network operator pays an interconnection charge to the terminating operator and the charge applies on a reciprocal basis. The *ex ante* regulation in Hong Kong for symmetric Type I interconnection has been applied with a light touch. The network operators have an obligation to interconnect with each other under their licence conditions. There is no specific provision in the legislation to require the dominant operator to publish Reference Interconnection Offers for interconnection.

OFTA has issued guidelines, setting out the charging principles for interconnection and the methodology for calculating the interconnection charges. Also, OFTA has adopted international best-practice that the interconnection charges should be based on the Long Run Average Incremental Costs. With guidance from these OFTA guidelines, the operators are expected to negotiate interconnection agreements on a commercial basis. OFTA may make a determination on the terms and conditions of interconnection if commercial negotiations fail.

The need for imposing the *ex ante* regulation was to address the imbalance in bargaining power between the incumbent and new entrants in interconnection. Without the obligations and OFTA guidelines, the incumbent probably did not have the incentive to interconnect with a new entrant when the entrant had a zero, or small, customer base.

Meanwhile, more than ten years have passed since Hong Kong opened the local fixed network market to competition. The networks that have entered the market have grown in size, and the relative bargaining positions of the network operators have become more balanced. Both the incumbent and the new entrants need to interconnect with each other in order to meet the expectation of their customers. Therefore, there is considerable countervailing power to work against the exercising of any market power in terminating calls.

For this reason, there is increasingly less need for OFTA to intervene in the negotiations; in fact, since liberalisation commenced, there have been just two occasions when OFTA stepped in to determine the interconnection charges between the incumbent and one new entrant. And since 2003, OFTA's intervention hasn't been needed at all in this type of interconnection between two fixed networks.

At this stage, OFTA does not consider that it can completely withdraw *ex ante* regulation on Type I interconnection. In principle, in a competitive market, operators should be free to negotiate terms and conditions for interconnection among themselves. OFTA's guidelines, and the power to determine as a backup, could be viewed as a form of intervention affecting the outcome of negotiations.

However, the appropriate balance must be struck between avoiding undue interference with the market process and minimising uncertainties that increase investment risks and raise entry barriers. Some basic ground rules would still be required; without them, there would be considerable uncertainty and protracted negotiations. So OFTA should, for the time being, continue to impose the obligation to interconnect and lay down the basic rules of interconnection. However, should the operators negotiate and agree to adopt different terms for interconnection, then OFTA should not interfere with their commercial agreements. Regulation would remain as a backup to resolve disputes which cannot be resolved through commercial negotiations.

Asymmetric arrangements

In Hong Kong, asymmetric Type I interconnection arrangements exist between fixed networks and mobile networks, and between networks and services. In the past, these arrangements have demanded more regulatory attention.

The first type of asymmetric interconnection arrangements exists between fixed networks and mobile networks. A Mobile Party Paying (MPP) arrangement is implemented at the retail level for mobile services; the mobile users pay for the calls they make and receive. At the interconnection level, for a call originating from a fixed network to be terminated on a mobile network – and for a call in the reverse direction – the mobile network operator has to pay either an origination charge or a termination charge, as the case may be, to the fixed network operator. (As far as OFTA is aware, this asymmetric arrangement is rather unique in the world; it is the legacy of treating mobile services as being 'value-added' services, when they first appeared in the market.)

As the costs arising from the interconnection charges are borne by the mobile customers, and would not directly affect the decisions of the fixed line customers in choosing their fixed network suppliers, the fixed network operators would not be subjected to competitive pressure to lower the origination or termination charges. Therefore, the market may not be working effectively to restrain the origination/termination charges. For this reason, the level of these charges of the incumbent has traditionally been regulated by OFTA.

OFTA uses a cost model to work out the origination/termination charges and agrees with the incumbent operator to publish a tariff based on OFTA's calculations. OFTA has a backup power to make a determination if the incumbent operator does not implement the charges it has worked out.

The current asymmetric arrangement is obsolete and would not be sustainable in the environment of 'fixed mobile convergence' (FMC), as user terminals served by the same network operator may be fixed at some times and moving at others. Interconnection rules based on the distinction on whether the user is fixed or mobile would become unenforceable. OFTA has therefore initiated a review on the appropriate interconnection charging arrangement in the FMC environment.

The first question to be addressed in the review is whether any market failure is expected when the existing regulation imposing the asymmetric arrangement is withdrawn. If not, it should indeed be withdrawn. The next question is whether some guidance to the industry on the basic rules of interconnection would still be necessary, in order to provide more certainty and predictability of the regulatory environment after the existing regulation is withdrawn.

If this guidance is warranted, in evaluating options of the basic rules, OFTA aims to identify arrangements that would facilitate the development of convergence services and encourage operators to negotiate efficient interconnection agreements among themselves without regulatory intervention. The arrangements should, in principle, simulate the outcome of commercial negotiations between operators of comparable bargaining power. A symmetric arrangement would better satisfy these criteria than the existing asymmetric one. This potential change is on the OFTA agenda of consultation with industry in 2006.

The second type of asymmetric arrangement applies to interconnection between networks and value-added services, such as internet access services and international telephone services. The service providers have to pay an interconnection charge to the network operator, to pay for the network resources consumed to connect the services to their users. The levels of the interconnection charges of the incumbent operator have traditionally been regulated by OFTA. A review has been initiated to look at whether market forces are able to constrain the level of these charges; the extent of the constraints differs according to the types of services.

In the case of origination charges for internet access services, the high penetration rate of broadband in Hong Kong would exert considerable pressure on those charges. Dial-up access is used mainly as backup. If the origination charge becomes excessive, users would simply stop using the dial-up access. It is therefore likely that the origination charge for internet access services may be deregulated.

When it comes to the origination of calls from fixed network users to international telephone service providers, and for the termination of incoming international calls to the called parties on the fixed networks, the competitive constraints on the fixed network operators to set the interconnection charges are not so obvious.

For outgoing international calls, the calling parties may, arguably, use (in addition to fixed phones) mobile phones and IP telephone services, but this ability to substitute is still to be assessed. For termination of incoming international calls, the calls can be terminated only by the networks to which the called parties are connected. As part of the review currently in progress, OFTA will examine whether market forces can act effectively to constrain the interconnection charges for call origination and termination.

Next generation networks

As in many other countries, operators in Hong Kong are busy implementing, or planning transition to, next generation networks (NGN). The equivalents of Type I and Type II interconnections to NGN should, in principle, still exist. Up to now, Type I interconnection between IP networks has been based on commercial agreements which are working well, and give no reason for the regulator to intervene. As explained in the section on Type II interconnections, regulated Type II interconnection will be phased out from the middle of 2008 and is unlikely to be re-introduced for NGN. However, one area that may need regulatory attention is interconnection and access to NGN infrastructure for the provision of services, applications and content.

In the NGN environment, intelligence and functionalities for the provision of services embedded in the switches of traditional networks will be moved to the network edges, to reside in the users' devices and servers of services and applications providers. This will enable the separation of facilities operation and services provision. Providers who do not own and operate physical infrastructure will also be able to offer services, applications and content in addition to the facilities operators.

The benefits of NGN in unbundling services provision and facilities operation would not be realised unless NGN operators opened the access to their network infrastructure to competing service providers. Inevitably, the number of infrastructure operators will be limited by economic and technical factors. Therefore, if the facilities operators do not allow 'open access' to their network infrastructure, oligopoly would develop in the market for services, applications and content. This open access to the NGN would enable competition for services, applications and content over the NGN infrastructure, bringing diversity, choice and innovation to the users. There will also be more opportunities generated for increasing numbers to participate in the market, including small and medium-sized enterprises which do not have the resources to invest in network infrastructure.

However, it is also necessary to preserve the commercial incentive for investment and deployment of advanced network infrastructure based on new technologies. Without such an infrastructure, the market for the services, applications and content would not materialise – after all, NGN involves substantial and possibly risky investment. The return on investment may depend on revenue from services, applications and content provided by the NGN operators themselves over the infrastructure. They may not have the commercial incentive to open their networks to competing services-based operators as they do not want to see revenue diverting to third-party providers. Regulatory intervention to mandate open access at cost-based prices could potentially dilute the incentive to invest.

The issue, then, is to what extent should *ex ante* regulation be employed to achieve this open access? In this regard, OFTA considers that there is no need to create new principles for NGN. The principles applicable to all economic regulation should be applied. *Ex ante* regulation should be employed to deal with market failures that cannot be overcome over time without regulatory intervention.

As is well known, the architecture of the NGN comprises four layers: namely (from bottom to top), the access layer; the transport (core network) layer; the control layer and the applications layer. In the Hong Kong environment, competing access networks have been constructed based on DSL-enabled local loops, fibre-to-the-building and hybrid-fibre-coaxial cable technologies. Four 3G networks are in operation and broadband wireless access (BWA) is planned to be licensed soon.

We can therefore expect that there will be multiple access networks connected to the core networks. There is also no entry barrier to the core network market. With alternative networks emerging, it would be unlikely to see a bottleneck at the access and core network layers. *Ex ante* regulatory intervention at these two layers is therefore unlikely to be justifiable, with the exception of some remote areas where the incumbent's local loops may remain as 'essential facilities' in the foreseeable future. As regards interconnection and access at the upper levels, the control and application layers involve new interfaces that may not have equivalents in the traditional networks. Whether new bottlenecks would emerge at these layers remains to be seen; however, if they do, regulatory attention might be necessary.

Although regulated access may not be warranted for infrastructure that has no bottleneck issues, experience in Hong Kong tells us that the mere existence of multiple access networks in competition does not automatically lead to open access. Whether the infrastructure operators would adopt a walled-garden or open access policy would depend on whether they viewed the provision of access to third-party service providers as a potential revenue threat to their own services, applications and content – or, conversely, an enhancement to their revenue from increased usage of their network infrastructure.

So far in Hong Kong, the business model commonly adopted by 3G and broadband fixed network operators is to bundle the broadband access with the provision of their own services, applications and content such as IPTV, IP telephony services and 3G content.

In the case of IPTV services, there are three operators in Hong Kong. So far, there has been one commercial agreement concluded between an IPTV service provider and the operator of a broadband network for delivery of the IPTV services; but in this case, the latter does not operate any IPTV services. On the other hand, two other broadband access network operators – who also act as content aggregators and IPTV service providers over their own networks – have yet to open their networks to third-party service providers.

In the case of IP telephony services, the broadband customer access network operators provide services over their own infrastructure to the customers. The users can also access, through their broadband connections, services provided as applications on the public internet. These applications may be provided on the internet by operators within Hong Kong, or by overseas operators. So far, there have been no cases of the broadband customer access network operators offering the access services to IP telephony service providers under a commercial agreement.

In the case of 3G content, the four 3G network operators have been aggregating and supplying content through their own portals. No commercial agreement for opening up the network to third-party content providers has been concluded. Third-party content may be accessed through the internet, although a combination of the data conveyance charges, the relatively more cumbersome steps involved and the suitability of the internet content for display on handsets with small screens probably discourages such access.

The 3G network licences contain obligations for the network operators to open 30 per cent of the network capacity to third-party content providers or Virtual Mobile Network Operators. However, OFTA does not intend to intervene until third-party content providers have made reasonable attempts to conclude commercial deals for access, but without success. So far, no request for intervention has been received by OFTA.

Although no substantial open access has been achieved at present, users are able to reach services, applications and content supplied over the internet. Such delivery modes would in the future exert competitive pressure on the operators of NGN infrastructure to open their networks to third-party service providers. So long as the NGN users can freely access applications on the internet without blockage or deliberate degradation of performance, service providers would be able to deliver their services to their customers through the internet, at least on a 'best-effort' basis. With an open internet, any walled garden policy of the NGN operators cannot be totally effective.

In the supply of applications over the internet, and in the absence of commercial relationships with the operators of the NGN, the delivery of the packets for the applications over the core and access networks of the NGN would be on a best-effort basis only. The service providers would have the incentive to seek direct access to the NGN under commercial agreement, in order to secure better control of the quality and reliability of the delivery services. NGN operators would also have the incentive to offer direct access with better grades of quality and security than best-effort delivery services, for which the NGN operators would charge a premium.

So there should be reasonable prospects for commercial relationships to be concluded between the NGN operators and the third-party service providers. Relationships forged this way would be more sustainable in the long-term than one imposed by regulation.

In the longer-term, it would be possible for the service providers to set up transmission paths over the public internet with guaranteed end-to-end quality of service for certain types of traffic, for which the service providers or the end-customers would be expected to pay.

There are no grounds to object to the service providers, or the end-customers, being charged for differentiated quality of delivery services over the internet – provided these charges are levied for specific levels of service open to all websites or service providers, on a non-discriminatory basis. The NGN operators should have the commercial incentive to interconnect with other operators to deliver traffic under specific service level agreements and share the additional revenue.

Taken as a whole, it is premature to conclude that *ex ante* regulation is necessary to achieve open access in the NGN environment. The NGN market is still new and emerging, and business models and relationships are evolving. Premature regulatory intervention could stifle both investment and innovation. Because multiple broadband access networks exist in the Hong Kong environment, there are real opportunities for market forces to bring about open access.

We should allow more time to see whether services-based operators would be able to obtain access to the NGN on a commercial basis. Without the imposition of *ex ante* regulation, the practices of the NGN operators in the market will continue to be constrained by competition law on an *ex post* basis.

Conclusion

Hong Kong's high population and building densities have meant that OFTA has been able to go further down the deregulation road than countries where the 'last-mile' to the users remains an economic bottleneck for the foreseeable future. OFTA has been able to withdraw some existing *ex ante* regulation on the traditional networks, such as setting a termination date for mandatory Type II interconnection (the equivalent of unbundling of local loops in other jurisdictions). Beyond that date (in mid-2008), competitors in the market will generally be facilities-based and not dependent on regulated access to the incumbent's network.

The extent of deregulation has only been possible because of the level of facilities-based competition which, although attributable to Hong Kong's population and densities, may also have been helped by certain regulatory policies for the rollout of competing infrastructure.

The first is the policy to promote competition in the market: competition drives the entrants to consider self-building their customer access networks in order to differentiate their services from those of the incumbent.

The second is the deliberate policy to contain the scope of Type II interconnection (unbundling of local loops) so that an optimum balance was struck between accelerating competition through access to the incumbent's last-mile and maintaining the incentive for infrastructure investment.

The third are the legislative and administrative measures to facilitate network rollout, such as the statutory right of access into apartment buildings for installation of equipment and wiring systems, and the proactive involvement of the regulator in facilitating road opening and building access.

While facilities-based competition is regarded by many regulators as the desirable goal, if it is applied on its own in the NGN environment it may lead to oligopoly in the market for services, applications and content if the network infrastructure is not opened to third-party service providers. This would then lose the benefits of the unbundling of facilities operation and services provision in the NGN environment. Therefore, OFTA considers that it should promote open access in order to enable a thriving market for services, applications and content to develop over the NGN infrastructure.

However, OFTA does not intend to apply prematurely *ex ante* regulation to the NGN. There is a need to preserve commercial incentive for infrastructure investment. In any case, in an environment where multiple networks can reach the customer premises, *ex ante* regulation would not be justifiable unless a bottleneck issue could be established in the NGN. OFTA would allow time for business models and business relationships to evolve before considering *ex ante* regulation.

By ensuring that NGN users have unrestricted access to the services, applications and content on the public internet, there may well be sufficient competitive pressure on the NGN operators to offer open access to their networks to third-party service providers; if not, the competing products would simply be accessed through the internet. Direct access to the NGN would enable the service providers to control the level of quality and reliability of service delivery to their customers, while the NGN operators received additional streams of revenue.

So, there is scope for market forces to be allowed to work in order to achieve open access. Should this access fail to materialise, due to bottlenecks in the NGN that were unlikely to be resolved by the market in a reasonable period, OFTA would consider *ex ante* regulation.

In this chapter, I have focused on economic regulation, but social and technical regulation may continue to be required in the NGN environment if the market cannot deliver the desired policy objectives on its own, and if the benefit achieved through regulation exceeds its cost. Social regulation may be required to achieve universal service and to protect public consumer interests, while technical regulation may be needed to facilitate interconnection and interoperability between networks.

Acknowledgements

Special thanks to those who contributed to the editing and production of this book:

Andy Bailey, Peter Dunkley, Robin Hull.

Thanks to those who contributed to the Approaches to Regulation project especially in the provision of helpful suggestions for topics and authors:

Chris Adams, Monica Arino, Azeem Azhar, Steven Barnett, Alex Blowers, Martin Cave, Pietro Crocioni, Julia Fraser, David Gillick, Ian Hargreaves, Lyn Maddock, David Mahoney, Chris Marsden, Gratham Mather, Dominic Morris, Anil Patel, Peter Philips, Stewart Purvis, Dougal Scott, Emily Seymour, Len Waverman, Sean Williams.